Lecture Notes in Computer Science 15748

Founding Editors

Gerhard Goos
Juris Hartmanis

AF173658

The series Lecture Notes in Computer Science (LNCS), including its subseries Lecture Notes in Artificial Intelligence (LNAI) and Lecture Notes in Bioinformatics (LNBI), has established itself as a medium for the publication of new developments in computer science and information technology research, teaching, and education.

LNCS enjoys close cooperation with the computer science R & D community, the series counts many renowned academics among its volume editors and paper authors, and collaborates with prestigious societies. Its mission is to serve this international community by providing an invaluable service, mainly focused on the publication of conference and workshop proceedings and postproceedings. LNCS commenced publication in 1973.

Manuel Egele · Veelasha Moonsamy ·
Daniel Gruss · Michele Carminati
Editors

Detection of Intrusions and Malware, and Vulnerability Assessment

22nd International Conference, DIMVA 2025
Graz, Austria, July 9–11, 2025
Proceedings, Part II

 Springer

Editors
Manuel Egele 🆔
Boston University College of Engineering
Boston, MA, USA

Veelasha Moonsamy 🆔
Ruhr University Bochum
Bochum, Germany

Daniel Gruss 🆔
Graz University of Technology
Graz, Austria

Michele Carminati 🆔
Politecnico di Milano
Milan, Italy

ISSN 0302-9743   ISSN 1611-3349 (electronic)
Lecture Notes in Computer Science
ISBN 978-3-031-97622-3   ISBN 978-3-031-97623-0 (eBook)
https://doi.org/10.1007/978-3-031-97623-0

This Springer imprint is published by the registered company Springer Nature Switzerland AG
The registered company address is: Gewerbestrasse 11, 6330 Cham, Switzerland

If disposing of this product, please recycle the paper.

Preface

On behalf of the Program Committee, we are pleased to present the proceedings of the 22nd Conference on Detection of Intrusions and Malware & Vulnerability Assessment (DIMVA 2025).

Over the past two decades, DIMVA has become a recognized venue for cutting-edge security research, attracting high-quality submissions and promoting collaboration among academia, industry, and government. DIMVA is organized by the Special Interest Group on Security, Intrusion Detection, and Response (SIDAR) of the German Informatics Society (GI).

This year, we received 103 valid submissions (11 desk-rejected), and accepted 25 full papers, resulting in a competitive acceptance rate of 24%. Each paper underwent a rigorous double-blind peer review process, with every submission reviewed by at least three experts. Each Program Committee member was assigned to review up to 7 papers. For the first time, the DIMVA proceedings include 11 poster papers selected through a single-blind review. For the third consecutive year, DIMVA followed a dual-deadline submission model. The acceptance rate was balanced across the two cycles: in the first cycle, 7 out of 40 submissions were accepted (2 directly and 5 with shepherding), while in the second cycle, 18 out of 63 submissions were accepted (8 directly and 10 with shepherding).

We extend our sincere gratitude to the Program Committee members for their tireless efforts in reviewing papers in both cycles, engaging in in-depth online discussions, and shepherding papers to completion. Across the reviewing phases, PC members exchanged over 900 comments.

We would also like to thank the Organizing Committee for their dedication and hard work in preparing this edition of DIMVA. We are grateful to our sponsors, including Genua and Springer, and our host institutions for their support.

Finally, we thank all authors for submitting and presenting their work, and all attendees for their participation. Your engagement and commitment continue to make DIMVA an impactful event. We look forward to your future contributions.

DIMVA 2025 was held in the Aula of the historic main building of Graz University of Technology. The venue offered a unique blend of historical atmosphere and academic prestige, creating an inspiring setting for insightful discussions and new collaborations. In addition to paper presentations, the program featured an engaging keynote session and a newly introduced poster session in the picturesque landscape of Southern Styria, providing attendees with the opportunity to discuss early-stage work.

Together, the program offered an excellent platform for discussion, collaboration, and social interaction in a relaxed setting in South-East Styria.

May 2025   Manuel Egele
                                                            Veelasha Moonsamy
                                                                   Daniel Gruss
                                                             Michele Carminati

Organization

General Chair

Daniel Gruss   Graz University of Technology, Austria

Program Committee Chairs

Veelasha Moonsamy   Ruhr University Bochum, Germany
Manuel Egele   Boston University, USA

Publication Chair

Michele Carminati   Politecnico di Milano, Italy

Poster Chairs

Tarini Saka   Ruhr University Bochum, Germany
Flavio Toffalini   Ruhr University Bochum, Germany

Steering Committee Chairs

Ulrich Flegel   Infineon Technologies, Germany
Michael Meier   University of Bonn and Fraunhofer FKIE,
                                     Germany

Steering Committee

Magnus Almgren   Chalmers University of Technology, Sweden
Sébastien Bardin   CEA, France
Leyla Bilge   Gen Digital, France
Gregory Blanc   Télécom SudParis, France
Herbert Bos   Vrije Universiteit Amsterdam, Netherlands
Danilo M. Bruschi   Università degli Studi di Milano, Italy

Roland Büschkes	RWE AG, Germany
Juan Caballero	IMDEA Software Institute, Spain
Lorenzo Cavallaro	King's College London, UK
Hervé Debar	Télécom SudParis, France
Sven Dietrich	City University of New York, USA
Mathias Fischer	Universität Hamburg, Germany
Giorgio Giacinto	University of Cagliari, Italy
Cristiano Giuffrida	Vrije Universiteit Amsterdam, Netherlands
Daniel Gruss	TU Graz, Austria
Bernhard Hämmerli	Acris GmbH and HSLU Lucerne, Switzerland
Thorsten Holz	CISPA Helmholtz Center for Information Security, Germany
Marko Jahnke	CSIRT, German Federal Authority, Germany
Klaus Julisch	Deloitte, Switzerland
Christian Kreibich	ICSI, USA
Christopher Kruegel	UC Santa Barbara, USA
Pavel Laskov	University of Liechtenstein, Liechtenstein
Federico Maggi	Amazon Web Services, USA
Clémentine Maurice	CNRS, IRISA, France
Nuno Neves	University of Lisbon, Portugal
Roberto Perdisci	University of Georgia and Georgia Institute of Technology, USA
Michalis Polychronakis	Stony Brook University, USA
Konrad Rieck	TU Braunschweig, Germany
Jean-Pierre Seifert	Technical University Berlin, Germany
Robin Sommer	Corelight, USA
Urko Zurutuza	Mondragon University, Spain

Program Committee

Advait Patel	Broadcom, USA
Amin Kharraz	Florida International University, USA
Andrea Continella	University of Twente, The Netherlands
Andrea Lanzi	University of Milan, Italy
Andrea Mambretti	IBM Research Europe - Zurich, Switzerland
Anita Nikolich	University of Illinois-Urbana Champaign, USA
Bart Coppens	Ghent University, Belgium
Behzad Ousat	Florida International University, USA
Daniel Gruss	Graz University of Technology, Austria
Daniel Plohmann	Fraunhofer FKIE, Germany
Daniele Antonioli	EURECOM, France

Daniele Cono D'Elia	Sapienza University of Rome, Italy
David Klein	Technische Universität Braunschweig, Germany
Eleonora Losiouk	University of Padua, Italy
Emilio Coppa	LUISS University, Italy
Fabio Pagani	Binarly, Italy
Fabio Pierazzi	King's College London, UK
Flavio Toffalini	Ruhr-Universität Bochum, Germany
Hervé Debar	Télécom SudParis, France
Ilya Grishchenko	University of California, Santa Barbara, USA
Jan Wichelmann	Universität zu Lübeck, Germany
Johanna Ullrich	University of Vienna, Austria
Johannes Kinder	LMU Munich, Germany
Juan Caballero	IMDEA Software Institute, Spain
Juan Tapiador	Universidad Carlos III de Madrid, Spain
Kaan Onarlioglu	Akamai, USA
Kimberly Tam	University of Plymouth/Alan Turing Institute, UK
Konrad Rieck	TU Berlin, Germany
Mannat Kaur	Max Planck Institute for Informatics, Germany
Manuel Egele	Boston University, USA
Marco Cova	VMware, UK
Martina Lindorfer	TU Wien, Austria
Mathias Fischer	University of Hamburg, Germany
Michael Meier	University of Bonn and Fraunhofer FKIE, Germany
Michael Schwarz	CISPA Helmholtz Center for Information Security, Germany
Michalis Polychronakis	Stony Brook University, USA
Michele Carminati	Politecnico di Milano, Italy
Moritz Schloegel	Arizona State University, USA
Prashast Srivastava	Columbia University, USA
Ricardo J. Rodríguez	Universidad de Zaragoza, Spain
Roland Yap	National University of Singapore, Singapore
Seungwon Shin	KAIST, South Korea
Silvia Sebastián	CISPA Helmholtz Center for Information Security, Germany
Simon Koch	TU Braunschweig, Germany
Stefano Zanero	Politecnico di Milano, Italy
Stijn Volckaert	KU Leuven, Belgium
Sven Dietrich	City University of New York, USA
Tapti Palit	UC Davis, USA
Tiago Heinrich	Max-Planck-Institut für Informatik, Germany
Urko Zurutuza	Mondragon Unibertsitatea, Spain

Vasileios Kemerlis	Brown University, USA
Vasilios Mavroudis	Alan Turing Institute, UK
Veelasha Moonsamy	Ruhr University Bochum, Germany
Vera Rimmer	KU Leuven, Belgium
Vinod Yegneswaran	SRI International, USA
Yang Zhang	CISPA Helmholtz Center for Information Security, Germany
Yinzhi Cao	Johns Hopkins University, USA

Contents – Part II

OS and Network

Resilient Systems

Contents – Part I

AI/ML and Security

Towards Explainable Drift Detection and Early Retrain in ML-Based Malware Detection Pipelines

Jayesh Tripathi[1], Heitor Gomes[2], and Marcus Botacin[1(✉)]

[1] Texas A&M University (TAMU), College Station, USA
{jtjayesh98,botacin}@tamu.edu
[2] Victoria University of Wellington, Wellington, New Zealand
heitor.gomes@vuw.ac.nz

Abstract. The current largest challenge in ML-based malware detection is maintaining high detection rates while samples evolve. Although multiple works have proposed drift detectors and retraining-aware pipelines that work with reasonable efficiency, none of these detectors and pipelines are currently explainable, which limits our understanding of the threats' evolution and the detector's efficiency. Despite previous works that presented taxonomies of concept drift events, no practical solution for explainable drift detection in malware pipelines existed until this work. Our insight to change this scenario is to split the classifier knowledge into two: (1) the knowledge about the frontier between Malware (M) and Goodware (G); and (2) the knowledge about the concept of the (M and G) classes. Thus, we can understand whether the concept or the classification frontier changed by measuring the variations in these two domains. We make this approach practical by deploying a pipeline with meta-classifiers to measure these sub-classes of the main malware detector. We demonstrate via 5K+ experiment runs the viability of our solution by (1) illustrating how it explains every drift point of the DREBIN and AndroZoo datasets and (2) how an explainable drift detector makes online retraining to achieve higher rates and requires fewer retraining points.

Keywords: Malware Detection · Concept Drift · Explainable AI

1 Introduction

Computer systems are targeted daily by a myriad of malware samples, such that we cannot keep up with this high volume of attacks without the help of automated tools, which are currently largely based on Machine Learning (ML). Although ML can be very effective, designing and deploying an efficient ML pipeline for malware detection has many challenges, as extensively discussed in related works [10]. The main challenge for ML-based malware detection nowadays is the degradation that the models face due to the attacker's new Tools, Tactics, and Procedures (TTPs), which require frequent classifier retrains. Constantly retraining the classifier is not efficient; thus, precise triggers for the retraining process are warranted. Currently, drift detection algorithms

© The Author(s), under exclusive license to Springer Nature Switzerland AG 2025
M. Egele et al. (Eds.): DIMVA 2025, LNCS 15748, pp. 3–24, 2025.
https://doi.org/10.1007/978-3-031-97623-0_1

are the best triggers for a retraining task. Drift detectors work by observing the distribution of the predicted labels in comparison to ground-truth labels (e.g., provided by malware sandboxes or human analysts) and report significant growths in the prediction error rate, which is used as a proxy for the distribution change.

The drawback of current drift detection algorithms is that, by the nature of their design, they do not explain the causes of the drift but just report the drift detection based on the observed error rates. Although it is enough to keep the pipeline operating at reasonable rates, this limits other applications, such as benchmarking and dataset characterization. Benchmarking is limited because it is impossible to know what the drift detector is recognizing as different in a stream. Therefore, it is hard to have a ground truth for eventual false positive reports. It also makes it hard to fairly compare two drift detectors as they detect different phenomena. In this scenario, identifying the best drift detectors for a given scenario has become a trial-and-error task rather than a scientific task. For similar reasons, the nonexplainability also limits the understanding of real scenarios, which limits incident response. Even if a drift detector points out that an in-the-wild stream of malware has changed its distribution, analysts cannot know what is present in the new samples that was not present in the previous ones, which limits remediation and prevention actions.

Ideally, Drift-aware pipelines should provide a clear view of what has changed from one period to another. Recent works in drift detection took their first steps in this direction. A proposed pipeline [11] innovates by retraining not only the classifiers but also the feature extractors when drift is detected. Thus, by comparing the vocabulary in different epochs, it is possible to know the malware features that gained and lost importance. This technique was used to identify, for example, when SMS sending permissions became less prevalent in Android malware because the OS stopped allowing apps with this permission in the app store. Despite this advance, this work still did not explain how the feature changes caused the models to change. A step in this direction was given by works that tried to model probabilistically and statistically the chances of a sample belonging to a distribution or not [23]. Again, despite being a significant advance, this approach still provides an incomplete treatment of the drift problem, even after its complement [6], because although identifying the best retraining moments, it still does not explain the drift points and changing features. Thus, the current scenario is: *Drift detectors that explain features do not explain models and vice-versa.*

We propose changing this scenario with a **practical** solution to explain drift events. This solution is based on the key insight that the classification process is composed of two different aspects: (**1**) the frontier between the classes and (**2**) the concept of the classes. This differentiation highlights the fact that sometimes drift is reported because (**i**) the frontier is misplaced and sometimes because (**ii**) the concept itself changed. We propose that if we measure these two aspects, we can explain drift events by assigning them to one of these two classes (frontier or concept change).

Unlike conformal evaluation [6], which also breaks down the problem in two, we do not rely on explicitly math modeling (e.g., via averages and variances) the problem, but on using their own ML classifiers (meta-models) for the task. This allows one to deploy our solution using the same ordinary drift detectors used in the main classifier (e.g., DDM, ADWIN, and so on) while still benefiting from the non-linearity of ML models to learn the concepts. We propose that detecting drift in the meta-models exposes the change in the frontier and the concepts. To test our hypothesis, we deploy the classifiers under test with a main drift detector in addition to **(1)** drift detectors in their classes, and **(2)** drift detectors in a second layer deployment of one-class versions of the main classifiers, specifically trained to recognize the concepts known by the main classifier.

We tested our solution via 5K+ experiments with the DREBIN and Androzoo datasets with different settings (e.g., detectors, imbalances, policies, etc.), and we found that:

- Our approach explains all classification points, including true drift points, but also false positive drift detections and bad frontiers caused by limited training data.
- Explaining detection and retraining are the two faces of the same coin, as recognizing true concept changes allows making online retraining procedures faster (early retrain), more effective (achieving higher rates), and more efficient (requiring fewer retrains) compared to traditional retraining.

In sum, our contributions are as follows.

- Proposing viewing concept drift as two separate problems: one is establishing the decision boundary, and another is learning the concept for the classes.
- Proposing using meta-classifiers as practical solutions to measure changes in the decision boundaries and the in-class concepts to explain drift detection decisions and suggesting early retraining.
- Evaluating our proposals on the DREBIN and AndroZoo datasets to demonstrate the practical viability of explaining concept drift events in malware detectors.

2   Analyzing Concept Drift

Understanding Classification Errors. When we train a binary classifier for a task like separating malware and goodware, we teach the classifier to divide the feature space with a decision frontier. The samples on one side of the frontier are assigned to one class, and the same is true for the other class. Figure 1 exemplifies it with a horizontal frontier in a 2D feature space for didactic purposes. It is key to note that the classifier might not learn the actual concept of the samples belonging to each class, i.e., what they have in common (the circle), but just a frontier that separates one from another.

In the ideal scenario, the frontier should adapt perfectly to the concept, but this is not what often happens in practice. This gap opens a space for positive and negative effects. On the positive side, some newer samples might still be classified in the right class even if they do not match the learned concept. To that, it is enough to be farther to the wrong class frontier. On the negative side, False Positives (FPs) might occur if the samples still belong to the same original concept but the initial frontier was improperly defined. This happens. For example, when the sampling process to define the training set was biased or did not consider enough representative samples. Figure 2 illustrates this case. Although the newer samples still belong to the same concept (the wider circle), they are flagged in the wrong class because the frontier is improperly defined. Ideally, the classifier should learn the actual concept to avoid these cases. In practice, the establishment of a new frontier (the diagonal line) is what is achieved with classifier retraining.

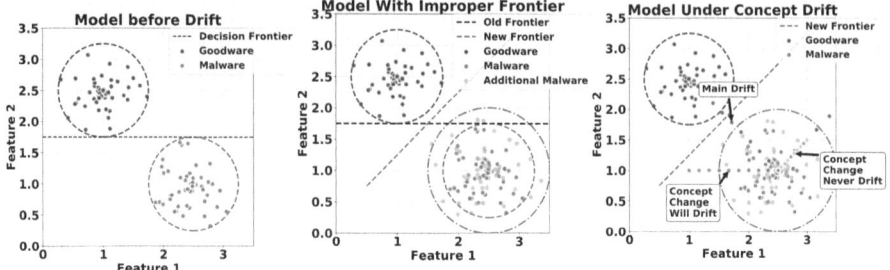

Fig. 1. Initial Training.   Fig. 2. Additional Data.   Fig. 3. Multiple Drifts.

If the classifier learned the concept, FPs due to an improper frontier would not happen, but the classifiers would still make an increasing number of incorrect predictions over time. This is caused by the concept drift/evolution effects, i.e., the malware samples changing their characteristics significantly to avoid detection such that the learned concepts do not apply anymore. In these events, the newer samples tend to slowly deviate from the learned concept to multiple directions, as illustrated in Fig. 3. When the samples drift towards the frontier, they might cross the boundaries and cause FPs.

Detecting vs. Explaining Drift. Constantly retraining the classifier to change the frontier is costly and inefficient. Thus, in practice, MLSec operators benefit from the loose coupling between concept and frontier to let systems operate while producing correct results–even outside the borders of the ideal concept–and only retrain the classifier when the frontier is affecting the correctness of results. The moment to change the frontier is indicated by drift detectors. The major limitation of this approach is that current drift detection algorithms and architecture do not explain their result, i.e., they do not draw a big picture of the phenomenon that is happening, as previously discussed. By observing only the main result, current solutions can't even tell if the result is wrong because the frontier is misplaced or because an actual concept change happened. Achieving greater explainability is possible by observing the multiple drift scenarios and mitigating the detector's blind spots, as follows.

Traditional Drift Detectors operate by only measuring the error rate in the classifier frontier. The underlying idea is that if the concept changes, more samples will cross the frontier, and thus, the error rate will increase. We can interpret this type of detector operation as measuring if the samples of our example are crossing the horizontal frontier from Fig. 1 to Fig. 2. The **first** blind spot is that samples cross the frontier not only when the concept changes but also when the frontier itself is misplaced. By only looking at the frontier, it is not possible to tell the cause of the drift alert. The **second** blind spot is that this type of detector is only triggered when the result is already affected, and this happens at a late drift stage. The concept might have started to change early and remained in the gap between the concept and the frontier before crossing it. In this case, the classifier did not benefit from this situation to start proactively retraining the classifier. The **third** blind spot is that drift detectors based on the frontier might be affected by class imbalance (depending on their internal construction, but we will remain agnostic here). If the number of samples crossing the frontier is relatively small, it is not "worthy" retraining the classifier. However, this decision is problematic in very imbalanced scenarios, where the majority class might always be correctly predicted, whereas the minority one might always be wrong. The **fourth** blind spot is that, due to practical storage and processing limits, the drift detectors do not observe the entire history of predictions but only a window, which makes them susceptible to FPs. If a sequence of wrong predictions (e.g., a few from each class) sequentially appears by simple randomness, the drift alarm will be triggered. Current drift detectors have no way to identify their own FPs. The **fifth** blind spot is that traditional drift detectors do not differentiate which classes are drifting but only observe if the frontier was crossed by a significant amount of samples. Therefore, many drift detectors are not even able to explain which class is problematic.

Class-Aware Drift Detectors is a potential solution for mitigating the blind spot of not telling which class crossed the line. By measuring the error rate in each class, it is possible to know which classes (malware, goodware, or both) have their samples crossed the lines. In the didactic example from Fig. 2, the malware class is causing the drift identification event. It also solves the blind spot of not identifying the FP cases because if none of the classes have their samples crossing the frontier, an eventual drift identification event can only be caused by a rare sequence of wrong predictions within the limited window. However, since it keeps monitoring only the decision frontier, it does not solve the blind spot of only detecting the drift at a late stage, and it is not able to explain if the frontier was misplaced or if the concept actually changed.

Concept-Aware Detectors is our proposal for an ideal model of a detector that learns the concept of the samples (the circles in Fig. 3) independently of the frontier. In this model, the samples are not only assigned to classes but also classified as belonging or not belonging to known concepts. This way, it is possible to know if a sample is part of a known concept regardless of where the frontier is placed, which allows differentiating the reasons for drift. If a drift detector is placed in the concept detector, we can identify if the concept actually changed. If there is no drift in the (ideal) concept classifier but a drift in the (real-world) main classifier, it means that the initial frontier was misplaced.

The concept-aware drift detector acts in tandem with the main classifying by helping to explain its drift and non-drift events. This interaction also allows for anticipating eventual drift occurrences. When the concept is actually changing, three cases might be identified, as illustrated in Fig. 3: **first**, the concept might change in a direction that does not go towards the frontier (the parallel line to the frontier in Fig. 3). In this case, the detection benefits from the concept-frontier gap to keep classifying the new samples correctly without the need for a retrain; **Second**, the new concept goes towards the frontier, but it has not crossed yet (the horizontal drift line in Fig. 3). In this case, the classifier would benefit from an early retrain, as crossing the frontier is imminent; **Third**, the new concept went towards the frontier and crossed it (the diagonal drift line in Fig. 3), which is the explanation for a true drift detection.

Based on the above discussion, we propose the following taxonomy on the type of information that the drift detection algorithm and/or architecture can provide and the case it identifies:

- **Type 1:** <u>Main Classifier Drift.</u> It detects whether a significant number of samples of any class crossed the detection frontier or not within a sampling window to the point of already harming the final classification result.
- **Type 2:** <u>Sub-Class Drift.</u> It detects whether a significant number of samples of a specific class crossed the detection frontier or not within a sampling window to the point of being noticeable but without guarantees that it affects the final classification result (contingent upon Type 1 detection).
- **Type 3:** <u>Concept Change.</u> It detects if a significant number of samples of a specific class do not match the previous knowledge the classifier had about that class, regardless of the correct class assignment (Type 1 and 2 events). The implications of the concept change causing drift or not are contingent on the following cases:

  - **Case A:** <u>Concept change without drift risk.</u> If the concept changes in a direction that does not go toward the decision frontier, it cannot cause drift events.
  - **Case B:** <u>Concept change with imminent drift risk.</u> If the concept changes towards the decision frontier (Type 2), it will eventually cause drift when crossing the frontier (Type 1). This point is a candidate for early retraining.
  - **Case C:** <u>Current Drift due to concept change.</u> If the concept changes towards the frontier (Type 2) and crosses it (Type 1), concept drift is detected late.

These types of detectors are cumulative, i.e., a detector type 3 implies the deployment of type 1 and type 2 detectors. Type 3 detectors currently do not exist. The development of a practical type 3 architecture is this work's goal, as following detailed.

A Practical Solution for Explaining Drift Events. A first idea for improving the explainability of drift detectors is to redesign the drift detection algorithms to be class- and concept-aware. This, however, imposes limits on the types of algorithms that can be used and significantly impacts the design of the malware detectors that become coupled to the used drift detectors. We aim to remain agnostic to the type of drift detection algorithm, and instead of algorithm redesign, we propose an external monitoring architecture that implements the class and concept-aware concepts independently of the detectors used.

We propose a 2-layer architecture where the first extends the Type 1 drift detectors already existing in current malware detection architectures to become Type 2 by instantiating copies of the main drift detector and directly linking them to the prediction of each class. The second layer consists of replicating the main classifier in 1-class settings to focus them on learning the class concept rather than the decision frontier, thus turning them into Type 3 detectors.

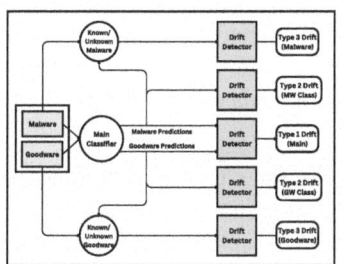

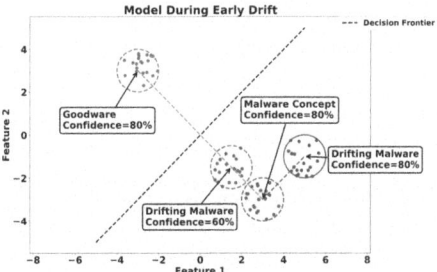

Fig. 4. Drift-Explainable Architecture.   **Fig. 5.** Direction-Change Drift Detection.

In the proposed architecture, illustrated by Fig. 4, the training process starts as usual, with malware and goodware ground-truth samples being provided to the main classifier (classifier under test). However, an additional step is introduced to train the second-layer classifier. To teach the classifier the concept of malware, for instance, the labels correctly predicted as malware after the training of the main classifier as provided as positive labels to the secondary classifier (*Known Malware*, classified correctly). The malware samples misclassified by the main classifier as goodware are provided to the second layer classifier as *Unknown Malware* samples. Thus, this second-layer classifier learns what the main classifier knows as malware and not its difference to goodware. The same process is repeated to train the goodware class. The prediction step also starts as usual, with the unknown sample being classified by the main classifier, whose predictions are later checked by the main drift detector. However, the labels of each class are also forwarded to individual detectors. The sample is also classified in parallel by the second layer classifier, having its ground-truth class as a reference, to identify if it fits the previous classifier knowledge about that class, independently of the main classifier prediction.

The proposed architecture is capable of detecting Type 3 drift, which is enough to point out concept change events, but it is still not enough to anticipate concept drift detection occurrences. to that, an additional step is required to understand the direction of the concept change. For a practical implementation, we approximate the identification of a drift direction by the Eq. 1: $Avg_{concept} - Avg_{drift} > k$,   given $k > 0$

The rationale for that is that the confidence level of the main classifier tells how close the samples are to the frontier. The lower the confidence, the closer to the frontier. If the concept is changing towards a lower confidence region, it is going closer to the frontier, and it is likely to cause a concept drift alarm. We here identify if the concept is going towards the frontier via the average confidence of the new concept cluster compared to the average confidence of the original concept cluster. It is key to highlight the difference between this and previous work's approach, as we do not simply trust in the overall change in the confidence level for all samples, but we rely on the second-layer model to cluster the samples whose confidence will be averaged.

Concept changes in different directions are illustrated in Fig. 5, which shows two concepts (malware and goodware) originally positioned at 80% confidence level. When the concept changes laterally, over the same diagonal line, it triggers a Type 3 detection, but the new concept is not a candidate for early retrain, as it does not go toward the frontier, which is indicated by the fact that the average confidence remained the same (it is the same in all points over the diagonal line parallel to the main frontier). In turn, if the concept changes towards the perpendicular line to the frontier, it will trigger a Type 3 alarm, and it is a candidate for early retrain, as it tends to cross the frontier, which is indicated by the reduced confidence level (60%).

The Relation Between the Drift Event Types in the Classification Stream. During the operation and/or the evaluation of a malware detection pipeline, the multiple drift detectors of our proposed architecture will trigger simultaneously, depending on the drift event that is happening. We summarized all valid states in Table 1, based on the previously introduced drift taxonomy.

The simplest situations are the ones when the detectors agree. When no drift detector fires, the operation is normal. When all detectors fire, there is a clear drift caused by concept change. The most challenging cases are when the detectors disagree. If only the Type 3 drift detector fires, it indicates an early drift case. Then, we need to check if the change is towards the frontier (Type 1) or not, if wanting to decide on an early retrain. If the Type 2 detector fires without a Type 3 one firing, this indicates that the frontier is the problem. If the Type 1 fires without the others, this indicates a false positive. Some detector-state combinations are not possible. For instance, if only the Type 3 detector fires, it can only be in case 1 or 2, but never in case 3 (crossing the frontier), as it requires the Type 1 drift to also fire. These states are our falseability points. We used them to validate our approach. If these cases appear in our test, this indicates that our hypothesis/theory does not hold.

Table 1. Explaining Drift Events. Information types for each combination of triggered detectors. Representing Triggered Detectors ($\checkmark$) and Possible ($\triangle$) and Not-Applicable ($\varnothing$) cases. Omitting Impossible cases.

Main Type			Cases			Conclusion
Type 1	Type 2	Type 3	Case A	Case B	Case C	
				$\varnothing$		Normal Operation
		$\checkmark$	$\triangle$			Early Concept Change with no impact on frontier
		$\checkmark$		$\triangle$		Early Concept Change with imminent impact on frontier
	$\checkmark$			$\varnothing$		Bad Frontier detected without concept change hold by imbalance in main class
	$\checkmark$	$\checkmark$		$\triangle$		Bad Frontier detected with concept change hold by imbalance in main class
$\checkmark$				$\varnothing$		False Positive Drift Detection
$\checkmark$		$\checkmark$	$\triangle$			False Positive with concept change in non-impactful direction
$\checkmark$	$\checkmark$			$\varnothing$		Bad Frontier detected without concept change, with impact in the main class
$\checkmark$	$\checkmark$	$\checkmark$			$\triangle$	Concept Change with Immediate Impact and Identification

3   Evaluation

3.1   Drift Detection Explained by Examples

Consider the Android malware detection case a representative example of the malware detection problem. For this demonstration, we considered the DREBIN [4] dataset, given its popularity and known drift cases. We also considered a simple model that classifies the APK's permissions by modeling them as a binary vector (as in DREBIN), which allows us to easily understand when features entered and left the concept, thus highlighting the concept change phenomenon. Whereas we initially enumerated all permissions in the dataset to create a feature vector of the ideal size, the feature vector is only filled on demand as the permissions appear in the data stream. In our experiment, the dataset was temporally ordered, as recommended by the best practices in the field [10,25] to avoid the data snooping pitfall [3]. Since DREBIN is very imbalanced [14] and could affect the drift results, we undersampled each temporal bin to the 50–50 proportion. Our goal in this demonstration is not to perform a real-world characterization (which is done in the next step) but to highlight the studied drift events.

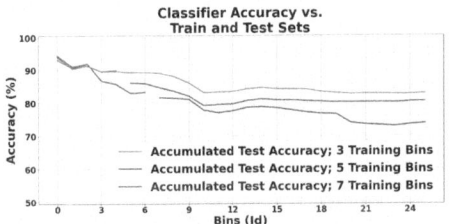

Fig. 6. Concept drift in practice. The classification accuracy decreases regardless of the initial training set size/period.

Fig. 7. Drift tendency vs. instantaneous detection. Drift points reported by the ADWIN algorithm.

Observing Drift in the Main Classifier Output. Drift is an attacker-induced phenomenon that can be noticed in any reasonably long temporal stream observation. Figure 6 illustrates this phenomenon for the evaluated DREBIN dataset setting. It shows the curves derived from the initial training with samples from the first 3, 5, and 7 bins (epochs). In all cases, the accuracy significantly decreased, from more than 90% to less than 80%.

Concept Drift is a Temporal Trend, but Drift Detection Algorithms Observe the Instantaneous Error Rate. When the drift effect significantly influences the classifier output, it is detected by a drift detection algorithm. Figure 7 shows how the Drift Detection Algorithm (DDM) [16] detects multiple drift points in our tested DREBIN setting. It is key to notice that the drift effect is a tendency, which is shown by the curve of the accumulated accuracy results. The drift detection, however, happens in the samples received within a window (the bins), which exhibit higher variations in the instantaneous accuracy values. In this work, we measure drift in the instantaneous results, but we display them mapped to the accumulated ones to highlight the tendencies.

Different Drift Detectors Detect Different Drift Points. In addition to DDM, we also tested the EDDM [5] and ADWIN [9] drift detectors in multiple settings. We consider the policies of (i) never resetting the detector in the entire stream; (ii) resetting upon detections; and (iii) resetting at every epoch. Figure 8 illustrates how each detection algorithms detect a different number of points and at different epochs according to their internal working and parameters. We notice that although the stream is the same, the first time each detectors identify the drift occurrence is different and depends on a different policy. This characteristic of current drift detectors makes it hard to compare them and to explain their detections. Although we tested all settings in all experiments, in the rest of this paper, we only show results for the detector that presented the best results.

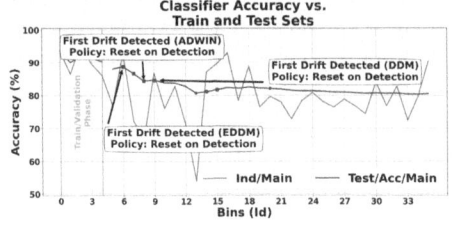

Fig. 8. Comparing algorithms and policies. Each one detects a different number of drift points/events and at different times.

Fig. 9. Separating detection rates per class reveals that the drift in the MW class causes the global performance degradation.

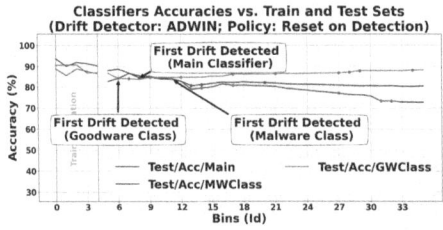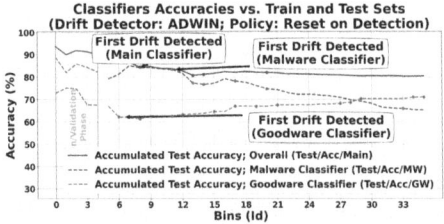

Fig. 10. Main class vs. sub-classes. A different number of drift points is identified in each class and at different epochs.

Fig. 11. Drift in the classes self-recognition rates. Drifts are represented both for the MW and GW meta-classifiers.

Class-Aware Detectors Provide an Initial Explanation Level. The previously presented results with Type 1 drift detectors do not allow identifying which class caused the drift identification. By adopting a Type 2 detector, we can easily identify it. Figure 9 shows the overall and per-class accuracies in the tested DREBIN settings. It is clear that the classifier detection rate is the average of two cases: **(i)** higher goodware and **(ii)** lower malware detection rates. It shows both that: **(i)** goodware is easier to classify than malware; and **(ii)** the drift in the malware is responsible for the classifier performance degradation.

Sub-classes Drift Differently than the Main Classifier. We proposed not only to measure the accuracy of individual classes but actually detect drift in the subclasses. Figure 10 shows that each class presents a different number of drift points and that the drifts in the main classifiers are not simply the sum of the drifts in the classes. It also shows that the first drift point for each sub-class is identified at different moments. This happens when the same classifier is applied to all classes, which reinforces that each class has its own dynamics that should be individually monitored and understood.

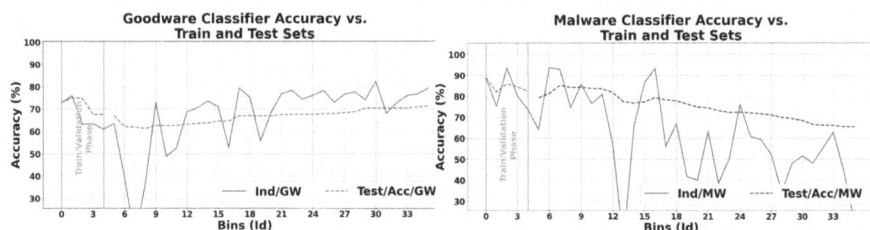

Fig. 13. MW class self-recognition rate.

Fig. 12. GW class self-recognition rate.

Concepts are really different than frontiers. In the previous experiments, we reached the limit of the Type 2 drift detector. It explained the classes that caused drift and the moments each one drifted, but it still does not explain the reason. To achieve that, we introduced the Type 3 detector with the concept of the classes. The remaining question was if the 2-nd layer classifier would learn something different than the main classifier. Figures 12 and 13 shows the accuracy for the self-recognition of the goodware and malware classes, respectively. It is possible to see both that: **(i)** the dynamics of these classifiers are different from the individual classes of the main classifier (Fig. 9); and **(ii)** each one of the classes has its own dynamics. The observed results are compatible with the overall results that the malware class is causing drift because the classifier is losing its ability to recognize malware over time.

A Drift in the Concept is Really Different than a Drift in the Frontier. A similar concern was if the concepts would drift and if these drift points would provide additional information than the drift points in the main classes. Figure 11 shows the drift points for the subclasses in comparison to the drift in the main classifier. It is possible to see that, once again, the drift dynamics for each concept/class are different among themselves and from the main classifier. The first time each concept drifts is noticeably different. Moreover, compared with the drift points for the subclasses of the main classifier (Fig. 10), the drift points are significantly different, which confirms the hypothesis that the second-layer classifier provides a different type of information about the drift events, which enables explaining them.

A Type-3 Detector Enables Explaining the Entire Classifier Operation and All Drift Events. Our key proposition is that looking at all types of drift detectors, we could explain the drift events, which was achieved in practice. Figure 14 shows the explanation of all points based on the detector's information combined as in Table 1, except for normal operation points, when nothing is plotted. It explains that the initial detection drop is due to the misplaced frontier and not because of actual concept change. This happens because the DREBIN dataset has an initial distribution very imbalanced towards goodware, and our artificial undersampling approach introduced a limited learning of the decision boundary. In turn, it further points out that later drift points are caused by real concept changes, which is indicated by the drift in the concept class. In all cases, the drift point in the main classifier is explained by a drift point also in another curve. No impossible case was observed, thus confirming the approach's correctness.

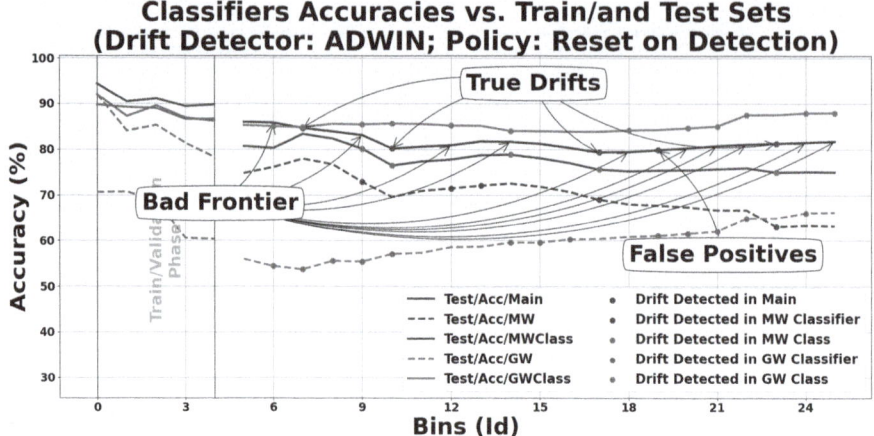

Fig. 14. Explaining all operational points and all drift occurrences. Omitting points of normal operation.

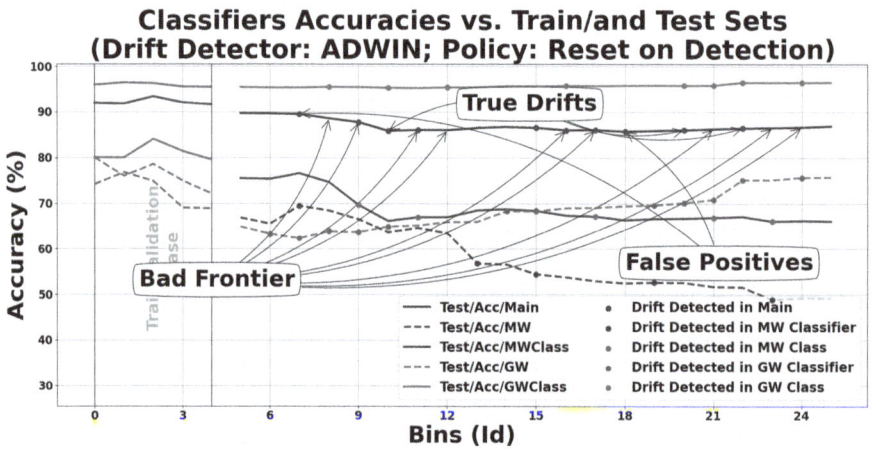

Fig. 15. Explaining all drift detection points (2:1 balance). We observe fewer frontier problems and more False Positives.

Type-3 Drift Detectors Reveal the Impact of Imbalanced Datasets in the Drift Detection. We attributed the previously identified frontier problems to the limited learning ability caused by the undersampling procedure. If our theory is correct, increasing the learning ability should result in fewer frontier problems being detected by our solution. We put this hypothesis to the test by increasing the proportion of goodware to malware (dataset balance) in an experiment.

Figure 15 shows the results for the 2:1 dataset imbalance. Our solution explained all drift and non-drift points, as expected. It also reported fewer frontier problems in this scenario, thus confirming our hypothesis that Type-3 drift detectors allow the identification of not only real drift cases but also frontier problems.

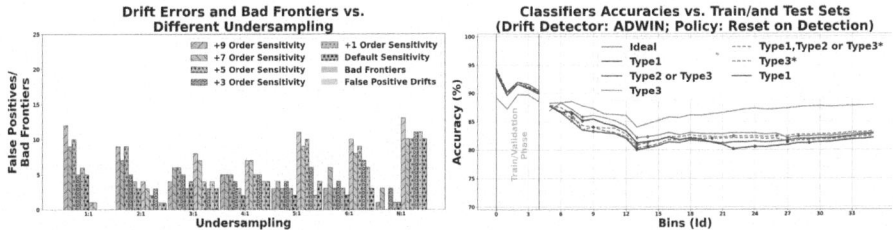

Fig. 16. Drift Detector Calibration. False Positives and bad frontiers are explained by the proposed approach.

Fig. 17. Early retraining on concept changes leads to improved accuracy than retraining only upon main class drift.

Type-3 drift Detectors Reveal True False Positives in Type-1 Drift Detectors. A phenomenon first observed in the previous experiment, when increasing the dataset imbalance, is that a drawback of having a more defined frontier causes False Positives (FP) to appear. In fact, we further discovered that the more we increase the imbalance– from fully balanced (1:1) to the natural imbalanced (N:1)–the more FPs are observed, as reported in Fig. 16.

We claim that our Type-3 solution can identify FPs in the Type-1 drift detectors. If our theory is correct and the reported points are true false positives, calibrating the drift detectors' sensitivity should decrease the FPs without affecting the frontier reports. In turn, if our theory is wrong and the reported FPs are true drift points, decreasing the drift detector sensibility should lead to more frontier problems as the samples evolve. We put our theory to the test by conducting experiments with different dataset imbalances. Figure 16 summarizes the results for the number of frontier problems and reported FPs. They confirm our hypothesis that our Type-3 solution reported true FPs. As we calibrated the drift detector sensitivity in different orders of magnitude, fewer FPs were reported. Moreover, the frontier problems decreased the more data the model had to learn.

Explainability Allows Early Retrain. The previously presented results are based on the continuous run of the initial model. In most pipelines, however, the model is retrained upon the drift alarms. Our main goal in this work was not to make drift detectors faster or more precise but more explainable. However, if you have a better situational awareness–i.e., you know what is happening, thus when concepts are changing–you have a better trigger for retraining procedures. Therefore, we conclude that explainability and precise early retraining are <u>the two faces of the same coin</u>.

This fact is clearly illustrated in Fig. 17, which shows the achieved accuracy when retraining models upon the trigger of the best and worst combinations of drift alarms (only Type 1, 2, or 3, or all combinations of Type 1, 2, and 3). The results are compared to the ideal case where the model was trained with the entire dataset, and no drift exists by definition/calibration. Retraining has a positive effect in all cases compared to not retraining. However, the best results are achieved when retraining upon Type 3 drift detectors and not when retraining upon a drift in the main classifier. This confirms the hypothesis that Type 3 detectors can identify concept change before the samples cross the decision boundary (Type 1 and 2), when it is already harming the detection results.

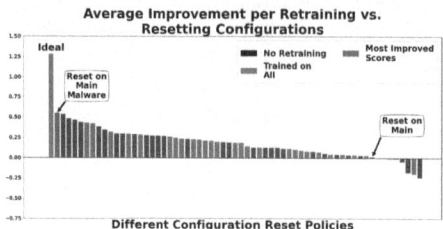
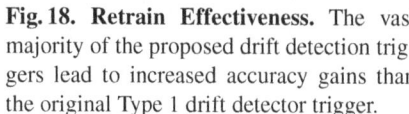
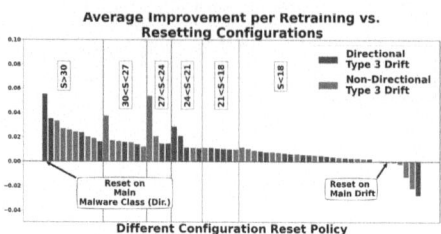

Fig. 18. Retrain Effectiveness. The vast majority of the proposed drift detection triggers lead to increased accuracy gains than the original Type 1 drift detector trigger.

Fig. 19. Retrain Efficiency. The best cost-benefit between the amount of retrains and accuracy increase is achieved by identifying concept changes.

Figure 18 presents the distribution of accuracy gains for all combinations of retrain triggers. It is noticeable that retraining upon a drift identification in the vast majority of the proposed Type 2 and Type 3 detectors leads to increased gains than only retraining upon a Type 1 drift detection event.

Explainability Makes Retraining More Efficient. Retraining procedures should be efficient in addition to effective, i.e., they should not only increase the overall detection rate but do it without spending excessive resources. A procedure that retrains the classifier every epoch is effective but not resource-efficient. We evaluated the efficiency of all retraining policies (the set of drift detectors that trigger a retrain) via their cost-benefit, i.e., how much they increase the total accuracy compared to the number of triggered retrains. Figure 18 demonstrated that the proposed directional approach was the most effective, but would it also be the most efficient?

Figure 19 shows the cost-benefit distribution for the multiple training strategies, i.e., their average accuracy increase per number of retrains. The results are broken down by their 10% increase intervals. This is required because, whereas a lower number of retrains is on average very efficient, it does not allow scaling the accuracy at higher levels. In turn, to raise the accuracy to cover all corner conditions, a series of additional retrains steps are required, which naturally decreases the efficiency. This is known as the 80:20 distribution or Pareto problem–when most of the gains come initially and a series of extra steps are required for the additional minor gains. This phenomenon is observed by the fact that the most efficient strategy is a non-directional drift strategy, but it did not achieve the highest increase levels. In turn, our proposed direction-aware retraining strategy was not only the most effective one but also the most efficient. Overall, directional strategies were the most effective for 3 out of the 5 ranges of accuracy gains.

3.2 Results Generalization

After showcasing our solution's potential via the above examples, we extended our analysis to evaluate how it generalizes. To do so, we repeated the previous experiments in a wide range of settings (e.g., different drift detectors, drift detection policies, and dataset imbalances) and assessed the outcomes' coherence. In total, we performed

5000 different runs of our solution over the DREBIN dataset. No run produced an "impossible" case, thus reinforcing our claims on the approach's correctness. Overall, the obtained large-scale results corroborated the previous experiments' results, which allows to explain the different effects datasets are subject to.

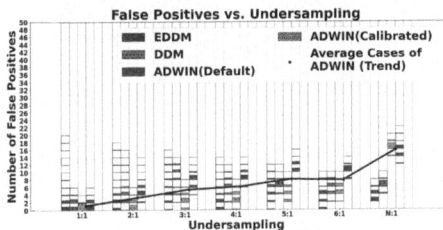

Fig. 20. DREBIN: FP Results Distribution for different imbalances. FPs grow with the imbalance for most detectors.

Fig. 21. ANDROZOO: FP Results Distribution for different imbalances. FPs grow with the imbalance for most detectors.

False Positives Identification at Scale. Figure 20 shows the FP distribution of different detector settings (e.g., detection policies) for different imbalances. For each possible FP value (y-axis), the stacked bars show the proportion of experiments that presented (hatched) or not (empty) that FP rate. The bars are stacked, covering all possible FP rates. We filled the bar representing the setting with the highest proportion to highlight patterns. In the case of ties, all settings were highlighted. We broke down the results by detectors.

The most prevalent results in each column shows that in general the more imbalanced the dataset, the more FPs happen, as previously showcased for the individual case. This happens for all detectors, but it is more pronounced for the default configuration of ADWIN (4th column). ADWIN calibration mitigates the problem (column 3) for most configurations (remember that each bar represents hundreds of runs). Retraining on drift detection (last column) also significantly mitigates the problem, but via a different mechanism: not because it is calibrated, but because of its continuous adjustments.

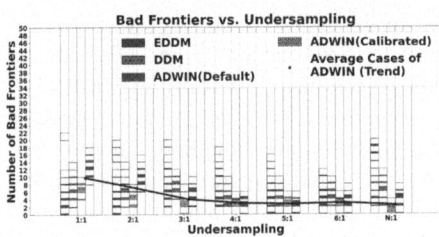 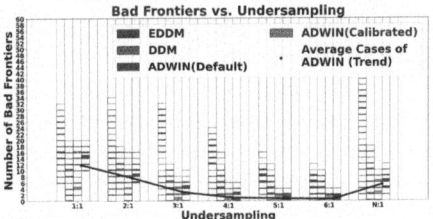

Fig. 22. DREBIN: Detectors' Results Distribution. Frontier problems for different undersamplings. The bigger the undersampling, the more frontier problems.

Fig. 23. ANDROZOO: Detectors' Results Distribution. Frontier problems for different undersamplings. The bigger the undersampling, the more frontier problems.

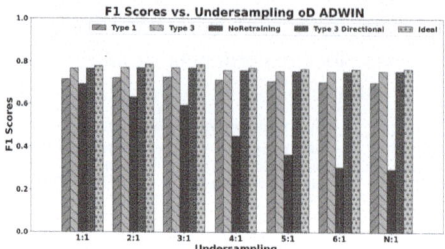

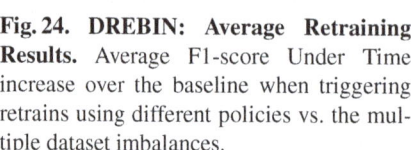

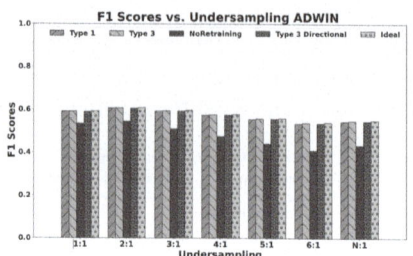

Fig. 24. DREBIN: Average Retraining Results. Average F1-score Under Time increase over the baseline when triggering retrains using different policies vs. the multiple dataset imbalances.

Fig. 25. ANDROZOO: Average Retraining Results. The imbalance effect is less pronounced in this dataset, but the Type-3 retraining strategy is still the superior one in all scenarios.

Bad Frontiers Identification at Scale. Figure 22 shows the distribution of bad frontier points reported by our solution in different dataset imbalances. In the overall case, as well as for the particular case previously demonstrated, the more undersampled the dataset is from the original distribution, the more frontier problems appear due to the limited data available to learn from. This effect is observed for all drift detection settings, and it is mitigated by calibration and/or retraining. This result shows not only that the effect is real but also that our solution can explain it.

Retraining. Once we have demonstrated how drift events are explained in the general case by our proposed solution, we shift our attention to remonstrating how an explainable drift detection also favors more effective and efficient retraining processes. We computed the retraining statistics for the thousands of runs previously presented and analyzed the behavior of different drift detection strategies.

Figure 24 shows the average F1-score for the multiple detectors configurations in the tests with multiple, different dataset imbalances. It is clear that performing no retraining (i.e., having no drift detection) is the worst case of all. The greater the dataset imbalance, the more noticeable the impact becomes as more samples are considered, thus accumulating the effect of wrong predictions. In this sense, using even a traditional (Type 1) drift detector is a good mitigation strategy. Class-aware (Type 2) drift detectors present gains over Type 1 because detecting when one class is drifting instead of waiting for the impact in the two classes at the same time allows early retraining, so the classifier spends less time making wrong predictions. In this same line, our proposed explainable drift detector (Type 3) presents even greater performance gains, as it allows early retraining when the concepts change, which reduces even more the window in which the detector makes wrong predictions. For all imbalances, our solution was the one that got closer to the ideal scenario–where retraining was not needed because it was trained with the complete dataset, thus not presenting drift by default.

Androzoo Generalization. While the imbalances and size of the DREBIN dataset highlight multiple positive aspects of our proposed drift explanation, they could also potentially hide negative aspects. Therefore, we repeated the previous experiments with Androzoo, a much larger and more balanced dataset, to demonstrate that our solution also operates properly under these different biases. As for the DREBIN dataset, our proposed solution explained all event points for the Androzoo one, and no invalid point was observed. More than that, the trends observed for the DREBIN dataset were also observed in Androzoo when different imbalances were enforced. Figure 21 shows the average number of FPs identified in the multiple Androzoo experiments for different dataset imbalances when considering ADWIN, the best detector in our tests (This time we omitted the results for the other detectors to not pollute the plot with redundant trends–they follow the same pattern as for DREBIN). As for DREBIN, the number of FPs increases with the imbalance. Also, similarly to DREBIN, the number of bad frontiers decreases with the imbalance, as shown in Fig. 23. Finally, the overall performance (F1 under time) also increases for AndroZoo (Fig. 25), as for DREBIN.

Runtime Performance Overhead Analysis. We conclude our evaluation with the analysis of the runtime performance overhead introduced by our solution. As a trade-off for making our solution agnostic to the main classifier, we made our solution bigger: as it requires 2 additional models, it is expected the total performance requirement to also increase by 2 times. We consider this an acceptable trade-off for one aiming to explain drifts. We measured the actual solution overhead via the total experiment runtime (training, predictions, and retraining) in our machines. While the baseline values might change from machine to machine, the relative overhead should be generalized consistently.

Table 2. Runtime Performance Overhead for DREBIN and AndroZoo. The cost of individual retrains is reduced, but the total execution time cost increases.

DREBIN				AndroZoo	
Models (#)	Retrain Policy	Total Time / Retrains (#)	Cost / Overhead / Normalized	Total Time / Retrains (#)	Cost / Overhead / Normalized
1	Type 1	11.65 s / 5	2.73 s / 0x / 0x	470.4 s / 5	94 s / 0x / 0x
3	Type 1	53.53 s / 5	10.7 s / 3.92x / 3.92x	1688.6 / 5	337.7 s / 3.6x / 3.6x
3	Type 2	105.77 s / 13	8.13 s / 7.74x / 2.98x	3563.9 s / 15	237.6 s / 7.5x / 2.5x
3	Type 3	102.70 s / 13	7.90 s / 7.52x / 2.89x	4161.3 s / 19	219 s / 8.84x / 2.32x

Table 2 shows the total execution time and overhead values normalized by the number of retrains for both datasets (different problem sizes). The first to second rows show the increase in the metrics scores caused by adding the external monitor to the system, i.e., by tripling the number of classifiers/models, but without actually using their information for classification. Although this scenario is not realistic in practice, it works as ground truth for isolating variables and measuring the individual impact of the architecture itself. The third and fourth rows exemplify cases that actually consider the results of the added classifiers, as used in practice, which allows measuring the impact of policies.

The results for the two datasets follow the same trend: The total execution time significantly increases when new the architecture is added (first to second line). It more than triples the total execution time, as not only the prediction and retraining time are tripled but there are also additional costs to compute drift distances and directions. The major overhead, however, comes from the additional number of retrains triggered by the new policies, which are required to increase the long-term detection accuracy. This overhead is non-linear, but exponential, because each new retrain includes all data from previous epochs, which makes each new retrain slower. Despite this effect, the normalize time per retrain remains constant (and lower than the triple from the original architecture addition). In fact, the more efficient the policy, the smaller the cost of an individual retrain. For instance, Type 3 triggers retrains in the most appropriate moments, such that it can achieve the same accuracy as Type 2, with the same number of retrains, but by training in a much earlier scenario, by predicting the drift occurrence, thus having to process less data. In sum, our results allow us to conclude that adding meta-classifiers to explain drift detector significantly increases the absolute time taken to process the samples, but it makes the individual retrains much more efficient.

4 Discussion

Wider Implications. Whereas exemplified in the malware detection domain, where we have domain expertise, the hereby presented findings apply to any domain that can be modeled as a binary classification problem under the same constraints. Thus, we expect our proposed architecture to explain drift occurrences in varied tasks.

Feedback for Operators. We expect operators of malware detection pipelines (e.g., MLSecOps) to benefit from our solution not only to understand the drift points but also to understand the pipeline operation at any point. Our solution's ability to identify improper frontiers might assist operators in identifying when the classifier was trained with a limited amount of data or a non-representative dataset and is not operating well in practice, which is common during the initial phases of malware detection pipeline deployments.

Limitations. We propose a practical solution for explaining drift points. We do not provide mathematical guarantees of the optimality of the retraining results. Future works should provide it via additional math formulations on top of our developments.

Future Work. This work is a first step towards explaining drift detection points, but our work is not exhaustive. Further developments should expand our contribution from the binary classification domain to the multi-class domain. It is also key to make analysis more fine-grained and robust. As cases of particular research interest to applying this approach in real scenarios, we point out the following open research problems and future research directions:

- **Label Delays.** In real deployments, the true labels used for drift detection are not immediately available but arrive delayed, which causes a delay in drift detection. Whereas here we assumed an ideal scenario, as most of the literature, it is key to develop strategies to handle drift delays to foster the solution adoption. It is key to be mindful that, in practice, the explanations provided by our solution would be delayed by the same amount of time as the label arrival delay.

- **Virtual Drifts.** In practical scenarios, it is common that the ground truth labels used for drift detection might flip over time from one class to another. These flips might cause improper drift detection (virtual drift points). Whereas this work, as most in the literature, assumed an ideal scenario of no label flips, the case of label flips should be addressed in real-world deployments. For such, strategies for identifying false drift reports should be developed.
- **Intra-Class Drifts.** The Sample's concept drifts not only across classes but also inside the same class as they evolve. A single concept approach, as most in the literature, is blind to such modifications. To be able to spot such cases, we should not only learn a single malware concept but also split the learned malware concept into multiple ones and temporally evaluate their drift dynamics.

Making Retraining as Practical as Explanation. Our approach makes the explanation of drift events practical. We consider that requiring additional processing for this task is acceptable since explanation and validation tasks are usually performed in controlled environments without processing constraints. In turn, whereas we demonstrated that it would foster early retraining, further studies are warranted to obtain the best cost-benefit for its deployment in practice. A lighter version of the approach can be developed, such as considering only the concept's centroid as the average confidence.

Expansion for Heterogeneous Architectures. Our experiments considered the first and second-layer classifiers as having the same classifier. This is an experimental choice to isolate variables and allows us to claim that any observed effect is due to the concepts, not due to the architecture. However, it is possible to hypothesize future deployments with heterogeneous architectures, where the second-layer classifier is even more powerful than the first layer to better learn the concept from limited data amounts.

The Data Coupling Invariant: Note that although the model can be decoupled in their architectures, they always should have a strong decoupling in their data, i.e., the first and second layer models should be trained with the same dataset and present the maximum performance on them. If the models are decoupled in data, the second layer model might report a drift point that does not exist in the primary model.

5 Related Works

Proposals for New Drift Detectors are the most common in the literature. They vary from eliminating the need for labels [30] to new models and architectures [2,24] for increased detection rates. However, these works do not focus their developments on explaining their drift results, which is our focus in this work.

Model Monitoring is the MLSecOps task this paper targets. Previous works proposed architectures to model a given model operation [7,8], but these works only approached the problem quantitatively, providing a final score of the system operation. In our qualitative approach, we used scores to explain the model operation at each moment.

Classifying Drift Events is a common contribution in the literature [21]. Previous works proposed characterizing drift events in multiple dimensions [28], such as the introduction of a new class, the recurrence of the drift events, the impact extent, and so on. Whereas characterizing drift in all its dimensions, the malware detection problem offers some advantages. such as its limitation to a binary classification problem, with no introduction of new classes, which allows for relaxing some constraints from the general case. In this case, we adopt a more practical classification [17] that considers real and virtual drift occurrences. We extend these concepts to drift in the malware detector class or frontier.

The limit of the existing explainable solutions is to only explain parts of the drift phenomenon, such as only local regions [18], only the wrong features [27], or only feature-frequency change [20]. We here present explanations that cover all these scenarios. In addition, we present tests with real and not synthetic data [22].

The Applications Benefiting from Drift Explanation described in the literature range from data mining [1] to COVID identification [15]. Whereas drift is a fundamental phenomenon in malware detection, few drift detection architectures are designed for the task, and the few existing ones [26] do not explain the phenomenon beyond feature change. Ideally, we would like to perform forensic procedures of malware detectors [13] but to explain the binary classification and not how new malware families appear.

Security-Focused Drift Detectors are limited in either providing information only about the features that changed without explaining the model changes [11] or by determining model changes without explaining the features used in the models [23]. Conformal evaluation [6], the closest idea to ours, also split detection in two (confidence and credibility), but unlike our proposal, these information are not used to explain the classifier situation at each moment, but only to early retrain the classifier. 2-level architectures are becoming more common over time. A recent proposal also split the problem into two [19] but adopted a mathematical approach based on the latent space conformity to evaluate drift points. Our proposal uses ML classifiers to model such deviations to explain the points in addition to triggering a retrain on them.

Boundary-based vs. Distance-based Explanations. Drift explanations are typically classified in these two classes, each one presenting pros and cons. Our approach fits into the first category as it aims to explain frontier-crossing events. An exemplary work in the second category is CADE [29], which uses contrastive learning to measure the difference between goodware and malware classes. We consider that boundary-based approaches are more interesting solutions to the scenario presented here as they allow explaining every drift point, including false positive drift reports, whereas distance-based approaches only report true concept changes.

Explaining Detection vs. Explaining Drifts. The explanation of ML-based malware detection models is a growing topic, but most of the existing solutions focus on explaining the detection model itself. Popular metrics for it include, for instance, the fidelity of the model [12], but these metrics do not explain the malware evolution. In this sense, the here proposed drift explanation is a step ahead to complement (and not replace) previous explanation initiatives.

6   Conclusion

This work investigated the problem of explaining concept drift detection occurrences in malware detection pipelines. We proposed the idea of splitting the concept drift phenomenon into (**1**) classifier's frontier changes and (**2**) concept changes to explain better the drift causes. We evaluated the viability of our proposal via experiments with the DREBIN and AndroZoo datasets. We discovered that our approach not only explains all drift events, identifying true and false drift reports, but also that the reliance on an explainable method increases the detector performance by allowing the retraining to occur in the most promising points (true drift events).

Acknowledgments. We thank the anonymous reviewers and shepherd for all the helpful insights. Marcus Botacin thanks NSF for the support via the CNS 2327427 grant.

Reproducibility.. All code developed in this search is available at https://github.com/Botacin-s-Lab/Concept.Drift.Explanation

References

1. Adams, J.N., van Zelst, S.J., Quack, L., Hausmann, K., van der Aalst, W., Rose, T.: A framework for explainable concept drift detection in process mining. In: Polyvyanyy, A., Wynn, M.T., Van Looy, A., Reichert, M. (eds.) BPM 2021. LNCS, vol. 12875, pp. 400–416. Springer, Cham (2021). https://doi.org/10.1007/978-3-030-85469-0_25
2. Andresini, G., Pendlebury, F., Pierazzi, F., Loglisci, C., Appice, A., Cavallaro, L.: Insomnia: towards concept-drift robustness in network intrusion detection. In: Proceedings of the 14th ACM Workshop on Artificial Intelligence and Security, pp. 111–122 (2021)
3. Arp, D., et al.: Dos and don'ts of machine learning in computer security. In: 31st USENIX Security Symposium (USENIX Security 22), pp. 3971–3988. USENIX Association, Boston, MA (2022). https://www.usenix.org/conference/usenixsecurity22/presentation/arp
4. Arp, D., Spreitzenbarth, M., Hubner, M., Gascon, H., Rieck, K.: Drebin: effective and explainable detection of android malware in your pocket. In: NDSS. The Internet Society (2014). http://dblp.uni-trier.de/db/conf/ndss/ndss2014.html#ArpSHGR14
5. Baena-Garcıa, M., del Campo-Ávila, J., Fidalgo, R., Bifet, A., Gavalda, R., Morales-Bueno, R.: Early drift detection method. In: Fourth International Workshop on Knowledge Discovery from Data Streams, vol. 6, pp. 77–86. Citeseer (2006)
6. Barbero, F., Pendlebury, F., Pierazzi, F., Cavallaro, L.: Transcending transcend: revisiting malware classification in the presence of concept drift. In: 2022 IEEE Symposium on Security and Privacy (SP), pp. 805–823 (2022). https://doi.org/10.1109/SP46214.2022.9833659
7. Bhaskhar, N., Rubin, D.L., Lee-Messer, C.: An explainable and actionable mistrust scoring framework for model monitoring. IEEE Trans. Artif. Intell. (2023)
8. Bhatt, U., et al.: Explainable machine learning in deployment. In: Proceedings of the 2020 Conference on Fairness, Accountability, and Transparency, pp. 648–657 (2020)
9. Bifet, A., Gavalda, R.: Learning from time-changing data with adaptive windowing. In: Proceedings of the 2007 SIAM International Conference on Data Mining, pp. 443–448. SIAM (2007)
10. Ceschin, F., et al.: Machine learning (in) security: a stream of problems. Digital Threats **5**(1) (2024). https://doi.org/10.1145/3617897
11. Ceschin, F., Botacin, M., Gomes, H.M., Pinagé, F., Oliveira, L.S., Grégio, A.: Fast & furious: on the modelling of malware detection as an evolving data stream. Expert Syst. Appl. **212**, 118590 (2023)
12. Chen, L., Yagemann, C., Downing, E.: To believe or not to believe: validating explanation fidelity for dynamic malware analysis. In: CVPR Workshops, pp. 48–52 (2019)

13. Chow, T., Kan, Z., Linhardt, L., Cavallaro, L., Arp, D., Pierazzi, F.: Drift forensics of malware classifiers. In: Proceedings of the 16th ACM Workshop on Artificial Intelligence and Security, pp. 197–207 (2023)
14. Daoudi, N., Allix, K., Bissyandé, T.F., Klein, J.: A deep dive inside Drebin: an explorative analysis beyond android malware detection scores. ACM Trans. Priv. Secur. **25**(2) (2022). https://doi.org/10.1145/3503463
15. Duckworth, C., et al.: Using explainable machine learning to characterise data drift and detect emergent health risks for emergency department admissions during COVID-19. Sci. Rep. **11**(1), 23017 (2021)
16. Gama, J., Medas, P., Castillo, G., Rodrigues, P.: Learning with drift detection. In: Bazzan, A., Labidi, S. (eds.) SBIA 2004. LNCS (LNAI), vol. 3171, pp. 286–295. Springer, Heidelberg (2004). https://doi.org/10.1007/978-3-540-28645-5_29
17. Gomes, H.M., Grzenda, M., Mello, R., Read, J., Le Nguyen, M.H., Bifet, A.: A survey on semi-supervised learning for delayed partially labelled data streams. ACM Comput. Surv. **55**(4) (2022). https://doi.org/10.1145/3523055
18. Haug, J., Braun, A., Zürn, S., Kasneci, G.: Change detection for local explainability in evolving data streams. In: Proceedings of the 31st ACM International Conference on Information & Knowledge Management, pp. 706–716 (2022)
19. He, Y., Lei, J., Qin, Z., Ren, K.: Going proactive and explanatory against malware concept drift (2024)
20. Hinder, F., Vaquet, V., Brinkrolf, J., Hammer, B.: Model-based explanations of concept drift. Neurocomputing **555**, 126640 (2023)
21. Hu, H., Kantardzic, M., Sethi, T.S.: No free lunch theorem for concept drift detection in streaming data classification: a review. Wiley Interdiscip. Rev. Data Min. Knowl. Discov. **10**(2), e1327 (2020)
22. Jacob, V., Song, F., Stiegler, A., Rad, B., Diao, Y., Tatbul, N.: Exathlon: a benchmark for explainable anomaly detection over time series. arXiv preprint arXiv:2010.05073 (2020)
23. Jordaney, R., et al.: Transcend: detecting concept drift in malware classification models. In: 26th USENIX Security Symposium (USENIX Security 17), pp. 625–642. USENIX Association, Vancouver, BC (2017). https://www.usenix.org/conference/usenixsecurity17/technical-sessions/presentation/jordaney
24. Panda, P., Kancheti, S.S., Balasubramanian, V.N., Sinha, G.: Interpretable model drift detection. In: Proceedings of the 7th Joint International Conference on Data Science & Management of Data (11th ACM IKDD CODS and 29th COMAD), pp. 1–9 (2024)
25. Pendlebury, F., Pierazzi, F., Jordaney, R., Kinder, J., Cavallaro, L.: TESSERACT: eliminating experimental bias in malware classification across space and time. In: 28th USENIX Security Symposium (USENIX Security 19), pp. 729–746. USENIX Association, Santa Clara, CA (2019). https://www.usenix.org/conference/usenixsecurity19/presentation/pendlebury
26. Shaer, I., Shami, A.: Thwarting cybersecurity attacks with explainable concept drift. arXiv preprint arXiv:2403.13023 (2024)
27. Vishnampet, R., Shenoy, R., Chen, J., Gupta, A.: Root causing prediction anomalies using explainable AI. arXiv preprint arXiv:2403.02439 (2024)
28. Webb, G.I., Hyde, R., Cao, H., Nguyen, H.L., Petitjean, F.: Characterizing concept drift. Data Min. Knowl. Disc. **30**(4), 964–994 (2016)
29. Yang, L., et al.: CADE: detecting and explaining concept drift samples for security applications. In: 30th USENIX Security Symposium (USENIX Security 21), pp. 2327–2344. USENIX Association (2021). https://www.usenix.org/conference/usenixsecurity21/presentation/yang-limin
30. Zheng, S., et al.: Labelless concept drift detection and explanation. In: NeurIPS 2019 Workshop on Robust AI in Financial Services: Data, Fairness, Explainability, Trustworthiness, and Privacy (2019)

InferONNX: Practical and Privacy-Preserving Machine Learning Inference Using Trusted Execution Environments

Konstantina Papafragkaki[1,2]([✉]) and Giorgos Vasiliadis[1,3]

[1] Institute of Computer Science, Foundation for Research and Technology - Hellas, Heraklion, Greece
{papafrkon,gvasil}@ics.forth.gr
[2] Computer Science Department, University of Crete, Heraklion, Greece
[3] Department of Management Science and Technology, Hellenic Mediterranean University, Agios Nikolaos, Greece

Abstract. Machine learning is increasingly applied in critical domains where sensitive data is involved. When models are deployed on untrusted devices, this raises significant privacy concerns for both model providers and end-users. Trusted Execution Environments (TEEs), which offer hardware-based protection for data during processing, can mitigate these concerns. However, their limited memory resources pose challenges for deploying traditional machine learning frameworks.

In this paper, we propose InferONNX, a lightweight machine learning inference service designed to run within Intel SGX. It embeds a high-level, portable, and framework-agnostic model format into the enclave, enabling easy execution of a wide range of machine learning and deep learning models. To address the memory limitations of Intel SGX, InferONNX employs two key strategies: a compact runtime with a small memory footprint, and model partitioning to reduce the memory required during inference. By executing model partitions instead of the full model, the system achieves $1.5\times$ to $4\times$ faster inference depending on the model size.

Keywords: Confidential computing · Trusted execution · Intel SGX · Machine learning · Inference

1 Introduction

Machine learning and deep learning models have made significant progress and are being used extensively across a wide range of application domains and business sectors, such as image classification [43], speech recognition [31], natural language processing [22], and healthcare diagnostics [37,42]. In terms of privacy though, machine learning and deep learning inference present challenges for both model providers and end-users [29]. Execution on untrusted end-user devices

M. Egele et al. (Eds.): DIMVA 2025, LNCS 15748, pp. 25–43, 2025.
https://doi.org/10.1007/978-3-031-97623-0_2

requires maintaining the confidentiality of proprietary models. In principle, a model can be a considerable investment and a valuable asset for a company or organization. Such models can be leaked in untrusted host scenarios, as end-users have full access to the hardware and the software installed on their devices. On the contrary, the execution on the model provider premises, via cloud-enabled services, raises serious privacy concerns for the end-users, as they are required in many cases to outsource private or sensitive data (such as images, voices, and text). Such data can be exposed to external threats due to vulnerabilities in these environments.

Overall, it is imperative to provide mechanisms that satisfy the confidentiality of both users' data and models, so as to enable privacy-preserving inference on untrusted hosts. A practical approach to tackle these needs is to utilize hardware-enabled trusted execution environments (TEE). These environments can be used to protect data while in use, a concept known as confidential computing. A TEE provided by hardware isolates programs or program fragments, and their data from potentially malicious operating systems, hypervisors, or any other privileged process. As a result, TEEs are becoming increasingly widespread in the machine learning domain [14,23,26], and provide a compelling paradigm for machine learning tasks, including inference. However, deploying complex machine learning workloads on TEEs has to address the limited memory resources of TEEs. This is a major constraint for machine learning workloads, as not only the models, but also the frameworks are quite large. For instance, PyTorch v3.9.11 is about 760 MB and TensorFlow v2.13.1 is 1.3 GB. These memory and security constraints not only affect runtime performance, but also introduce overheads during model initialization. As shown in Fig. 1, inference tasks pay increased overheads when models are loaded from disk (*cold-start*). The most straightforward way to mitigate cold-start latency is to keep models permanently in memory, a strategy adopted by the majority of previous work [24,27,38]. Unfortunately, this is not practical in typical scenarios where the size of the models exceeds the system memory capacity. Therefore, it is crucial to design mechanisms for efficiently loading models from disk.

This paper presents InferONNX, a lightweight machine learning inference service designed to run within Intel SGX. InferONNX's key insight is to embed the runtime environment of a high-level machine learning format, namely ONNX (Open Neural Network Exchange) [6], into Intel SGX. ONNX is chosen due to its high-level semantics, its wide popularity, and its small memory footprint: the resulting inference engine has a size of 46MB, consuming minimal enclave memory and thus leaving more memory resources available for the machine learning and deep learning models running atop InferONNX. To manage the limited memory resources of the machine, we further partition those models and store them on disk rather than keeping them in memory; the partitions are loaded from disk at runtime and executed sequentially.

The main contributions of this paper are:

– We design and implement InferONNX, a lightweight service for Intel SGX, that enables confidential inference using the ONNX language. InferONNX

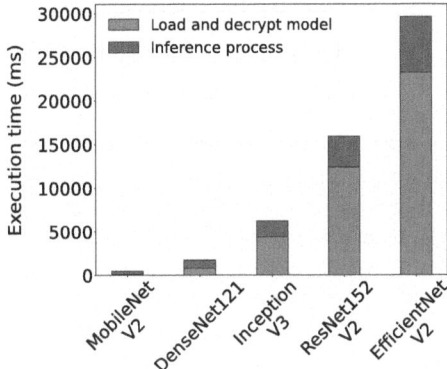

Fig. 1. Execution time breakdown for five popular machine learning models that span different sizes. The blue portion represents the time spent loading and decrypting the models from disk (cold start), while the red portion represents the execution of the inference process that produces the result (warm state). (Color figure online)

relies on disk-based storage for machine learning models to ensure scalability and manage memory constraints effectively through model partitioning (1.5×-4× reduced overhead compared to executing the full model).
– We show that InferONNX can handle machine learning models of varying sizes, from the lightweight SqueezeNet1.0 to the larger EfficientNet V2, with a maximum overhead of 3.65× compared to their unprotected counterparts.

2 Background

This section provides background on Intel SGX and the use of Library Operating Systems (libOSes) to support secure application deployment within Intel SGX enclaves.

2.1 Intel SGX

Intel SGX (Software Guard Extensions) introduces secure enclaves to protect the confidentiality and integrity of sensitive data and applications. These secure enclaves provide isolated regions within the CPU, ensuring that sensitive information is processed in isolation, even if the operating system, hypervisor or other parts of the system are compromised. Code and data within an enclave are stored in a protected region of memory called the Enclave Page Cache (EPC).

Since the EPC has limited capacity, Intel SGX includes a paging mechanism to manage applications requiring more memory. This mechanism encrypts EPC pages and transfers them to an untrusted DRAM buffer, maintaining security during this process. However, these operations involve encryption, decryption, and privileged instructions executed outside the enclave, resulting in performance overhead. Moreover, when EPC pages are reused, the Translation Lookaside Buffer (TLB) is cleared, introducing further delays.

The creation of an enclave begins with the ECREATE instruction, which initializes its control structure within the EPC. Memory setup and cryptographic measurements for remote attestation are managed by subsequent instructions like EADD and EINIT. Once prepared, the enclave code is executed through the EENTER instruction, which switches the processor to enclave mode. SGX also supports multi-threading within enclaves, with each thread's context maintained in a Thread Control Structure (TCS). The EEXIT instruction is used to terminate enclave execution and return control to the untrusted application.

2.2   Library OSes for Intel SGX

Deploying only part of an application in a Trusted Execution Environment (TEE) often requires manual code partitioning, along with recompilation and relinking of the entire application—even for components that remain outside the TEE. This static and tightly coupled development model limits support for applications that depend on runtime code extensibility. Moreover, additional complexities, such as handling cryptographic operations and enabling end-to-end encryption, further complicate the development process.

Library Operating Systems (libOSes) help address the challenges of deploying applications in TEEs by offering two key advantages: reducing the size of the Trusted Computing Base (TCB) and simplifying the development workflow. A smaller TCB improves overall system security by limiting the amount of trusted code, thereby reducing the potential for vulnerabilities. LibOSes achieve this by including only minimal, essential components—such as a shim C library—within the enclave, while leaving larger system libraries and runtimes outside the trusted boundary. At the same time, libOSes streamline secure application development by enabling the entire application stack—including code, libraries, and system functions—to run inside Intel SGX with minimal or no code changes. They transparently manage interactions with untrusted system components, such as I/O operations, making it easier to adapt existing applications to run securely within a TEE.

3   Design Objectives

The widespread adoption of machine learning and deep learning models has led to a significant expansion of software stacks designed to improve efficiency. Typically, frameworks, such as TensorFlow and PyTorch, are used for training and optimizing machine learning models in an iterative process. Once a machine learning model has been trained, it can be deployed for inference. However, inference with these frameworks brings unnecessary overheads mainly due to the bloated size of the target application. To overcome this, applications can manually use optimized operators that utilize carefully-designed assembly instructions [1–3], or directly embed the model's architecture and parameters into the source code. By doing so, the target application is lightweight and efficient, as it does not rely on any external library. The procedure can be performed either manually, by porting the models on hand, or automatically, by using domain-specific deep learning compilers, such as TVM [12]. Both approaches offer quite improved performance; however, manual porting requires substantial effort and is time-consuming. Compiler-based approaches, such as TVM, are also scalable in terms of model size and hardware heterogeneity, by producing appropriate executables that are also optimized for the target hardware. However, these benefits come at a cost when executing in an environment where the model is not fully trusted. In such scenarios, the compiled code operates outside the user's direct control, meaning the binary could potentially be tampered with, creating an opportunity for malicious code injection. This, in turn, could allow attackers to steal sensitive data, either directly or indirectly, for example through covert channels; even if the code is isolated within a secure enclave. One way to prevent this is to confine the untrusted compiled code into a trusted sandbox, or have a trusted inference engine that executes machine learning models on a unified format, such as the ONNX. Using the latter, the inference engine can execute any machine learning model, without requiring any instrumentation or code validation/verification. In addition, embedding an ONNX runtime inside a TEE tackles the challenges of transparency, dynamic extensibility, and run-time safety—by piggy-backing on the characteristics of a high-level open source format for machine learning and deep learning models, such as ONNX.

Our approach offers significant developer economy compared to low-level abstractions, because of the productivity benefits stemming from a high-level widely-adopted format. Applications can leverage the ONNX model ecosystem transparently, without having to develop them from scratch—such as image analysis, object detection, and natural language processing. Finally, since ONNX is a high-level, declarative format for machine learning and deep learning models, its execution is constrained, minimizing the risk of unintended behavior and helping protect data owners from potential leaks by untrusted model providers.

4   Design

This section outlines the design of InferONNX. We describe its client-server architecture, the use of enclaves for secure ONNX model execution, and the partitioning strategy used to manage models that exceed the memory limits of Intel SGX.

4.1   Client-Server Architecture

We design an inference server that enables an end-to-end secure environment between clients and model providers, so as to protect the sensitive data of clients and the intellectual property of model providers. As shown in Fig. 2, the server operates within hardware-based secure enclaves, isolating the execution environment and protecting sensitive data during processing. This data is encrypted and shielded from unauthorized access at the hardware level, ensuring that information within these enclaves remains secure.

To establish secure connections between the server and clients, we use the Transport Layer Security (TLS) protocol, which ensures encrypted communication, protecting the confidentiality and integrity of data exchanged over the network. Along with this, we implement access control by authenticating each client, ensuring that only clients with valid credentials can interact with the server. For clients within Intel SGX environments, Remote Attestation (RA) can be used to verify their authenticity. This process ensures that the client is genuine before any sensitive data is exchanged, establishing trust between the parties, and can also be optimized to reduce latency and enhance performance [10]. In contrast, clients running on conventional CPUs (without Intel SGX support) are assigned a unique private key, enabling the establishment of a mutually authenticated TLS session.

The client-server interaction consists of two primary modes:

- **Model uploading.** In this mode, the server receives models, in ONNX format, securely from model providers. The models are encrypted using the AES-256-GCM mode and stored on disk. Each model is associated with a unique identifier (ID), through a hash table that also contains its cryptographic metadata (encryption key and initialization vector). The clients can use these IDs to submit inference requests on the corresponding models.
- **Inference serving.** In this mode, clients submit inference requests that include the model's ID and the corresponding input data. The server uses the ID to locate the appropriate model, loads it from disk into the enclave, decrypts it, and then initiates the inference process. Inference is performed by the ONNX interpreter described in Sect. 4.2, and the resulting output is returned to the client.

4.2   ONNX Interpreter Enclaves

We deploy an interpreter within secure enclaves to execute machine learning and deep learning models in ONNX format. Unlike compiled execution, which requires pre-compilation of the models, the interpreter dynamically processes any model by reading its structure and executing the respective operations sequentially. This approach offers flexibility in handling a variety of model architectures, making it well-suited for environments where models may vary or need to be updated without recompilation. By executing models in a secure enclave, we ensure that sensitive data, such as inputs and model weights, remain protected throughout the execution process, leveraging the strong isolation capabilities of Intel SGX.

ONNX provides a universal format for machine learning and deep learning models, allowing models trained in different frameworks (such as TensorFlow, PyTorch, and Scikit-Learn) to be represented in a standardized form. ONNX represents models as computational graphs, where nodes correspond to operations (referred to as operators) and edges define the data flow between these

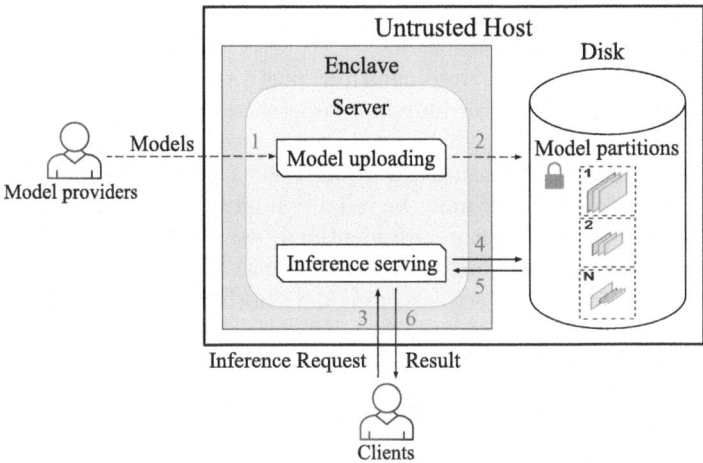

Fig. 2. Overview of InferONNX. The untrusted section (white) represents the untrusted device, while the trusted section (light blue) represents the enclave where the server operates. During *model uploading*, the model providers upload models to the server (step 1), which partitions, encrypts and stores them on disk (step 2). During *inference serving*, the clients submit inference requests to the server (step 3). The server performs inference by sequentially loading the relevant model partitions from disk into the trusted environment (steps 4-5), executing them in order, and then returning the final inference result to the client (step 6). (Color figure online)

operations. During execution, the ONNX interpreter reads the model's graph and iteratively processes each operator, applying the respective computations and data transformations. This sequential execution ensures that the operations are carried out in the correct order while maintaining the integrity of the model's intended functionality. The interpreter handles different types of operators by using pre-implemented functions.

4.3 Model Partitioning

As we experimentally verify in Sect. 6.4, executing entire models within the enclave incurs performance overhead, which we mitigate through model partitioning. Model partitioning divides the model architecture into smaller, sequential components that can be processed more efficiently within the memory constraints of hardware-based enclaves. Model partitioning can be classified into two categories: (i) *intra-operator partitioning*, where the computation and input data of a single operator are split into multiple segments—each processing a subset of the data and passing intermediate results between segments; and (ii) *inter-operator partitioning*, where groups of operators are treated as independent execution units. In this work, we focus on inter-operator partitioning. While intra-operator partitioning could be useful in scenarios where a single operator exceeds the enclave's EPC capacity, it may introduce additional overhead due to the need for intermediate data transfers. We leave the exploration of intra-operator partitioning to future work.

The models are divided into multiple partitions based on the EPC size and the computational cost of individual operators, similar to [17]. To identify memory-intensive (heavy-weight) operators, we profile each model on both Intel SGX and a standard CPU, using execution time as a proxy for memory usage—since memory consumption cannot be reliably inferred from the operator's input size alone. Once these operators are identified, we perform model partitioning. This process involves traversing the computational graph in reverse order—from the last operator to the first—to accommodate both simple and complex topologies. As we iterate through the operators, we accumulate their estimated sizes. If the combined size of the current and previous operators exceeds the EPC capacity, we define a partition from the starting operator to the previous one. The next partition starts at the current operator, and size accumulation resets. If an operator is marked as heavy-weight, it is treated as a standalone partition, and the accumulation restarts from zero.

This partitioning procedure ensures that all partitions meet the system's memory requirements. Once precomputed, model providers upload the partitions to the server, where they are used for inference. During *inference serving*, clients submit requests to run inference using these sub-models. After loading all of them from disk within the enclave, the partitions are executed in a pipeline, with the output of one partition passed as input to the subsequent partition. The result of the final partition corresponds to the inference result of the entire model, which is then returned to the clients.

5   Implementation

This section outlines the implementation of InferONNX, focusing on the base runtime built with Occlum, the TLS-based secure communication layer, and the trusted inference engine that executes models within Intel SGX enclaves.

5.1   Base Runtime

To simplify the development of InferONNX, we adopt Occlum [39], a libOS tailored for Intel SGX. Occlum enables secure execution of application processes entirely within the enclave. The Occlum runtime uses a packaged file system, known as the Occlum image, to configure and manage the enclave environment. This image includes the InferONNX binaries, configuration files, and required libraries, totaling approximately 59MB. At runtime, it is used to initialize the enclave and securely execute the InferONNX server alongside the deployed models.

In its current design, InferONNX runs a single Occlum image at a time, avoiding concurrent sessions and multi-client access. This serialized execution model reduces the attack surface and helps mitigate side-channel threats, such as timing and cache attacks, that are known to affect Intel SGX [9,11,19,44].

5.2   Secure Communications Layer

We integrate the MbedTLS library [16], which provides full support for TLSv1.2 and v1.3 and is optimized for resource-constrained environments. Each client request is handled over a dedicated TLS session, ensuring end-to-end encryption between the client and the InferONNX server. Upon receiving a request, the server authenticates the client and establishes a secure channel before proceeding with either *model uploading* or *inference serving*, depending on the request type. This communication model ensures the security of data in transit, maintaining both confidentiality and integrity throughout the client-server interaction.

5.3   Trusted Inference Engine

InferONNX performs inference operations, using Tract [32], a Rust-based inference engine that supports various model formats, including ONNX. Tract leverages Rust's built-in memory-safety features, and its lightweight design, at approximately 46MB, makes it particularly well-suited for memory-constrained environments, such as Intel SGX.

6   Evaluation

In this section, we evaluate the performance of InferONNX, focusing on the overheads introduced by Intel SGX and the impact of model partitioning.

6.1   Experimental Setup

Platform. We evaluate InferONNX on a single machine with an Intel Core i7-7700 3.60 GHz CPU and 32GB RAM, running Ubuntu 20.04 and supporting Intel SGX v1.0 [13]. In Intel SGX v1.0, each enclave is limited to 128MB, reduced to around 100MB due to metadata and system-reserved memory, with static memory initialization for the enclave's lifetime. Intel SGX v2.0, by contrast, supports scalable memory through dynamic allocation and thread creation, but this flexibility comes with a relaxed threat model, offering full protection against cyberattacks but only partial protection from physical attacks. While Intel SGX v2.0 could provide better performance and scalability in certain cases, the inference process in our system does not require frequent memory allocations, making Intel SGX v1.0's static memory model sufficient for our needs. Further exploration is needed to assess if an Intel SGX v2.0-based solution would offer meaningful performance gains and whether those gains would justify the security trade-offs.

Workload. We use models from image classification and general-purpose tasks, including SqueezeNet1.0, MobileNet V2, DenseNet121, EfficientNet Lite4, Inception V3, and ResNet101/152 V2. Most of these models were downloaded from the ONNX Model Zoo [4], while the ResNet models were obtained from a GitHub repository [7].

6.2 Memory Usage Profiling

We use the Valgrind Massif tool [5] to profile the memory usage of InferONNX. Massif tracks cumulative memory allocations throughout a process and captures real-time memory snapshots. The peak value among these snapshots reflects the highest memory usage observed during inference for a given model or partition. This peak is used to estimate the memory footprint of each operator during partitioning. In this analysis, we focus on 'heavy-weight' operators—those incurring a performance overhead of at least 12×—since partitioning offers little to no benefit for lighter operators.

Table 1. The sizes of various models (on disk) and the corresponding number of partitions needed for each of them to fit the EPC size.

Model	Disk Size (MB)	Partitions (#)
SqueezeNet1.0	4.8	3
MobileNet V2	13.5	6
DenseNet121	31.2	31
EfficientNet Lite4	49.5	36
Inception V3	90.9	23
ResNet101 V2	170	25
ResNet152 V2	230	34
EfficientNet V2	451	77

Table 1 presents details on the disk sizes of various models and the number of partitions required for each. We observe that models with similar disk sizes do not necessarily require the same number of partitions. This is due to the heterogeneous structure of machine learning models, where differences in layer composition, operator memory usage, and execution patterns significantly impact memory demands during inference. Consequently, partitioning decisions depend not only on disk size but also on the internal characteristics of each model.

Figure 3a illustrates the memory requirements of each model, highlighting the percentage of snapshots that exceed the EPC capacity (∼100MB). We observe that smaller models, such as SqueezeNet1.0 and MobileNet V2, consume less than 100MB of memory. However, since the Occlum image occupies around 59MB, only about 41MB of usable memory remains for models. This constraint requires partitioning for these models to fit within the available memory. As the model size increases, the memory demands grow accordingly, with the two largest models, ResNet152 V2 and EfficientNet V2, requiring around 470MB and 530MB of memory at the 97th percentile. To verify the effectiveness of the partitioning technique, we also show the memory requirements of each model's partitions during sequential execution in Fig. 3b. We observe that, after partitioning, all models remain under the 100MB range, confirming the technique's efficiency.

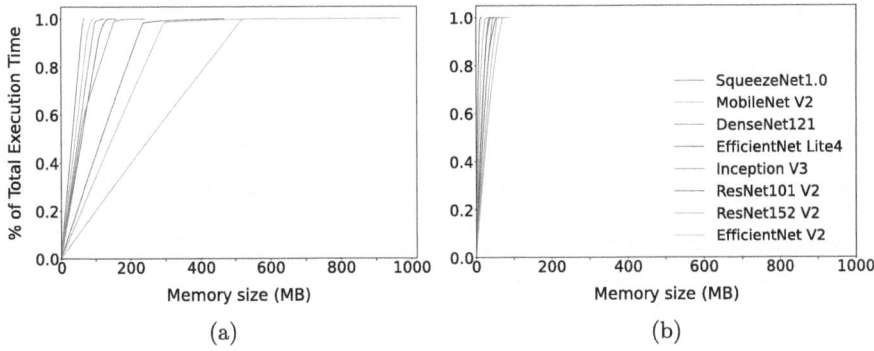

(a)   (b)

Fig. 3. Memory requirements of models when running as a whole (a), and when running in partitions (b). The average percentage of memory snapshots is shown, indicating whether they exceed the EPC capacity.

6.3   Baseline Performance

In this section, we evaluate the performance of machine learning inference on the CPU without Intel SGX, disk encryption, or TLS connections. The primary metric is inference time, measured in milliseconds (ms), capturing the duration from when a client sends a request to when the server returns the result over an unencrypted, plain connection. This baseline serves as a reference point to quantify and justify the overheads introduced by Intel SGX, secure disk access, and TLS in subsequent experiments.

As shown in Table 2, execution is faster when models are stored in memory. However, when models are loaded from disk, execution times increase by a factor of 1.27× to 2×. The smallest model, SqueezeNet1.0, experiences a more pronounced slowdown. A similar trend is observed with MobileNet V2, the second

Table 2. Base inference execution time (in milliseconds) on CPU without Intel SGX or TLS, with models either stored in memory or loaded from disk.

Models	Stored in memory	Loaded from disk
SqueezeNet1.0	34	70
MobileNet V2	162	247
DenseNet121	381	512
EfficientNet Lite4	511	647
Inception V3	597	771
ResNet101 V2	839	1093
ResNet152 V2	1073	1374
EfficientNet V2	1257	1754

smallest model, while larger models show a more moderate increase in execution time, ranging from 1.27× to 1.4×.

6.4   Performance of InferONNX During Full Model Execution

In this section, we analyze the overheads introduced by Intel SGX, along with the computational and I/O costs associated with loading and decrypting the full models from disk. To isolate these effects, we run InferONNX without using model partitioning; instead, the full model is loaded from disk on each inference request. This setup is designed to highlight the overhead of stressing the limited EPC memory of Intel SGX during end-to-end inference. For comparison, we evaluate two additional configurations: (i) InferONNX running outside the Intel SGX enclave to quantify SGX-related overheads, and (ii) InferONNX running with models preloaded into memory, eliminating disk access and decryption to expose only the computational cost within the enclave.

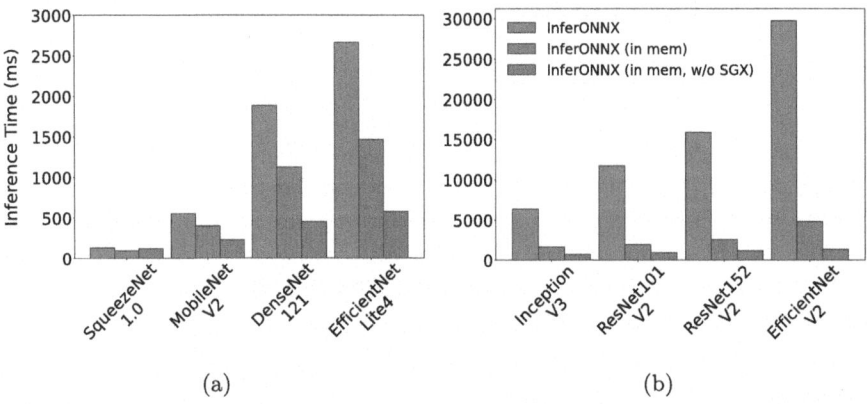

(a)   (b)

Fig. 4. Performance evaluation when running inference on small (a) and large (b) models, across three configurations: the approach where the full models are loaded from disk and decrypted, the in-memory approach, and the baseline where the execution is on CPU, without Intel SGX.

Figure 4a and 4b present the performance of InferONNX across small and large models respectively. We define small models as those whose memory requirements are within or only slightly above the EPC limit–such as SqueezeNet1.0, MobileNet V2, DenseNet121, and EfficientNet Lite4. The remaining models exceed this threshold and are classified as large. For small models, the impact of SGX-induced page swapping is minimal, as their memory demands fit within the EPC capacity. In contrast, large models, experience more pronounced performance degradation due to frequent page swapping, as illustrated in Fig. 4b. The largest model, EfficientNet V2—requiring approximately five times the EPC capacity—suffers the highest overhead, with a slowdown of

3.65×. Other large models incur overheads ranging from 2.15× to 2.38×. These performance penalties are primarily attributed to the overhead of encryption, decryption, and secure data transfers between the enclave and untrusted memory. Despite this, as shown in Table 3, the Instructions Per Cycle (IPC) remains stable across all models, ranging from 2.27 to 2.67. This consistency indicates that CPU efficiency is largely unaffected, and that the main bottleneck stems from SGX-induced page swapping.

Next, we compare the performance of InferONNX when models are preloaded into memory versus when they are loaded from disk. The results highlight the overhead introduced by disk I/O, as loading models from disk and performing decryption during inference leads to additional latency. For small models, this overhead results in a performance penalty of 1.37× to 1.82×. For large models, the overhead is even more significant, with performance degradation ranging from 3.84× to approximately 6.2×. These penalties are primarily due to the time spent on disk access, SGX-induced page swapping, and the decryption process during runtime.

Table 3. Instructions Per Cycle (IPC) values across different models.

Model	IPC
SqueezeNet1.0	2.62
MobileNet V2	2.66
DenseNet121	2.38
EfficientNet Lite4	2.67
Inception V3	2.35
ResNet101 V2	2.46
ResNet152 V2	2.58
EfficientNet V2	2.27

6.5   Performance of InferONNX with Model Partitioning

We now evaluate the performance of InferONNX with model partitioning. Due to Intel SGX's memory constraints, model partitioning is employed to enhance efficiency by loading large models from disk in smaller, manageable segments. To assess its impact, we compare the execution times of full models with those of their partitioned versions, loaded and executed sequentially.

As shown in Fig. 5a, small models, such as SqueezeNet1.0 and MobileNet V2, are not significantly affected by partitioning, as their execution times remain similar to when running the full model. However, as model size increases, the benefits of partitioning become more apparent. DenseNet121, with a size of 31.2MB, shows a 1.46× improvement, while EfficientNet Lite4, at approximately 50MB, achieves a 1.48× improvement. For large models, the impact of partitioning is even more pronounced, as shown in Fig. 5b. Partitioning leads to further improvements in execution times, with performance gains becoming more substantial as the model size increases. These improvements range from 2.64× for Inception V3 (90.9MB) to 4.04× for the largest model, EfficientNet V2 (451MB).

Overall, model partitioning can significantly improve execution efficiency, especially for larger models. However, its effectiveness is influenced by factors such as model size and the overhead introduced by disk accesses. As a result, clients must find the optimal balance between memory utilization and execution time, tailoring the partitioning strategy to meet the specific needs of their use case and system constraints.

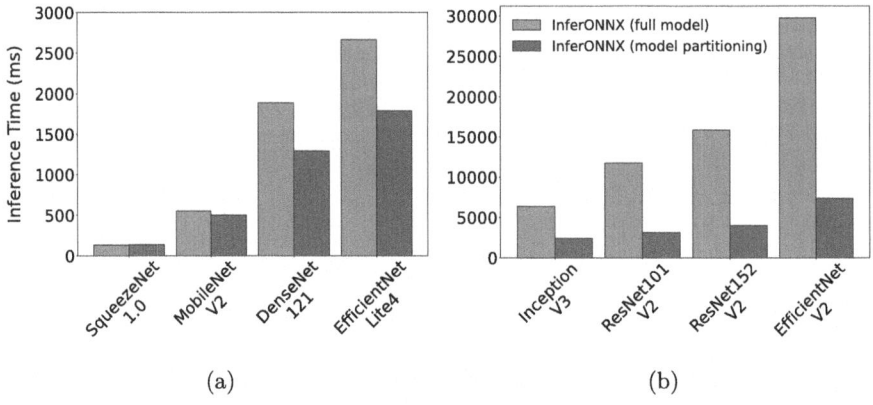

(a)   (b)

Fig. 5. Performance evaluation of InferONNX during inference on small (a) and large (b) models, comparing full model execution with model partitioning.

7   Related Work

In this section, we present existing approaches for confidential machine learning inference, including cryptographic techniques and TEEs.

Cryptographic approaches have long been a key solution for securing sensitive data. Rivest et al. [36] first introduced Homomorphic Encryption (HE), which enables computations on encrypted data without decrypting it, ensuring that third parties can handle data securely. Gentry [18] later developed Fully Homomorphic Encryption (FHE), which supports arbitrary computations on ciphertext, making it a powerful but computationally expensive solution for secure data processing. Multi-Party Computation (MPC) is another cryptographic approach, allowing multiple parties to perform a joint computation while keeping their input private. Some approaches combine these methods for improved performance. Bourse et al. [8] proposed Fast HE Discretized Neural Network (Fast HE DiNN), which leverages both HE and discrete neural networks to reduce the computation complexity of HE. Although this method sacrifices some accuracy, it improves efficiency in cases where DiNNs are used for training instead of discretizing during inference. Xue et al. [46] addressed limitations in HE by proposing a multi-key FHE scheme, which improves privacy protection for client data but does not guarantee model privacy. Several works, such as SecureML [33], MiniONN [30], and Chameleon [35] have leveraged MPC for secure inference. Gazelle [21] combined FHE and MPC to achieve better performance, though it requires two-party computation.

To overcome the limitations of cryptographic methods, hardware-based TEEs like Intel SGX [13], ARM TrustZone [45], and ARM CCA [25] are being used to create isolated execution environments for machine learning operations. One example is GuaranTEE [40], which builds on ARM CCA to provide a secure framework for protecting machine learning inference on edge devices. In contrast, our work utilizes Occlum, a libOS for Intel SGX, which simplifies the development and deployment of applications within Intel SGX. Fortanix Enclave Development Platform [15] is another widely used libOS for Intel SGX, offering strong cloud-native integration and support for multiple programming languages, particularly Rust, which offers memory safety features.

Intel SGX faces strict memory constraints, which necessitate further optimization techniques, such as model partitioning, to improve performance while staying within enclave memory limits. Slalom [41] presents a system for secure machine learning inference that partitions model execution between a trusted enclave (Intel SGX) and an untrusted CPU, enabling efficient execution while maintaining model confidentiality. TEESlice [28] highlights the limitations of existing post-training model partitioning approaches under knowledgeable adversaries and proposes a partition-before-training method that isolates privacy-sensitive weights within TEEs, while offloading the remaining, less sensitive weights to GPUs. Soter [38] partitions sensitive model layers, executing them inside Intel SGX enclaves, while the remaining layers run on GPUs to accelerate inference. While this approach reduces computation time, it introduces performance overhead due to frequent CPU-GPU context switching, especially for large models. Additionally, the integrity checks required to ensure correct results can increase latency by up to 1.27x in some cases. Approaches like Soter and TEESlice leverage GPU acceleration to address the memory constraints of Intel SGX, however they still require careful orchestration of data movement and memory usage to avoid performance bottlenecks and ensure security guarantees. SecureTF [34] provides a secure enclave-based runtime specifically tailored for TensorFlow models. MEDIA [27] partitions Deep Neural Networks (DNNs) into multiple partitions in an edge cloud environment, addressing the limitations posed by cyclic graphs. It optimizes inference by routing models to one of N servers, selecting the server that achieves the best inference time. Unlike our approach, which strictly partitions models based on EPC capacity, MEDIA prioritizes inference time, allowing some degree of exceeding the EPC limit as long as overall performance remains efficient. Finally, MLCapsule [20] enables client-side execution while keeping the model and computations confidential, allowing service providers to protect their intellectual property and business models. Similarly, our system supports this model using ONNX, a format that further confines execution to the client's environment.

8   Limitations

Model partitioning is a practical method for managing execution within the memory limitations of hardware enclaves. Nonetheless, the current approach has certain constraints. One such limitation arises when an individual operator in the model exceeds the capacity of the EPC. In these cases, partitioning has limited effect, as the large operator introduces execution overhead that cannot be avoided through inter-operator partitioning alone. The current design does not address such cases, and support for intra-operator partitioning remains an area for future investigation.

Another limitation concerns models in which partitions depend on intermediate state produced by preceding operators. Preserving and managing this state across partition boundaries introduces additional memory requirements. When executing model partitions sequentially, the cumulative memory footprint can exceed the EPC capacity. In such cases, the intended benefits of partitioning are diminished, as the overhead from enclave memory constraints—such as page swapping—remains significant, potentially affecting overall performance.

9   Conclusions

In this work, we propose InferONNX, a lightweight machine learning inference service designed to run within Intel SGX. Our approach enables model providers to securely deploy their models, allowing clients to perform inference on sensitive data while preserving both model confidentiality and data privacy. In addition, it tackles Intel SGX's memory constraints using two key mechanisms: a compact runtime with a minimal memory footprint and model partitioning that reduces memory usage during inference. Our evaluation shows that InferONNX reduces the overhead associated with full model execution by approximately $1.5\times$ to $4\times$.

Acknowledgments. We thank the shepherd and the anonymous reviewers for their helpful comments. This work was supported by dAIEdge funded by the European Commission under Grant 101120726. The source code is available at https://github.com/Konstantina155/InferONNX.

References

1. Anakin inference framework. https://github.com/PaddlePaddle/Anakin
2. Mobile AI Compute Engine (MACE) inference framework. https://github.com/XiaoMi/mace
3. NCNN inference framework. https://github.com/Tencent/ncnn

4. ONNX Model Zoo. https://onnx.ai/models/
5. Valgrind Massif: a heap profiler. https://valgrind.org/docs/manual/ms-manual. html
6. Bai, J., Lu, F., Zhang, K., et al.: ONNX: open neural network exchange (2019). https://github.com/onnx/onnx
7. Bao, B.: ONNX models. https://github.com/BowenBao/models-1
8. Bourse, F., Minelli, M., Minihold, M., Paillier, P.: Fast homomorphic evaluation of deep discretized neural networks. In: Shacham, H., Boldyreva, A. (eds.) CRYPTO 2018. LNCS, vol. 10993, pp. 483–512. Springer, Cham (2018). https://doi.org/10. 1007/978-3-319-96878-0_17
9. Brasser, F., Müller, U., Dmitrienko, A., Kostiainen, K., Capkun, S., Sadeghi, A.R.: Software grand exposure: SGX cache attacks are practical. In: Proceedings of the 11th USENIX Conference on Offensive Technologies (2017)
10. Chalkiadakis, N., Deyannis, D., Karnikis, D., Vasiliadis, G., Ioannidis, S.: The million dollar handshake: secure and attested communications in the cloud. In: 2020 IEEE 13th International Conference on Cloud Computing (CLOUD), pp. 63–70 (2020). https://doi.org/10.1109/CLOUD49709.2020.00022
11. Chen, G., Chen, S., Xiao, Y., Zhang, Y., Lin, Z., Lai, T.H.: SgxPectre: stealing intel secrets from SGX enclaves via speculative execution. In: 2019 IEEE European Symposium on Security and Privacy (EuroS&P), pp. 142–157 (2019)
12. Chen, T., et al.: TVM: an automated end-to-end optimizing compiler for deep learning. In: Proceedings of the 13th USENIX conference on Operating Systems Design and Implementation, pp. 579–594 (2018)
13. Costan, V., Devadas, S.: Intel SGX explained. In: IACR Cryptology ePrint Archive, pp. 1–118 (2016)
14. Duy, K.D., Noh, T., Huh, S., Lee, H.: Confidential machine learning computation in untrusted environments: a systems security perspective. IEEE Access **9**, 168656– 168677 (2021)
15. FortanixEDP: Fortanix enclave development platform. https://edp.fortanix.com/
16. Peskine, G., Pégourié-Gonnard, M., et al.: Mbed-TLS library. https://github.com/ Mbed-TLS/mbedtls
17. Gallego, A., Odyurt, U., Cheng, Y., Wang, Y., Zhao, Z.: Machine learning inference on serverless platforms using model decomposition. In: Proceedings of the IEEE/ACM 16th International Conference on Utility and Cloud Computing, pp. 1– 6. Association for Computing Machinery (2024)
18. Gentry, C.: Fully homomorphic encryption using ideal lattices. In: Proceedings of the Forty-First Annual ACM Symposium on Theory of Computing, pp. 169–178 (2009)
19. Götzfried, J., Eckert, M., Schinzel, S., Müller, T.: Cache attacks on intel SGX. In: Proceedings of the 10th European Workshop on Systems Security, pp. 1–6 (2017)
20. Hanzlik, L., et al.: MLCapsule: guarded offline deployment of machine learning as a service. In: 2021 IEEE/CVF Conference on Computer Vision and Pattern Recognition Workshops (CVPRW), pp. 3295–3304 (2021)
21. Juvekar, C., Vaikuntanathan, V., Chandrakasan, A.: GAZELLE: a low latency framework for secure neural network inference. In: Proceedings of the 27th USENIX Conference on Security Symposium, pp. 1651–1669 (2018)
22. Lauriola, I., Lavelli, A., Aiolli, F.: An introduction to deep learning in natural language processing: models, techniques, and tools. Neurocomputing **470**, 443– 456 (2022)

23. Lee, T., et al.: Occlumency: privacy-preserving remote deep-learning inference using SGX. In: The 25th Annual International Conference on Mobile Computing and Networking, pp. 1–17 (2019)

24. Li, F., Li, X., Gao, M.: Secure MLaaS with temper: trusted and efficient model partitioning and enclave reuse. In: Proceedings of the 39th Annual Computer Security Applications Conference, pp. 621–635 (2023)

25. Li, X., et al.: Design and verification of the arm confidential compute architecture. In: 16th USENIX Symposium on Operating Systems Design and Implementation (OSDI 22), pp. 465–484 (2022)

26. Li, Y., et al.: Lasagna: accelerating secure deep learning inference in SGX-enabled edge cloud. In: Proceedings of the ACM Symposium on Cloud Computing, pp. 533–545 (2021)

27. Li, Y., Zeng, D., Gu, L., Guo, S., Zomaya, A.Y.: DNN partitioning and assignment for distributed inference in SGX empowered edge cloud. In: 2024 IEEE 44th International Conference on Distributed Computing Systems (ICDCS), pp. 635–644 (2024)

28. Li, D., Zhang, Z., Yao, M., Cai, Y., Guo, Y., Chen, X.: TEESlice: protecting sensitive neural network models in trusted execution environments when attackers have pre-trained models. ACM Trans. Softw. Eng. Methodol. (2024)

29. Liu, B., Ding, M., Shaham, S., Rahayu, W., Farokhi, F., Lin, Z.: When machine learning meets privacy: a survey and outlook. ACM CSUR **54**(2), 1–36 (2021)

30. Liu, J., Juuti, M., Lu, Y., Asokan, N.: Oblivious neural network predictions via MiniONN transformations. In: Proceedings of the 2017 ACM SIGSAC Conference on Computer and Communications Security, pp. 619–631 (2017)

31. Malik, M., Malik, M.K., Mehmood, K., Makhdoom, I.: Automatic speech recognition: a survey. Multimedia Tools Appl. **80**, 9411–9457 (2021)

32. Poumeyrol, M., et al.: Tract inference engine. https://github.com/sonos/tract

33. Mohassel, P., Zhang, Y.: SecureML: a system for scalable privacy-preserving machine learning. In: 2017 IEEE Symposium on Security and Privacy (SC), pp. 19–38 (2017)

34. Quoc, D.L., Gregor, F., Arnautov, S., Kunkel, R., Bhatotia, P., Fetzer, C.: secureTF: a secure TensorFlow framework. In: Proceedings of the 21st International Middleware Conference, pp. 44–59 (2020)

35. Riazi, M.S., Weinert, C., Tkachenko, O., Songhori, E.M., Schneider, T., Koushanfar, F.: Chameleon: a hybrid secure computation framework for machine learning applications. In: Proceedings of the 2018 on Asia Conference on Computer and Communications Security, pp. 707–721 (2018)

36. Rivest, R.L., Adleman, L., Dertouzos, M.L.: On data banks and privacy homomorphisms. Found. Secur. Comput. **4**(11), 169–180 (1978)

37. Shamshirband, S., Fathi, M., Dehzangi, A., Chronopoulos, A.T., Alinejad-Rokny, H.: A review on deep learning approaches in healthcare systems: taxonomies, challenges, and open issues. J. Biomed. Inform. **113**, 103627 (2021)

38. Shen, T., et al.: SOTER: guarding black-box inference for general neural networks at the edge. In: Proceedings of the 2022 USENIX Annual Technical Conference, pp. 1651–1669 (2022)

39. Shen, Y., et al.: Occlum: secure and efficient multitasking inside a single enclave of intel SGX. In: Proceedings of the Twenty-Fifth International Conference on Architectural Support for Programming Languages and Operating Systems, pp. 955–970 (2020)

40. Siby, S., Abdollahi, S., Maheri, M., Kogias, M., Haddadi, H.: GuaranTEE: towards attestable and private ML with CCA. In: Proceedings of the 4th Workshop on Machine Learning and Systems, pp. 1–9 (2024)
41. Tramèr, F., Boneh, D.: Slalom: fast, verifiable and private execution of neural networks in trusted hardware. In: 7th International Conference on Learning Representations, ICLR 2019, New Orleans, LA, USA, May 6-9 (2019)
42. Tran, K.A., Kondrashova, O., Bradley, A., Williams, E.D., Pearson, J.V., Waddell, N.: Deep learning in cancer diagnosis, prognosis and treatment selection. Genome Med. **13**, 1–17 (2021)
43. Wang, P., Fan, E., Wang, P.: Comparative analysis of image classification algorithms based on traditional machine learning and deep learning. Pattern Recogn. Lett. **141**, 61–67 (2021)
44. Wang, W., et al.: Leaky Cauldron on the dark land: understanding memory side-channel hazards in SGX. In: Proceedings of the 2017 ACM SIGSAC Conference on Computer and Communications Security, pp. 2421–2434 (2017)
45. Winter, J.: Trusted computing building blocks for embedded Linux-based ARM trustzone platforms. In: Proceedings of the 3rd ACM Workshop on Scalable Trusted Computing, pp. 21–30 (2009)
46. Xue, H., et al.: Distributed large scale privacy-preserving deep mining. In: 2018 IEEE Third International Conference on Data Science in Cyberspace (DSC), pp. 418–422 (2018)

Hiding in Plain Sight: On the Robustness of AI-Generated Code Detection

Saman Pordanesh, Sufiyan Bukhari[ID], Benjamin Tan[ID], and Lorenzo De Carli[✉][ID]

University of Calgary, Calgary, AB T2N 1N4, Canada
{saman.pordanesh,sufiyanahmed.bukhari,benjamin.tan1, lorenzo.decarli}@ucalgary.ca

Abstract. AI code assistants, such as GitHub Copilot, are an increasingly popular coding aid, but they also present risks. Large language models (LLMs) upon which those assistants are built may generate insecure/incorrect code, either by accident or as a result of code poisoning attacks. In general, LLMs obfuscate the lineage of source code used for training. This is a problem, for example, in the context of supply chain security, where tracking provenance is of the utmost importance. While a number of recent approaches can flag AI-generated code based on a combination of lexical and syntactic features, such works have not been evaluated in realistic settings. First, we identify and operationalize a number of recently proposed AI code identification tools, measuring their baseline performance on datasets generated by state-of-the-art models. Then, we verify the robustness of such approaches to variations in training sets and prompting strategies. Results show that existing AI code detectors tend to be fragile and have limited accuracy in real-world scenarios.

1 Introduction

AI code assistants are quickly becoming ubiquitous within the software development cycle. Such tools can quickly generate source code on demand, either by completing developer-written code or in response to specific user requests. In doing so, they can greatly reduce the effort to write boilerplate code, tests, and similar components. This, in turn, reduces development time and frees programmers to concentrate on higher-level tasks, such as debugging and extending functionality. As such, these tools enjoy ever-increasing popularity [31].

While AI code assistants have the potential to be immensely useful, they also give rise to security concerns. Recent research points to the fact that code assistants may, in certain situations, generate code that is less secure than that written by humans [26,27]. Further, Large Language Models (LLMs), on which assistants are based, are trained on large datasets of unvetted Open-Source code. As such, they may learn to generate copyrighted [8] or incorrect code, or even be the target of code poisoning attacks [30,39,41]. This is potentially problematic, both in the context of direct use of the tools and when importing software dependencies that may be AI-generated. Overall, LLMs obfuscate provenance [5],

M. Egele et al. (Eds.): DIMVA 2025, LNCS 15748, pp. 44–64, 2025.
https://doi.org/10.1007/978-3-031-97623-0_3

as code generation is based on a training dataset, but the relation between training and output is not clearly maintained. Thus, being able to track *code provenance* back to a code assistant is of the utmost importance for security, correctness, and legal reasons. Indeed, it is not uncommon for software companies to limit or qualify the use of such tools by their employees [13,29]. For the reasons above, until the threat model surrounding AI-generated code is better understood, there is a need for tools that can highlight the presence of AI-generated code in the wild so that it can be appropriately reviewed if necessary.

Several recent works propose the design of classifiers that can identify the human or AI provenance of source code with high accuracy, at least under certain assumptions [5,17,22,32,35,40]. We believe such algorithms can fill an important gap and be useful for AI code detection and general code measurement studies. Unfortunately, the robustness of such tools has seen limited to no investigation. There are multiple threats to their accuracy. One is overfitting the training set: as multiple LLMs capable of generating code exist, there may be intrinsic differences in the code they generate. This may make a detector trained on one model underperform on code generated by another. Another issue is that differences in the characteristics of human and AI-generated code may depend on specific programming tasks. Indeed, past work observed that "[detector] performance considerably improves when the common patterns – those that may occur in data curated from the same domains – have been learned during the training" [22].

In this paper, we examine the robustness of recently proposed AI code detectors. We consider multiple classifiers [5,22,32,41]. First, we measure baseline classifier performance and examine the impact of classifier design parameters. Then, we evaluate the effect of the classifier training dataset (where applicable) and prompt variations on classifier effectiveness. Finally, we consider whether diversifying training sets can help improve classifier performance.

For evaluation accuracy, it is important to use cleanly labeled, diverse datasets. We use a set of 6K software samples from two source code datasets commonly used in this domain [15,16], each including programming task assignments and human-generated solutions. To maximize external validity, we generate corresponding AI solutions using three prominent models: OpenAI's GPT [24], Google's Gemini [28], and Anthropic's Claude [2]. Furthermore, there exist numerous AI code detectors based on different approaches. Most only have proof-of-concept implementations, and for some, a full implementation has not been released. We perform a literature review, selecting four representative detectors (see Sect. 4) [5,22,32,41]. We engineer all these approaches to perform classification on our dataset, implementing missing components when necessary[1].

Overall, zero-shot detectors tend to exhibit a significant drop in accuracy compared to the originally published results, even after parameter tuning. Classifiers based on train+evaluate ML pipelines tend to fare better, but this advantage dissipates when training and evaluation sets have different characteristics. These results suggest that the problem of AI-generated code detection is not trivially solvable, and more work is needed to produce effective detectors.

---

[1] Data and code package: https://osf.io/jahxs/?view_only=dff479fdad8c4b9cb2060e6a3c2c5e5a.

2   Background

2.1   AI Code Assistant

We use the term "AI code assistant" to refer to any AI-based tool that can generate code on demand. We include in our definition both tools that are directly integrated into software development IDEs, such as GitHub Copilot [11], and tools that offer some form of chat UI where users can post questions and retrieve code from the answers, such as OpenAI's ChatGPT [24]. These code assistants are generally based on LLMs. In extreme synthesis, those are transformer models with billions of parameters trained on petabytes of human-generated text. They have the ability to continue a prompt with a stream of words (tokens) likely to follow, thus generating, in many circumstances, "human-like" answers to queries. In the coding domain, those models are used to complete source code or to generate source code from scratch based on natural language descriptions.

The training dataset of prominent "foundation" LLMs includes many forms of text, including, in the case of models used for coding, a substantial amount of source code. Due to its size, text in the training data is often used as-is without vetting for correctness or risks. In many cases, the specifics of those training datasets are not made public. This has raised a number of concerns, including the possibility of bias and/or generation of technically incorrect information [38], and vulnerability of LLMs to poisoning attacks [30].

2.2   Supply Chain Security

Software supply chain security focuses on identifying risks and vulnerabilities in components used within software artifacts. Much modern software is built compositionally, importing open-source software (OSS) components within a project. These components provide ready-made, freely available implementations of functionality that may be tedious or complex to implement correctly (e.g., JSON parsers, messaging middleware, web UI frameworks, etc.). In the last 15 years, this approach has revolutionized software development, enabling companies to quickly and economically build infrastructure that previously would have needed to be entirely developed in-house.

However, this approach also comes with security drawbacks, as including a large amount of external modules can significantly extend the attack surface of an software artifact [36]. AI coding assistants further complicate the issue: as discussed in Sect. 1, they may obfuscate the provenance of code within imported modules, and may even introduce additional vulnerabilities. Thus, we consider AI code detection relevant to software supply chain security.

2.3   Threat Model

We consider a scenario where an AI code assistant generates vulnerable or malicious code that gets embedded in a software artifact. This may be accidental, e.g., due to defective examples in the training set, or purposefully caused by an attacker conducting dataset poisoning [30, 39]. To contain such attacks, an organization may opt to identify and track the provenance of AI-generated code to a specific model/training set. This may enable countermeasures such as only allowing the use of models in non-security-sensitive contexts or limiting code generation to specific, vetted models/training sets. To do so, we consider the problem of identifying AI-generated code and distinguishing it from human-generated code. While classifiers have been proposed for this problem [5, 22, 32, 41], as part of a more general trend towards distinguishing AI- and human-generated content [12, 21], these tools have only been evaluated in "clean lab" settings, typically on limited datasets and in the absence of confounding factors that exist in the real world (e.g., training and evaluation sets with different characteristics) and may degrade their performance. In this work, we aim to evaluate and compare the performance of different classifiers in the presence of such factors.

Prompt (abridged): "Utility function to calculate md5 hashes while being light on memory usage. By reading the fileobject piece by piece, we are able to process content that is larger than available memory"

(a) Prompt used to generate the function

```
def calculate_md5(fileobject, size=2**16):
    fileobject.seek(0)
    md5 = hashlib.md5()
    for data in iter(lambda: fileobject.read(size), b''):
        if not data: break
        if isinstance(data, six.text_type):
            data = data.encode('utf-8') # md5 needs a byte string
        md5.update(data)
    fileobject.seek(0) # rewind read head
    return md5.hexdigest()
```

```
def calculate_md5(fileobject):
    md5 = hashlib.md5()
    while chunk := fileobject.read(8192):
        md5.update(chunk)
    return md5.hexdigest()
```

(b) Human-generated example   (c) AI-generated example

Fig. 1. Example of prompt and human- and AI-generated code (CSN dataset).

3 Dataset

3.1 Prompt Corpora

To assess the effectiveness of classifiers, we required a diverse dataset containing code samples generated by both humans and AI. To control confounding factors, we sought to have AI generate code for tasks for which corresponding human implementations exist. To ensure and maintain the integrity of the dataset generation process, we established specific criteria for our dataset selection.

Comprehensive Problem Descriptions: Each human-written code sample must be accompanied by a detailed problem statement. This context is crucial for a fair comparison between human and AI-generated code. Without it, AI may produce incorrect or irrelevant code, potentially leading to biased analysis.

Sufficient Number of Code Samples: The dataset should contain a substantial number of code samples to enable large-scale analysis. A larger dataset leads to more reliable and robust conclusions.

Prompt (abridged): "Write a program which prints multiplication tables in the following format: 1x1=1 1x2=2 [...] [template for various languages follows]"

(a) Prompt used to generate the function

```
def main():
    for a in range(1,10):
        for b in range(1,10):
            print("{}x{}={}".format(a,b,a*b))
    return None
```

```
def print_multiplication_table():
    for i in range(1, 10):
        for j in range(1, 10):
            print(f"{i}x{j}={i*j}")

print_multiplication_table()
```

(b) Human-generated example   (c) AI-generated example

Fig. 2. Example of prompt and human- and AI-generated code (IBM dataset).

As the target language, we choose Python due to (i) the large amount of Python code available for model training, which makes widely available AI code assistants particularly suited to Python programming (Python routinely tops lists of most popular programming languages [34]); and (ii) the widespread use of Python in literature on AI-generated code detection [32,41].

CodeSearchNet. The CodeSearchNet Corpus [15] is a collection of 6M+ code functions from popular open-source GitHub repositories. They are written in a variety of programming languages, including Go, Java, JavaScript, PHP, Python, and Ruby. In the rest of the paper, we refer to this as the **CSN dataset**.

The dataset comprises approximately 2 million pairs of code functions and their corresponding documentation. Additionally, it includes around 4 million functions without associated documentation to facilitate model training and evaluation for training, validation, and testing.

The primary goal of this dataset is to support the CodeSearchNet challenge. This challenge focuses on the development of advanced code search techniques that can accurately retrieve code snippets based on natural language queries.

Prompt Set Review. Manual review of this dataset reveals potential issues. Documentation is oftentimes truncated, resulting in loss of contextual information. In some cases, the documentation fails to reflect requirements that are clear in the human implementation, and/or the documentation is outdated and no longer accurately reflects the original code. Figure 1 presents an example, together with a human solution and a sample AI solution generated with OpenAI GPT 4o. It is worth noting that while this dataset is popular in the AI code detection literature [32,41], these issues are virtually ignored. Ultimately, we decided to use this dataset as, despite its limitations, it has the advantage of comprising realistic software development tasks while noting that the information asymmetry between human developers and AI may introduce systematic differences.

IBM Project CodeNet. Project CodeNet [16], by IBM, is a massive dataset of 13.9 million code samples, each designed to solve one of 4,000 coding challenges.

```
Provide python code as a function for the below problem statement and
produce no other text. Do not include the function inside a docstring.

Problem Statement:

PROBLEM DESCRIPTION GOES HERE.
```

Fig. 3. Template used for AI code generation.

Sourced from AIZU Online Judge[2] and AtCoder[3], these samples cover over 50 programming languages, primarily C++, C, Python, and Java. Each sample includes detailed metadata like size, memory usage, execution time, and outcome (accepted or rejected with reason). Human code samples are pre-vetted by human judges. In the following discussion, we refer to this as the **IBM dataset**.

Over 90% of the problems in Project CodeNet are accompanied by detailed descriptions. Additionally, over half of the 4K problems in the dataset have at least one accepted code solution, providing a benchmark for evaluation.

Problem Set Review. Different from CodeSearchNet, samples in this dataset do not suffer from information asymmetry: human samples were developed using exactly the same information available for AI code generation. These descriptions outline the specific problem and expected input and output formats, often including sample input/output pairs. This rich contextual information provides a comprehensive understanding of the problem to be solved. Figure 2 presents an example, together with a human solution and a sample AI solution generated with OpenAI GPT 4o. However, this dataset suffers from a different limitation, as prompts veer towards artificial programming exercises rather than real-world tasks. We decided to still retain this dataset, as it has the advantage of providing clear, unambiguous tasks together with rigorously vetted human solutions.

3.2 Dataset Generation and Discussion

Human-Generated Samples. We identified potential prompts for the LLMs (coding task *problem descriptions* for the AI) in both CSN and IBM sets, requiring (i) the prompt to be in English (to eliminate language as a confounding factor); (ii) for the IBM prompt set, an accepted solution to be available; and (iii) the solution to be non-empty and valid Python (as some classifiers generate features based on the extracted Abstract Syntax Tree, which requires syntactic validity). Finally, we randomly sampled a set of approximately 3000 solutions from the acceptable samples for each dataset. As discussed in Sect. 3.3, we only retained human samples for which a corresponding machine-generated solution could be obtained from all models. This resulted in **3042** code/prompt samples for CSN and **2984** code/prompt samples for IBM.

---

[2] https://onlinejudge.u-aizu.ac.jp/home.

[3] https://atcoder.jp.

3.3   AI-Generated Samples

To construct code-generation prompts, we used the same problem descriptions present in the set of human samples (described above) for both the CSN and IBM datasets. We selected three state-of-the-art enterprise LLMs as our generative models for addressing code-related problems: GPT-4o [24], Gemini 1.5 Flash [28], and Claude 3.5 Sonnet [2], based on the most recent versions available as of mid-2024. Despite their strong performance, open-source models such as LLaMA 3.0 [20] were excluded. They are hosted on limited cloud infrastructures and are, therefore, less commonly used for day-to-day tasks like programming. All code generation processes using these models were conducted via their APIs.

Generation Process We created a two-part initial prompt template to guide the generative models. The first part provided specific instructions about the desired response format and content, while the second part incorporated a problem description from one of the prompt sets. The prompt was created by following OpenAI's suggested best practices [25] and iteratively refined to minimize invalid/incorrect code generation in initial informal experimentation. It also includes instructions to minimize the generation of extraneous non-code text. The prompt is given in Fig. 3. We further discuss the potential impacts of prompt structure on detection results in Sect. 5.3.

After querying the generative models, the generated samples were checked for non-emptiness and correctness. When checks failed, we re-tried code generation up to three times, after which we dropped the prompt from the evaluation set. Additionally, we observed that AI-generated samples occasionally only consist of documentation (i.e., docstrings); we removed all such documentation before checking for empty samples.

During generation, all LLMs failed to generate code for several problem description prompts; the set of failed prompts overlaps only partially. To ensure a fair evaluation across models, we only retain prompts/generated code for prompts from which all models could generate a valid code sample.

Table 1. Effect size for Wilcoxon pairwise comparison between length and complexity of human and AI samples (strong correlation highlighted)

	GPT		Gemini		Claude	
	CSN	IBM	CSN	IBM	CSN	IBM
Length	0.06	0.76	0.66	0.96	0.70	0.97
Cyclomatic Comp.	0.31	-0.20	0.48	-0.28	0.23	-0.26

Quantitative Datasets Review. Observing information asymmetry in Code-SearchNet raises the question of whether systematic differences between human-

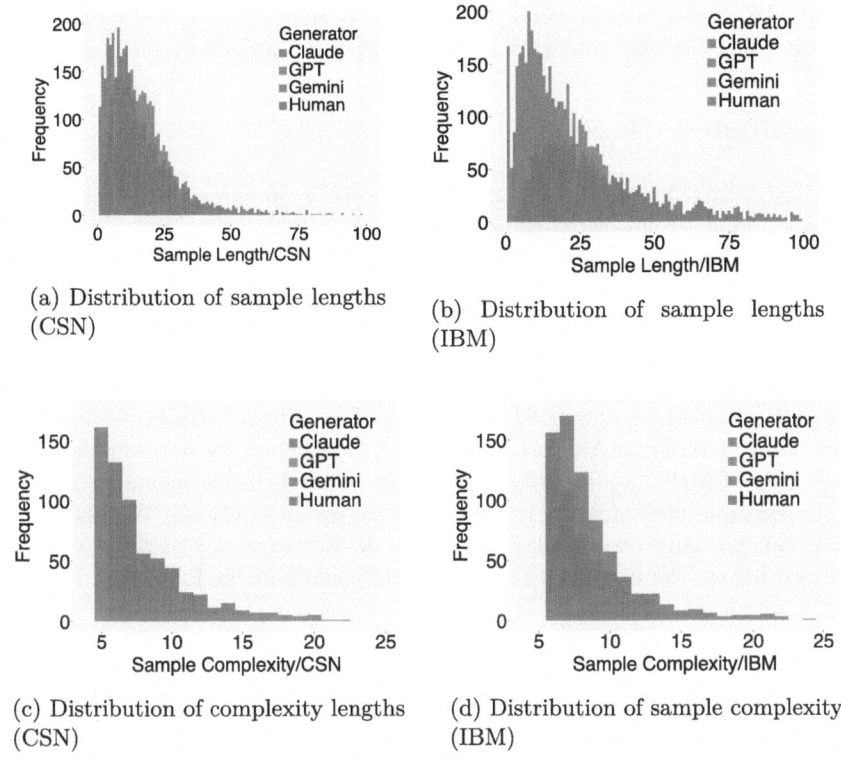

(a) Distribution of sample lengths (CSN)

(b) Distribution of sample lengths (IBM)

(c) Distribution of complexity lengths (CSN)

(d) Distribution of sample complexity (IBM)

Fig. 4. Characterization of basic dataset metrics.

and AI-generated code exist. To address it, we compute two basic measures between paired code samples from our dataset: lengths in lines of code (LOCs), and cyclomatic complexity [19]. The latter is defined as the number of independent paths through a source code artifact and is frequently used to quantitatively represent the level of complexity of a source code artifact.

Distribution of sample lengths from samples generated by human programmers, and the GPT, Gemini and Claude models on CodeSearchNet and IBM are depicted in Fig. 4(a) and (b), while complexity distributions are depicted in Fig. 4(c) and (d). The plots present a nuanced picture, showing small but noticeable distributional differences. Motivated by this, we conducted Wilcoxon signed-rank tests pairwise between human samples and each set of model-generated samples. All metrics are found to present statistically significant differences at $p <= 0.01$. Effect sizes are measured using Rank Biserial Correlation (RBC) (r). Following practices from the literature, we interpret r values above 0.6 as representing high correlation. Full results are given in Table 1. Overall, these results suggest macroscopic statistical differences between samples generated by humans and most sets of AI-generated samples. While this does not imply that individual samples may be correctly labeled, it suggests that there are indeed systemic effects at play that may make classification possible.

4   Evaluated Classifiers

To identify suitable classifiers from the analysis, we reviewed the literature on the detection of AI-generated code. We did so by reviewing publications in high-profile security venues, such as IEEE S&P, ACM CCS, and similar, searching the arXiv online archive, and using search engines (e.g., Google Scholar) to identify any missing publications. We only retained publications with partial or full code released and sufficient details to achieve a working implementation on our dataset with reasonable effort. We further filtered approaches for which we could not get accuracy and or F1 score consistently above 60%. For example, we were unable to replicate Yang et al. [40] 70–80% AUC on Python samples in our experiments. Finally, we filtered approaches that used similar techniques, retaining one example per category. For example, we did not evaluate Whodunit [17], as it is conceptually very similar to the one developed in our previous work [5], discussed below. We discussed the selected classifiers in the following.

4.1   Bukhari et al.

In our previous work [5] we evaluated a classifier based on syntactic and semantic features mutuated from the code stylometry feature set by Caliskan et al. [6]. Such features are computed from source code and an intermediate AST representation, and fed to a trained ML classifier. While the initial code release[4] uses a subset of the stylometric features by Caliskan et al., we were able to extend it with limited effort to the full feature set, thus more closely representing the potential of stylometry-based approaches. This classifier's performance are heavily dependent on which algorithm is used for classification (SVM, XGBoost, random forest etc.). Based on preliminary experiments, XGBoost returns the best result, and we use it for our experiments. For simplicity, in the rest of this paper we refer to this classifier as "Bukhari et al.".

4.2   GPTSniffer

GPTSniffer [22] is a CodeBERT-based classifier designed to detect source code generated by ChatGPT. By fine-tuning a pre-trained language model, it aims at identifiying patterns and anomalies specific to machine-generated code. This approach has a fully functional code release[5], only requiring to format our dataset appropriately. We further make minor alterations to the code to perform cross-validation, compute additional metrics, and add small quality-of-life improvements (e.g., saving the model after training for reuse).

---

[4] https://osf.io/46nva/?view_only=9110c4a94f0a4b4591f14fdd976deeca.
[5] https://github.com/MDEGroup/GPTSniffer.

4.3   Ye et al.

Ye et al. [41] introduce a zero-shot synthetic code detection technique through code rewriting. Specifically, this model measures how code properties change when a sample is partially rewritten by an AI code generation tool. This model has the least complete implementation, requiring a substantial effort to extend its code release[6] to a functional tool. We reimplemented missing components based on the description given in the paper. Further, this detector can be implemented in different ways, as it is heavily dependent on which model is internally used for rewriting, and how many times the code is regenerated. To provide a fair evaluation, we experimented extensively with different models (including OpenAI GPT 4o Mini and Gemini 1.5 Flash), and 4/8/16 rewrites, and retained the combination offering the best results.

Table 2. Performance of evaluated classifiers on CSN dataset using GPT-4o, Gemini Flash 1.5, and Claude Sonnet 3.5 for code generation. * denote zero-shot models which were evaluated on whole dataset; for other models, 4-fold cross-validation was used (**P:** Precision; **R:** Recall; **A:** Accuracy). Highest values in green, lowest in purple.

Approach	GPT 4o					Gemini					Claude				
	P	R	A	F1	AUC	P	R	A	F1	AUC	P	R	A	F1	AUC
Bukhari et al.	0.86	0.89	0.89	0.88	0.97	0.83	**0.83**	0.83	0.83	0.92	0.94	0.93	0.94	0.94	**0.99**
GPTSniffer	0.95	0.98	0.97	0.97	0.99	0.84	0.94	0.89	0.89	0.95	0.93	0.98	0.95	0.95	0.99
Ye et al.*	0.58	0.58	0.58	0.58	0.6	0.5	1.00	0.5	0.67	0.44	0.65	0.62	0.65	0.63	0.69
DetectCodeGPT*	0.61	0.85	0.85	0.71	0.72	0.51	0.97	0.74	**0.67**	0.63	**0.60**	0.85	0.79	0.7	0.72

4.4   DetectCodeGPT

DetectCodeGPT [32] applies zero-shot machine-learning to distinguish between machine-generated and human-written code. This approach works by introducing perturbations in code samples, and measuring how these alter code *naturalness*. Like GPTSniffer, this approach has a fairly complete implementation[7], only requiring reformatting our dataset and computing additional metrics.

5   Experimental Evaluation

5.1   Research Questions

In this section, we seek to answer the following research questions:

- *RQ1: What are the baseline performance of evaluated classifiers on distinguishing AI- and human-generated code?* Sect. 5.2 demonstrates that classifiers exhibit varying degrees of accuracy.

---

[6] https://anonymous.4open.science/r/code-detection-6B35/README.md.
[7] https://github.com/YerbaPage/DetectCodeGPT.

- *RQ2: How robust are these classifiers to variations in training set and code generation prompts?* Experiments in Sect. 5.3 show that, in many cases, model accuracy decreases drastically when training samples come from a different problem domain than evaluation samples.
- *RQ3: Can classifier performance be improved by diversifying training set?* Experiments in Sect. 5.4 suggest that diversifying training set has limited impact on performance.

5.2   Baseline Classifier Performance

In this section, we evaluate baseline performance of the four classifiers under examination. For "baseline performance", we intend executing the classifiers in the absence of confounding factors. These are primarily a concern in the case of training based classifiers (GPTSniffer and Bukhari et al.), as the accuracy of those may suffer when trained on samples whose characteristics differ from the evaluation set. Overall, this analysis also serves to establish baseline performance expectations on our dataset.

Table 3. Performance of evaluated classifiers on IBM dataset using GPT-4o, Gemini Flash 1.5, and Claude Sonnet 3.5 for code generation. * denote zero-shot models which were evaluated on whole dataset; for other models, 4-fold cross-validation was used (**P:** Precision; **R:** Recall; **A:** Accuracy). Highest values in green, lowest in purple.

Approach	GPT 4o					Gemini					Claude				
	P	R	A	F1	AUC	P	R	A	F1	AUC	P	R	A	F1	AUC
Bukhari et al.	0.94	0.95	0.97	0.95	1.00	0.94	0.95	0.97	0.95	0.99	0.95	0.96	0.98	0.96	**1.00**
GPTSniffer	**0.99**	**1.00**	**0.99**	**0.99**	1.00	**0.99**	**1.00**	**0.99**	**0.99**	1.00	**0.99**	**1.00**	**0.99**	**0.99**	1.00
Ye et al.*	0.63	**0.76**	0.65	0.69	0.71	0.58	0.12	0.52	0.20	0.46	0.67	**0.61**	0.66	**0.64**	0.72
DetectCodeGPT*	**0.51**	0.99	**0.51**	**0.67**	**0.51**	0.63	0.80	0.73	0.70	0.73	**0.53**	0.93	**0.60**	0.67	**0.59**

Tables 2 and 3 presents baseline results for the five classifiers under examination. We do not report variations across cross-validation folds as for all metrics the range is at most 0.03, and typically well below. Training-based models (Bukhari et al. and GPTSniffer) generally perform reasonably well, with F1 scores substantially above 0.8 and in line with published results. However, there is the question of whether such results hold up when training and evaluation set have different charateristics, as samples derived from different sources may be distributionally different. We evaluate this question in the next subsection.

Interestingly, zero-shot models performance measured on our datasets were significantly lower than published results. Ye et al. [41] report AUCs above 80% for many experimental scenarios on APPS [14] and MBPP [3] benchmarks. We note that this approach performance appear sensitive to specific implementation choices, such as the number of rewrites and the model used for rewriting. We further investigate the impact of these choices below.

Similarly, DetectCodeGPT [32], which is a zero-shot classifier, shows varied base-line results through different models and dataset in comparison with the original paper. While the original publication was also based on the CSN dataset, it used relatively small models (1-7B parameters) for AI code generation, which may explain the discrepancies with our results.

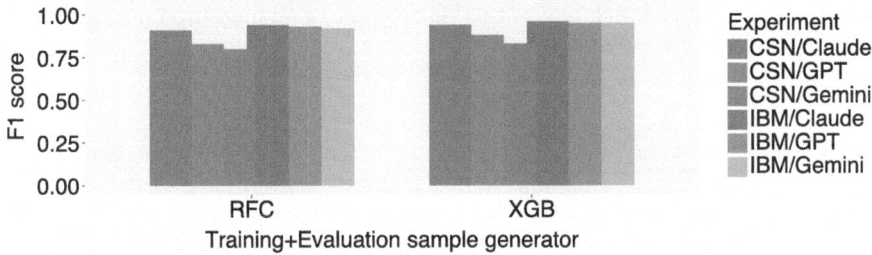

Fig. 5. F1 score for Bukhari et al. depending on classification algorithm used.

Bukhari et al. Design Space Analysis. Bukhari et al.'s approach works by (i) using AST-based analysis to transform each sample program in a feature vector; and (ii) training and using a Machine Learning classifier to label sample vectors as either human- or machine-generated. As such, it is sensitive to the particular algorithm used to train the classifier. We evaluate both XGBoost and Random Forest classifiers as those resulted in the best performance in their original paper [5]. F1 scores for both classifiers for all combinations of problem set/generator model (using cross-validation) are depicted in Fig. 5. XGBoost results in marginally better performance, and we use it for all other experiments.

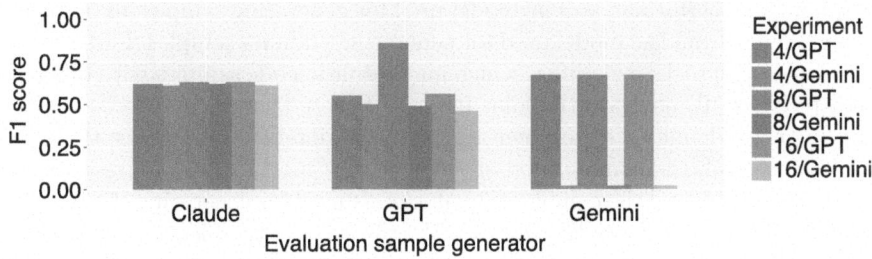

Fig. 6. F1 score for Ye et al. depending on number of rewrites/model used to generate rewrites (CSN prompt set only).

Ye et al. Design Space Analysis. As Ye et al.'s approach is dependent on specific parameters including (i) number of times the code is truncated and rewritten; and (ii) model used for rewriting (note, this is different from the model used to generate the AI portion of the dataset), we investigate sensitivity of the results on those choices. Figure 6 displays the resulting F1 scores on our whole dataset for the CSN prompt set (we omit the IBM prompt set for brevity). Results suggest no clear trend and for all other experiments we pick the combination of parameters which maximizes the average of all measured metrics across both prompt sets (8 rewrites w/ GPT).

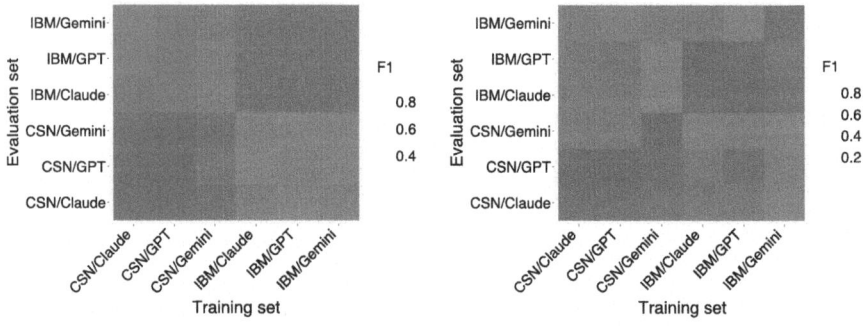

(a) Effect of training/evaluation set on F1 score for GPTSniffer classifier

(b) Effect of training/evaluation set on F1 score for Bukhari et al. classifier

Fig. 7. Training/evaluation set characteristics and F1 scores.

5.3   Factors Affecting Classification

Impact of Training and Evaluation Set. For classifiers which requires training, a relevant question is whether mismatch between training and evaluation set can impact performance. The model used for generating samples to be classified may differ from the model used for generating training samples. Further, the nature of coding tasks may differ, which may result in code samples with different characteristics. To evaluate the joint impact of these factors, we proceed as follows. First, we define an experiment as a combination of three factors: the *Classifier* being used, the combination of *Problem sets* used for training/evaluation (e.g., *CSN_IBM*), and the combination of *Models* used for generating the training and evaluation set (e.g., *GPT_Gemini*). We run a code classification experiment on all feasible combinations of factors, recording Precision and Recall, and we tabulate the results. Finally, we build a random forest regressor predicting each metric from the factors, and we extract Gini feature importance.

We resort to this approach as structured statistical tests would require a large number of repeated samples for each experiment, which are prohibitive to obtain due to the large and costly amount of computation required for individual experiments. While we verified the variation among repetitions to be minimal, we prefer the regression approach as it does not require repeated measures. We acknowledge that this approach does not allow us to generalize results beyond our set of experiments, but we believe that it is sufficient to identify general trends and qualitatively identify significant factors that may affect classification. As our goal here is not to achieve generalization but to maximize explanatory power, we perform hyper-parameter tuning.

The Precision regressor achieves good explanatory power ($R^2 = 0.78$), ranking Problem sets as the top feature ($Gini = 0.84$), followed by Models (0.10) and Classifier (0.07). The Recall regressor only achieves $R^2 = 0.46$ but ranks factors in the same order, respectively, with Gini scores 0.54, 0.25, 0.21. These results suggest that, regardless of the algorithm being used, training-based classifier performance is largely defined by differences between the characteristics of code used for training and classification, with the AI model used for code generation also playing a role.

To further investigate this observation, we plot heatmaps depicting how F1 scores vary based on the combination of training and evaluation set characteristics (i.e., the combination of problem set and model used for the generation of AI samples). Results for both GPTSniffer and Bukhari et al. are shown in Fig. 7. The plots clearly show how classifier performance degrades significantly when the training and evaluation sets differ, particularly in terms of the problem set.

Table 4. Base Prompt Variation Results: Each row presents the mean ± standard deviation for Precision, Recall, and F1 score across Base Prompts 1–5, showing the impact of different prompts on classifier performance. Highest values in green, lowest values in purple.

Approach	GPT 4o			Gemini			Claude		
	P	R	F1	P	R	F1	P	R	F1
Bukhari	0.84±0.01	0.79±0.02	0.82±0.01	0.84±0.01	0.79±0.02	0.82±0.01	0.84±0.01	0.79±0.02	0.82±0.01
GPTSniffer	0.92±0.01	0.99±0.00	0.96±0.00	0.92±0.01	0.99±0.00	0.96±0.00	0.92±0.01	0.99±0.00	0.96±0.00
Ye et al.	0.67±0.04	0.61±0.07	0.63±0.04	0.67±0.04	0.61±0.07	0.63±0.04	0.67±0.04	0.61±0.07	0.63±0.04
DetCodeGPT	0.52±0.04	0.95±0.08	0.67±0.01	0.52±0.04	0.95±0.08	0.67±0.01	0.52±0.04	0.95±0.08	0.67±0.01

Impact of Prompt Template. Each prompt used in our AI code generation experiments consists of two main components. The first part is the base prompt, which provides the model with instructions on how to generate code, what aspects to focus on, and the required output format. The second part is a detailed problem description from either the CSN or IBM dataset. During the code generation process, we kept the base prompt constant while iterating through problem descriptions from the datasets.

In this section, we investigate whether the choice of prompt template used for generation affects classification results. Our base prompt is discussed in Sect. 3.3 and given in Fig. 3. In this experiment, we designed five additional base prompts to assess their impact on classification performance. To design of the variations, we performed an extensive literature review[4,7,15,18,42] and syncretized five styles: *Validation-Centric, Minimalist, Self-Contained, Strict Output-Only*, and *Testing-Oriented*. Content and description of each style are provided in our data package (see Sect. 1). We evaluate the effect of each prompt template by randomly selecting 100 problems (evenly distributed among the CSN and IBM sets) and feeding them to each of the models used for generation, resulting in a dataset of 300 instances (we used 4-fold cross-validation for training-based classifiers, and we fed the whole dataset to zero-shot classifiers).

Results are presented in Table 4. For brevity's sake, for each prompt/model combination we list the average value for each metric and the standard deviation. Variations across different base prompts are relatively small, which suggests that the choice of base prompt has a minimal effect on the final generated code and, consequently, on the classification process.

Table 5. Multi-model, multi-problem training set: Each row presents Precision, Recall, and F1 score for a classifier trained on two out of three models, using both CSN and IBM problem sets.

Approach	GPT+Gem vs Cl			GPT+Cl vs Gem			Gem+Cl vs GPT		
	P	R	F1	P	R	F1	P	R	F1
Bukhari et al.	0.96	0.90	0.93	0.93	0.82	0.87	0.90	0.59	0.71
GPTSniffer	0.95	0.66	0.78	0.91	0.92	0.91	0.96	0.73	0.83

Table 6. Multi-model, single-problem training set: each row presents Precision, Recall and F1 score for a classifier trained on samples generated by all models on one problem set, and evaluated on samples generated by all models for a different problem set.

Approach	CSN vs IBM			IBM vs CSN		
	P	R	F1	P	R	F1
Bukhari et al.	0.63	0.43	0.51	0.27	0.84	0.41
GPTSniffer	0.50	0.99	0.67	0.51	0.99	0.67

5.4 Impact of Diversified Training Set

For models that require training, another relevant question is whether diversification aids classification performance. In particular, we ask: (i) does diversifying the training set helps with classifying samples from previously unseen models?; and (ii) does diversifying the training set helps with classifying samples from a previously unseen problem set?

To answer the first question, we run one-out experiments where classifiers are trained using combined samples from CSN and IBM generated by two models and evaluated on samples generated by the third (we split the problem set so each problem only appears in either training or evaluation). Results are shown in Table 5. These results suggest that diversifying the training set does not consistently improve detection performance on unseen models. For example, Bukhari et al. suffers from low recall (0.28 average) in detecting Gemini/CSN when trained on GPT/CSN or Claude/CSN only (see Fig. 7(b)). When trained on GPT+Claude/CSN+IBM, its recall jumps to 0.87. However, the same classifier exhibits an average recall of 0.71 in detecting GPT/CSN based on training on Gemini/CSN or Claude/CSN, while in this experiment, its recall falls to 0.59.

To answer the second question, we run four experiments where (i) we train each classifier on all samples generated from all models on the CSN problem set, and evaluate them on the IBM problem set; and (ii) vice versa. Results are presented in Table 6. Metrics in columns 2–4 were computed by training on the CSN problem set and evaluating on the IBM problem set; metrics in columns 5–7 were computed by training on IBM and evaluating on CSN. Results remain poor, suggesting that diversifying the dataset in terms of the generator model does not enable the classifiers to generalize across problem domains.

Overall, results show that the benefits of diversifying the training dataset may be limited, at least for the current generation of AI code detectors. Classifiers are able, to an extent, to generalize even from training data generated by a single model to detect code from another model - as long as the problem set remains the same. However, they are unable to transfer insights gleaned from one set of software development tasks, to a different set of tasks. We further discuss these observations in Sect. 6.

6 Discussion

Threats to Validity. We mitigate threats to internal validity by explicitly modeling confounding factors such as model used for generation, problem domain, and discrepancies between training and evaluation set. We strive to mitigate external validity threats by diversifying our set of models used for generation, and for considering two different coding problem datasets. We believe our selection of models to be representative; the popularity of GPT, Gemini and Claude is empirically confirmed by the fact that these are the three models supported by the popular GitHub Copilot plugin [9]. While our selection of problem sets is limited to CSN and IBM, results show that they are sufficient to identify limitations of existing models, and they both consist of high-quality prompts with directly or indirectly vetted human solutions. As for construct validity, some models are sensitive to design parameters. We pre-analyze the effect of such parameters to ensure each model is tested under the most favourable conditions.

Another limitation of our study lies in the scope of the analyzed code sample: we focused exclusively on comparing source code that was entirely authored by humans with source code that was entirely generated by AI. This binary distinction enabled clearer ground-truth labeling and more controlled experiments. However, it does not capture the scenario in which human-authored and AI-generated code is intertwined—such as when developers use AI assistants to suggest, complete, or modify code—highlighting the need for future work to study these mixed-authorship settings. Thus, our current approach is inherently geared towards a best-case classification scenario, where the inputs represent fully human or fully AI-generated code.

A final limitation is that our study exclusively focused on Python source code. While this decision removes language as confounding factor, it nonetheless restricts the generalizability of our findings to other programming languages. Characteristics such as syntactical, lexical, and style may vary significantly across languages and may influence the performance of code origin classifiers. As a result, further investigation is necessary to evaluate whether our methods and conclusions hold when applied to code written in other coding languages.

Generated Code Functionality Preservation. An orthogonal but relevant question is whether AI-generated code correctly implements the desired functionality (human code samples in our datasets are directly or indirectly pre-vetted for correctness). In a small-scale experiment evaluating the correctness of machine-generated code, we compared 25 human-written code samples from the IBM problem set with AI outputs from the three models under examination. The IBM problems typically include sample test inputs/outputs, which we used to generate per-problem unit tests. Claude achieved a 48% pass rate (12/25), Gemini 24% (6/25), while GPT attained 52% (13/25). Notably, all three models succeeded concurrently in only 24% of cases (6/25), whereas 40% (10/25) of samples failed across the board, often due to recurring issues such as timeouts, incorrect values, and missed edge cases. In these cases, human programmers are likely not to use AI-generated code "as is", but rather modify it to correct recurring errors, resulting in mixed-authorship samples discussed above.

Implications and Future Directions. Results presented in Sect. 5 present a complex picture, with some clearly identifiable insights. First, zero-shot models tend to perform poorly, with precision oftentimes below 70%, and recall below 80%. Models based on training or fine-tuning can achieve high performance, provided that the code submitted for classification comes from the same problem domain (i.e., problem set) used for training/fine-tuning. This observation is consistent with past work [22], and with the observation that human- and AI-generated code for the same problem set presents distinct statistical properties (ref. Section 3.3). The same-origin assumption may be reasonable in some contexts – for example, detecting AI code in implementations of well-defined critical functionality, such as implementations of specific ciphers.

In general, however, existing tools do not appear to be able to detect AI-generated code in the wild, under realistic conditions, with sufficient accuracy to be practical. Thus, it may be worth investigating alternative solutions such as watermarking [33]. In situations where compliance requirements can be enforced, it may also be possible to use developer tools to track the use of AI code assistants and tag AI-generated code as it is created [5].

In parallel, incorporating developer perspectives may prove invaluable. This would entail conducting a surveys on the experiences of developers with AI detection tools, developer views on the performance and security of machine-generated code, and the broader culture surrounding the use of AI-assisted programming.

7 Related Work

Supply Chain Security. Supply chain security issues emerge from current software design practices where code from various, potentially untrusted origins is incorporated into a project. This increases the risk of malicious, vulnerable, or other undesirable code being present in projects. Recent work includes identifying malicious packages, such as in PyPI [10] or npm [1]. Such solutions are, however, imperfect, suggesting that any externally sourced code should be tracked and examined. As AI-generated code potentially introduces new threat vectors into the supply chain (e.g., through the generation of insecure code), there is a need for automated processes that can complement existing human-driven software practices for provenance tracking and analysis; our work provides insight into the current state-of-the-art on this front.

Code Classification and Stylometry. Author identification using coding style dates back to the 1980s, with Oman et al. [23] pioneering the approach by analyzing the typographic and layout styles of programs to identify the authors of three Pascal algorithms from various computer science textbooks. More recently, Caliskan et al. [6] developed the Code Stylometry Feature Set (CSFS) for programmer de-anonymization, specifically for identifying human authors from a set of potential candidates. Caliskan's technique, used for source code attribution, is considered a closed-world machine learning task involving multi-class classification. CSFS offers a comprehensive code representation, categorized into three feature types: lexical features, which reflect programmer choices like keyword, identifier, and operator frequency; layout features, capturing the visual structure of the code, including indentation, line length, and comments; and syntactic features, derived from Abstract Syntax Trees (ASTs), which delve deeper into the code's structure by examining element types, nesting levels, and control flow.

Building upon the foundation laid by Caliskan et al.'s [6] Code Stylometry Feature Set (CSFS), researchers have explored various applications and extensions. Watson [37], for example, presented a method to de-anonymize source code contributors based on intrinsic programming style, building upon Caliskan-Islam et al.'s work but modifying the feature set and modeling strategy for improved scalability and feature-selection robustness. In our previous work [5], on the other hand, we have leveraged the CSFS for a binary classification task, distinguishing between AI-generated and human-written code. Nguyen et al. [22] developed GPTSniffer, a machine learning solution designed to detect source code potentially generated by ChatGPT. GPTSniffer's classification engine utilizes

CodeBERT, a pre-trained model specialized in code analysis and trained on the extensive CodeSearchNet dataset. Shi et al. [32] introduced DetectCodeGPT, a novel method for identifying machine-authored code. DetectCodeGPT modifies code and analyzes the responses of a pre-trained model. It focuses on stylistic tokens like whitespaces and newlines. By strategically inserting such tokens, DetectCodeGPT aims to highlight stylistic differences between human and machine-written code, making identification easier. Ye et al. [41] developed a zero-shot method for detecting synthetic (AI-generated) code. Their method is based on the principle that the similarity between the original code and versions rewritten by LLMs is indicative of whether the code is AI-generated. The process involves rewriting the code and subsequently comparing the original and rewritten versions for similarity.

8   Conclusion

We performed a comparative investigation of the performance of tools for detecting AI-generated code. We investigated multiple classifiers, taking into account both the impact of the model used for generating code, and of different programming tasks. Results suggest that classifiers can be effective under narrow assumptions, but are not yet sufficiently accurate to be used in the wild.

Acknowledgements. We thank the anonymous reviewers for their feedback. We also thank Brian Meta, Masroor Posh and Elizabeth Wyss for their help with this work. This work was supported by NSERC Alliance Grant #2341206 "Managing Risks of AI-generated Code in the Software +Supply Chain".

References

1. Sejfia, A., Schafer, M.: Practical automated detection of malicious NPM packages. In: ICSE (2022)
2. Anthropic: Introducing claude (2023). https://www.anthropic.com/news/introducing-claude
3. Austin, J., et al.: Program synthesis with large language models. arXiv preprint arXiv:2108.07732 (2021)
4. Brown, T.B., et al.: Language models are few-shot learners. arXiv preprint arXiv:2005.14165 (2020)
5. Bukhari, S., Tan, B., De Carli, L.: Distinguishing AI- and human-generated code: a case study. In: ACM CCS SCORED Workshop (2023)
6. Caliskan-Islam, A., et al.: De-anonymizing programmers via code stylometry. In: USENIX Security Symposium (2015)
7. Chen, M., et al.: Evaluating large language models trained on code. arXiv preprint arXiv:2107.03374 (2021)
8. Claburn, T.: GitHub and OpenAI fail to wriggle out of Copilot lawsuit (2023). https://www.theregister.com/2023/05/12/github_microsoft_openai_copilot/
9. Dohmke, T.: Bringing developer choice to copilot with anthropic's Claude 3.5 sonnet, Google's Gemini 1.5 pro, and OpenAI's O1-preview (2024). https://github.blog/news-insights/product-news/bringing-developer-choice-to-copilot/

10. Ly Vu, D., Newman, Z., Speed Meyers, J.: Bad snakes: understanding and improving python package index malware scanning. In: ICSE (2023)
11. Friedman, N.: Introducing GitHub Copilot: your AI pair programmer (2021). https://github.blog/2021-06-29-introducing-github-copilot-ai-pair-programmer/
12. GPTZero: AI detector - the original AI checker for ChatGPT & more. https://gptzero.me/
13. HackerNews: Ask HN: does your company ban GitHub Copilot? Hacker news (2023). https://news.ycombinator.com/item?id=34914810
14. Hendrycks, D., et al.: Measuring coding challenge competence with apps. arXiv preprint arXiv:2105.09938 (2021)
15. Husain, H., Wu, H.H., Gazit, T., Allamanis, M., Brockschmidt, M.: Codesearchnet challenge: evaluating the state of semantic code search. arXiv preprint arXiv:1909.09436 (2020)
16. IBM Research: github/project_codenet (2025). https://github.com/IBM/Project_CodeNet
17. Idialu, O.J., Mathews, N.S., Maipradit, R., Atlee, J.M., Nagappan, M.: Whodunit: classifying code as human authored or GPT-4 generated – a case study on CodeChef problems. In: MSR (2024)
18. Mathews, N.S., Nagappan, M.: Test-driven development and LLM-based code generation. In: ASE (2024)
19. McCabe, T.J.: A complexity measure. IEEE Trans. Softw. Eng. **SE-2**(4) (1976)
20. Meta AI: Introducing Meta Llama 3: the most capable openly available LLM to date (2025). https://ai.meta.com/blog/meta-llama-3/
21. Mitchell, E., Lee, Y., Khazatsky, A., Manning, C.D., Finn, C.: DetectGPT: zero-shot machine-generated text detection using probability curvature. arXiv preprint arXiv:2301.11305 (2023)
22. Nguyen, P.T., Di Rocco, J., Di Sipio, C., Rubei, R., Di Ruscio, D., Di Penta, M.: GPTSniffer: a CodeBERT-based classifier to detect source code written by ChatGPT. J. Syst. Softw., 112059 (2024)
23. Oman, P.W., Cook, C.R.: Programming style authorship analysis. In: CSC (1989)
24. OpenAI: Introducing ChatGPT (2022). https://openai.com/index/chatgpt/
25. OpenAI: Best practices for prompt engineering with the OpenAI API, OpenAI Help Center (2025). https://help.openai.com/en/articles/6654000-best-practices-for-prompt-engineering-with-the-openai-api
26. Pearce, H., Ahmad, B., Tan, B., Dolan-Gavitt, B., Karri, R.: Asleep at the keyboard? Assessing the security of GitHub Copilot's code contributions. In: IEEE S&P (2022)
27. Perry, N., Srivastava, M., Kumar, D., Boneh, D.: Do users write more insecure code with AI assistants? In: ACM CCS (2023)
28. Pichai, S., Hassabis, D.: Introducing Gemini: our largest and most capable AI model (2023). https://blog.google/technology/ai/google-gemini-ai/
29. Roberto Torres: Apple restricts ChatGPT, GitHub Copilot use over data worries: report (2023). https://www.ciodive.com/news/apple-chatgpt-openai-copilot-generative-AI/650816/
30. Schuster, R., Song, C., Tromer, E., Shmatikov, V.: You autocomplete me: Poisoning vulnerabilities in neural code completion. In: USENIX Security Symposium (2021)
31. Shani, I., GitHub Staff: survey reveals AI's impact on the developer experience (2023). https://github.blog/2023-06-13-survey-reveals-ais-impact-on-the-developer-experience/
32. Shi, Y., Zhang, H., Wan, C., Gu, X.: Between lines of code: unraveling the distinct patterns of machine and human programmers. In: ICSE (2025)

33. Suresh, T., Ugare, S., Singh, G., Misailovic, S.: Is the watermarking of LLM-generated code robust? arXiv preprint arXiv:2403.17983 (2025)
34. TIOBE: TIOBE index (2025). https://www.tiobe.com/tiobe-index/
35. Tufano, R., Mastropaolo, A., Pepe, F., Dabić, O., Di Penta, M., Bavota, G.: Unveiling ChatGPT's usage in open source projects: a mining-based study. In: MSR (2024)
36. Vaidya, R.K., De Carli, L., Davidson, D., Rastogi, V.: Security issues in language-based software ecosystems. arXiv preprint arXiv:1903.02613 (2021)
37. Watson, D.: Source code stylometry and authorship attribution for open source, Ph.D. thesis, University of Waterloo (2019)
38. Weidinger, L., et al.: Taxonomy of risks posed by language models. In: ACM FAccT (2022)
39. Yan, S., et al.: An LLM-assisted easy-to-trigger backdoor attack on code completion models: injecting disguised vulnerabilities against strong detection. In: USENIX Security Symposium (2024)
40. Yang, X., Zhang, K., Chen, H., Petzold, L., Wang, W.Y., Cheng, W.: Zero-shot detection of machine-generated codes. arXiv preprint arXiv:2310.05103 (2023)
41. Ye, T., et al.: Uncovering LLM-generated code: a zero-shot synthetic code detector via code rewriting. arXiv preprint arXiv:2405.16133 (2024)
42. Zhou, Y., et al.: Large language models are human-level prompt engineers. arXiv preprint arXiv:2211.01910 (2023)

FlexGE: Towards Secure and Flexible Model Partition for Deep Neural Networks

Xiaolong Wu$^{(\boxtimes)}$, Aravind Kumar Machiry, Yung-Hsiang Lu, and Dave Jing Tian

Purdue University, West Lafayette, IN, USA
{wu1565,amachiry,yunglu,daveti}@purdue.edu

Abstract. Proprietary deep neural network (DNN) models are being deployed in the cloud nowadays. With the increased usage of AI accelerators in the cloud, there is a growing need for privacy protection for outsourced deep learning computations. Existing works use a Trusted Execution Environment (TEE) to shield DNN partitions, which puts a subset of the DNN model in TEEs and offloads the rest of the computation on GPUs. However, these solutions use fixed security primitives and model partition policy, which precludes per-model specialization to balance the security and performance requirements. In this paper, we present a novel on-demand model inference system, FlexGE, that partitions the DNN model between TEE and GPU accelerator with programmable partition policies and protection primitives based on the user's configuration. FlexGE achieves this by tailoring the protection profile as well as the model partition policy and partitioning the model at *build time* as opposed to design time. We implement FlexGE using Darknet and GEVisor, and evaluate it on five popular DNNs. Our evaluation shows that FlexGE is flexible and outperforms the state-of-the-art in terms of security and performance.

Keywords: Model Partition · GPU TEE · DNN

1 Introduction

Deep neural networks (DNNs) are widely used and often require excessive computational resources. Meanwhile, cloud computing makes deep learning more accessible, flexible, and cost-effective while allowing developers to build deep learning algorithms faster. Artificial Intelligence as a Service (AIaaS) [26] in the cloud uses pre-trained models and enables vendors to reduce the risk and hardware investment of their customers. At the same time, the rapid increase in the complexity of the software stack in the cloud expands the attack surface for machine learning applications. This raises privacy concerns about DNN computations in untrusted environments, in particular, for DNN models outsourced by a client to a remote cloud server. Attackers are financially motivated to steal these models derived from expensive training with a significant engineering effort. As a result, leakage of such proprietary models can cause severe financial loss and security issues.

To make matters worse, existing proprietary models are found to be not well protected, especially since GPU as a Service (GPUaaS) [12] is prominent in the cloud.

© The Author(s), under exclusive license to Springer Nature Switzerland AG 2025
M. Egele et al. (Eds.): DIMVA 2025, LNCS 15748, pp. 65–85, 2025.
https://doi.org/10.1007/978-3-031-97623-0_4

Using GPUs to accelerate DNN makes model privacy protection even more challenging. Nevertheless, directly applying TEEs to protect entire DNN models presents significant challenges, as most commercial GPUs (e.g., V100 and A100) lack TEE functionality. While some high-end GPUs, such as those based on the Nvidia Hopper architecture [1], support confidential computing, their cost remains prohibitively high for typical model users. For instance, an Nvidia H100 GPU is over 15 times more expensive than a GeForce RTX 4090[1]. Therefore, researchers propose Trusted Execution Environment (TEE) shielded DNN partition [46,47,54], which puts a subset of privacy-sensitive and critical components of the DNN model in TEEs and offloads the rest computation on GPUs. However, these approaches suffer from several drawbacks. First, they preclude per-model specialization to balance the security and performance requirements. The rigid use of security primitives in these techniques permanently locks the design into a fixed combination of security primitives that is likely to result in suboptimal security/performance in many scenarios. Second, they fix the model partition policy for each model, losing the flexibility to explore the advantages of different partition policies. As a result, when the protection offered by a hardware security primitive breaks down (e.g., physical attacks [28]), or the protection offered by a security primitive is too expensive (e.g., the homomorphic encryption may counteract the performance benefit of GPU acceleration), or the model partition policy is not optimal, it is difficult and costly to decide how it should be replaced.

When multiple protection mechanisms or model partitioning policies are available for a given model, selecting the most suitable configuration depends on various factors and is best deferred until deployment time. This leads to one important research question: *Is it possible to switch between different protection primitives and model partition policies at deployment time, avoiding the lock-in model partition that characterizes the status quo?* Our answer is that protection profile as well as model partition policy can cost-efficiently be tailored towards a specific DNN model at *build time*, as opposed to design time.

To verify this idea, we design *FlexGE*, the first system framework that supports different model partition policies and protection primitives and enables flexible fine-grained model partition at *build time* via delegating partial computations to different back-ends, potentially in different protection domains, with different protection mechanisms. The challenge is how to instantiate protection primitives for each model partition, what partition granularity to use between different partitions, and what software hardening mechanisms should be applied to mitigate the potential vulnerabilities of hardware primitives. To that aim, we abstract the common operations required when partitioning arbitrary models behind a generic API that is used to retrofit an existing DNN model into FlexGE. This API reduces the manual porting effort of the existing DNN model by partitioning weight matrix data using annotations. These annotations, alongside other abstract source-level constructs, are replaced at build time by FlexGE to instantiate a given configuration.

There exists a significant gap in security and performance between TEE and GPU. This large disparity presents a substantial trade-off space with potential for optimiza-

---

[1] At February 2025, the price of an H100 GPU is about $25,000, while the price of a GeForce RTX 4090 is less than $1600.

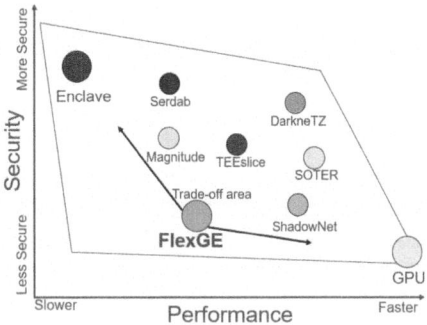

Fig. 1. Design space of DNN model partition.

tion. The configuration space enabled by model partition, illustrated in Fig. 1, is large and almost impossible for a non-expert user to explore manually. This leads to the second research question we explore: *how to guide a typical user to navigate the vast configuration space unlocked by FlexGE?* To answer this, we propose a quantitative analysis framework that formally defines the security-performance trade-off across various model partitioning policies and includes an automatic algorithm for selecting the optimal configuration. Existing works only consider a model-level partition policy that does not reflect the effect of system protection profile variation (i.e., strong protection primitives offer higher security but can reduce performance, while weaker protection primitives deliver better performance at the cost of reduced security). We address the security-performance separation gap and argue that model partitioning should incorporate not only model-level partitioning policies but also system-level protection primitives. However, this design introduces an additional challenge: while a weak protection profile can reduce performance overhead, it also risks an extreme scenario where a single compromised protection primitive could lead to a complete system crash. Therefore, we propose *Configuration Space Layout Randomization (CSLR)*, a co-design method for model and system co-configuration, hardening both the partition and the protection primitives. As the workload has been partitioned and delivered to different back-ends, FlexGE makes it impossible for the attackers to collect and piece together all the information to complete the attack, as it requires the attackers to breach all back-ends with different protection primitives. This design makes the system robust and resilient to future attacks, assuming the partitioning mechanism fails.

We have developed a FlexGE prototype that integrates Intel Software Guard Extensions (SGX) to support CUDA kernel I/O encryption and VM/EPT-based GPU I/O protection through a hypervisor, extending the security boundary of SGX from the CPU to the GPU, as well as two hardening mechanisms (CFI [2] and ASLR [41]). Our evaluation of several deep neural networks demonstrates the potential security versus performance tradeoff space unlocked by FlexGE, e.g., exploring over 80 configurations for AlexNet. Finally, we demonstrate that under an equivalent requirement, FlexGE outperforms the state-of-the-art in terms of security and performance.

The contributions of this paper are as follows.

- We design FlexGE, a model inference system for DNNs that supports arbitrary partitioning policies between TEE and GPU, allowing flexible delegation of partial computations to various back-ends with different protection mechanisms (Sect. 4).
- We propose a novel *build time* model partitioning mechanism that enables flexible configuration and fine-grained partitioning of models (Sect. 4.2).
- We propose a novel *Configuration Space Layout Randomization (CSLR)* mechanism to enhance the security of the model inference system (Sect. 4.3).
- We propose *Quantitative Metrics* for characterizing both security and performance, and the corresponding algorithm to identify the optimal configuration for each of the partition policies (Sect. 5).
- Our evaluation demonstrates FlexGE's security and flexibility for real-world usage on a diverse range of DNN architectures (Sect. 7).

2 Background

2.1 Deep Neural Network

Convolutional neural network (CNN) [35] is a class of deep neural networks that typically consists of an input and an output layer with a sequence of linear and non-linear layers stacked in between. The linear layers include convolutional layers and fully connected layers; the non-linear layers include activation and pooling layers.

Convolutional Layer. The parameters of a convolutional layer consist of a set of learnable filters. Each filter is characterized by the width, height, and depth of the receptive field. The depth must be equal to the number of channels of the input feature map. Let h, b, d represent the height, width, and depth of the filter ω, respectively, and (x, y) refer to the coordinates in the 2D output feature map. Formally, the convolution operation on a given image I with filter ω can be described as follows:

$$CONV(I, \omega)_{x,y} = \sum_{i=1}^{h}\sum_{j=1}^{b}\sum_{k=1}^{d} \omega_{i,j,k} I_{x+i-1, y+j-1, k} \tag{1}$$

Let X and Y denote the input and output, respectively, and $W = [\omega_1, \ldots, \omega_n]^T$ be the convolution filter. The corresponding convolutional layer is thus given by:

$$Y = Conv(X, W^T) \tag{2}$$

Fully Connected Layer. The dense layer connects every input node to every output node. It can be implemented as a convolutional layer with both filter height and width: 1. For example, a dense layer connecting n input to m output can be viewed as a convolutional layer that has m filters of size $(1, 1, n)$.

Residual Neural Network (a.k.a. Residual Network, ResNet) [14] is another deep learning model extended from CNN for addressing the vanishing gradient problem of CNN to some extent, in which the network skips connections that perform identity mappings, merged with the layer outputs by addition.

Table 1. Comparison between existing works and FlexGE.

	Model Privacy	Cloud/Device	Backend Protection	Partition Policy	Flexible Configuration	Fine Grained
Slalom [47]	-	Cloud	Obfuscation	Linear Layers	-	-
DarkneTZ [29]	✓	Device	-	Deep layer	-	-
Serdab [10]	✓	Device	-	Shadow layer	-	-
Magnitude [16]	✓	Device	-	Large-Mag. Weights	-	✓
SOTER [43]	✓	Device	-	Intermediate Layers	-	-
ShadowNet [46]	✓	Device	Obfuscation	Non-Linear Layers	-	-
TEEslice [54]	✓	Device	Obfuscation	Slice	-	✓
FlexGE	✓	Cloud (or Device)	EPT and Cryptographic (Extensible)	All	✓	✓

Residual Layer. Suppose the output of linear operation of layer $1+2$ is z^{l+2}, and the ReLU [30] function is g. For a 2 layer skip residual layer, instead of $a^{l+2} = g(z^{l+2})$, the output of layer $1+2$ is:

$$a^{l+2} = g(z^{l+2} + a^l) \tag{3}$$

2.2 Intel SGX

Intel Software Guard Extensions (SGX) [32] is a Trusted Execution Environment (TEE) that ensures the confidentiality and integrity of user code and data. SGX allows a process to allocate a protected memory region, i.e., an enclave, within its address space. Intel SGX affords the enclave hardware protections against CPU-based attacks but is not designed to secure the communication between the enclave and external devices attached to the system; that is, SGX's security boundary is only within the CPU. One of the main reasons for this is that all external device communication is traditionally handled by the OS. In particular, the device drivers (loaded onto the OS) create and maintain a memory-mapped I/O channel between a program and the intended device(s).

2.3 GPU Software Stack

The GPU device driver is responsible for the creation, deletion, and upkeep of a communication channel with the GPU. Gdev [19], an open-source GPU stack consists of an implementation of the CUDA driver API and `libdrm` and `nouveau`, which implement the user- and kernel-space GPU device driver.

3 Motivation

Based on the model partition policy, existing TEE-based DNN model partition research can be classified into five categories in Table 1, including shielding the deep layer into TEE, such as DarkneTZ [29], shielding the shallow layer into TEE, such as Serdab [10], putting intermediate layers to enclave like SOTER [43], putting fine-grained sub-layers into the enclave, such as Magnitude [16] and TEEslice [54]), and shielding no-linear layers into the enclave, e.g., ShadowNet [46].

Slalom [47] outsources the linear layers to the GPU for acceleration with masked inputs while keeping the other layers inside SGX. However, Slalom protects the user

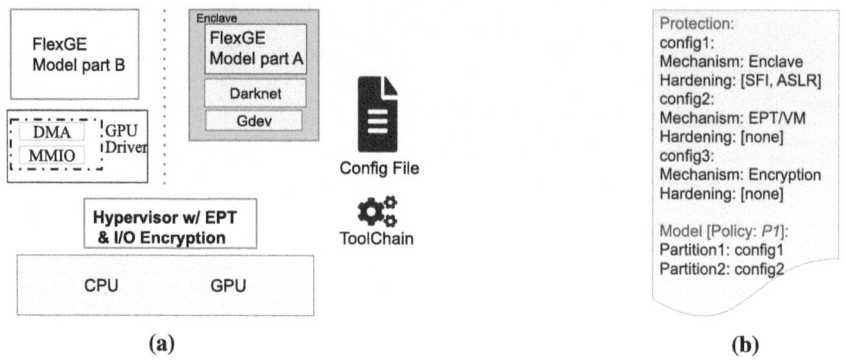

Fig. 2. (a) FlexGE Architecture. (b) Configuration file.

input privacy but not the model weights from the untrusted cloud server. State-of-the-art (i.e., TEEslice [54]) proposes a training-before-partition strategy, which involves expensive training on a private model.

We make the following key observation from Table 1: existing DNN model partition policies are fixed at the design time. This motivates us to design a flexible DNN model partition framework corresponding to different users' diverse security and performance requirements. FlexGE seeks to enable users to easily and securely switch between TEE and GPU with different protection primitives and partition policies at deployment time.

4   Design

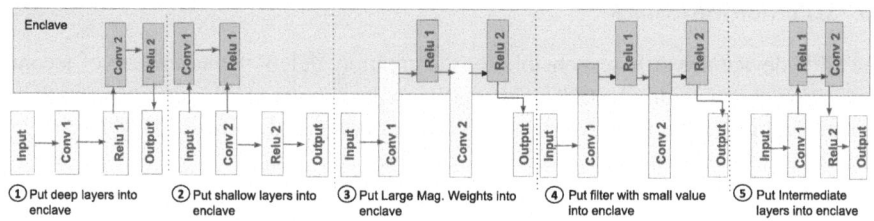

Fig. 3. Model partition policy

We now present an overview of the design of FlexGE in Fig. 2(a). FlexGE is composed of SGX-based CPU TEE and GPU, and two GPU I/O protection backends between CPU TEE and GPU through a hypervisor and GPU runtime (i.e., Gdev): EPT/VM and CUDA kernel Encryption. The encryption backend provides strong but expensive I/O channel protection, while the EPT-based backend provides weaker but more performant protection. The CPU TEE (i.e., enclave) and EPT/VM backend can be hardened via techniques such as Control-Flow Integrity (CFI) and address space layout random-ization (ASLR). The security configuration is provided at build time in a configuration file provided by the user including different protection configurations of various pro-tection primitives and model configuration with selected partition policy and protection configurations, and FlexGE's tool-chain produces a DNN model partition implementa-tion with the desired security characteristics. An example of such a configuration file is given in Fig. 2(b).

FlexGE's tool-chain enables calling external security functions via abstract gates and data sharing between the enclave and GPU via abstract code annotations. Gates and annotations form an API used to partition a DNN model in FlexGE. A given partition configuration in Fig. 2(b) is automatically replaced by our toolchain with a particular implementation at build time.

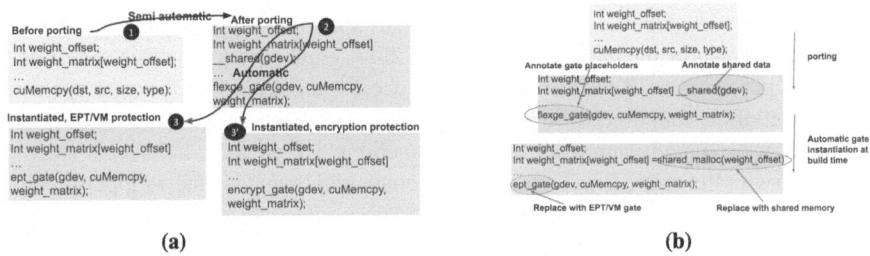

(a)   (b)

Fig. 4. (a) FlexGE code transformations. First, users manually annotate partitioned data with shared (Gdev) annotation, and gate placeholders are automatically inserted. At build time, API primitives are automatically replaced with the chosen mechanism. (b) EPT Backend Instantiation.

4.1 Fine-Grained Model Partition Policy

The goal of FlexGE's model partition policy is to support all the policies in existing works. We omit the approach of ShadowNet [46] because it does not require configu-rations. For sliced-based partition [54], we instead propose a fine-grained filter-based partition policy to represent it. Totally, we provide five partition policies (P) for a DNN model including layer-based and fine-grained filter-based partition, as shown in Fig. 3.

① According to the layer depth and closeness to the output layer in the TEE, two deep-est layers (Conv2 andReLU2) are put into the enclave (P1).
② According to the layer depth and closeness to the input layer in the TEE, two shal-lowest layers (Conv1 and ReLU1) are put into the enclave (P2).
③ According to the absolute weight value, partial convolution layers with large-magnitude weights and ReLU layers are put into the enclave (P3).
④ According to the filter value, ReLU1 and Conv2 with the filter of small values put into the enclave (P4).
⑤ Putting randomly chosen intermediate layers in the TEE. In Fig. 3, ReLU1 and Conv2 as the random-selected layers are put into the enclave (P5).

4.2   Build-Time Model Partition Mechanism

API and Build-Time Instantiation. The GPU I/O backend security functions are made through abstract call gates that are instantiated at build time. *Shared data* (e.g., weight matrix) between CPU TEE and GPU is marked using compiler annotations, used at build time to instantiate a given DNN model partition policy. FlexGE performs replacements using source-to-source transformations, which gives it a better performance advantage over heavyweight runtime abstraction interfaces.

Call Gates. In FlexGE, security function calls (EPT/VM, and encryption) are represented in the source code by abstract call gates. At build time, as part of the transformation phase, abstract call gates are replaced with a specific implementation. For instance, when the DNN model is configured to be in the GPU with EPT/VM backend, the call gate performs VM enter call. Figure 4(a) presents an example of gates from the porting (step ❷) to the replacement by the toolchain (❸ and ❸').

Porting existing Gdev code to FlexGE consists of marking call gates, which can be automated: Knowing the system's control flow graph, static analysis determines whether a procedure call performs GPU I/O access and, if so, performs a syntactic replacement of the function call with a call gate instead. However, in our current implementation, we manually annotate the Gdev functions that perform I/O access (MMIO and DMA) with the abstract call gate. The toolchain will then generate wrappers enclosing the implementations of the functions in the appropriate call gates.

EPT/VM Backend. The EPT/VM backend is based on a small hypervisor. FlexGE's hypervisor provides GPU I/O protection, including MMIO and DMA, with an EPT mechanism. The hypervisor ensures that the DMA and command buffers are inaccessible to an attacker who attempts to use the CPU to access these memory regions with EPT trapping. In particular, the hypervisor maintains memory region mapping tables containing the virtual and physical address pairs of both MMIO and DMA memory regions per enclave within a reserved memory region and traps access to these regions for access control. Compared with the encryption backend, the EPT-based backend provides weaker but more performant protection. FlexGE provides two variants of EPT backend: EPT and EPT_2M with huge page optimization [50].

Data Sharing. The EPT backend relies on shared memory areas to share model weights across VMs in EPT and EPT_2M backends. Areas are always mapped at the same address in the VMs so that pointers to/in shared model matrix structures remain valid. Each VM manages its own portion of the shared memory area to avoid the need for complex multithreaded bookkeeping.

Figure 4(b) shows the instantiating procedure of an EPT protection backend. Using the programmable API, the user would first annotate shared data and add gate placeholders. Then, at build time, FlexGE would replace the annotated shared data with a shared memory location and replace the gate placeholder with an EPT/VM gate.

Encryption Backend. While FlexGE could use EPT to prevent unauthorized accesses to DMA from privileged software with better performance, it cannot stop attackers from attaching probes to the I/O bus and snooping the traffic to steal the code and data. Fundamentally, an EPT/VM protection would have to downgrade the threat model by excluding potential physical attacks.

Instead, we design a *crypto CUDA kernel* to augment GPU with *in-device* Diffie-Hellman (DH) key exchange, encryption, and decryption, enabling a crypto-secured DMA communication channel between enclaves and GPU. A user enclave and the GPU first performs local attestation to verify each other. Once they establish the trust through attestation, the crypto kernel is loaded from the enclave to the GPU using MMIO, which is protected by the hypervisor. Once loaded, the crypto kernel within the GPU launches a DH key exchange to establish a shared secret key within the GPU memory and the enclave. A physical bus snooping attack could observe our crypto kernel and even the DH key exchange but not the shared secret established after. From this point, enclaves can encrypt the code and data before exposing them in DMA. GPU will decrypt them inside the device and encrypt the results again before writing them into DMA. Assuming an authenticated encryption scheme, e.g., AES-GCM, we will not need a hypervisor anymore since any tampering will fail the decryption and thus be detected.

Software Hardening. The flexible DNN model secure protection provided by FlexGE allows enabling/disabling software hardening (SH) such as CFI, etc. Isolating DNN model layers with SH from layers without it allows the former to maintain the guarantees offered by SH. This flexibility allows for alleviating the performance impact of SH by enabling it only for sub-layers of a DNN model. Our prototype currently uses address space layout randomization (ASLR) and CFI. ASLR hides the memory layouts from adversaries by randomly placing code and data in runtime, which makes it hard for the victim code or data to find the location so that control-flow hijack or data-flow manipulation attacks are prevented. FlexGE employs fine-grained randomization by splitting the code section of SGX enclave into a set of randomization units [41]. For CFI, our enforcement is performed for all control transfer instructions for the program in the SGX enclave, including indirect branches as well as return instructions. In the case of indirect branches, we add masking operations to the destination so that it only points to one of the randomization unit's entry points. In the case of a return instruction, we replace it with two equivalent instructions, *pop reg* and *jmp reg*, where *reg* can be any available register. Then, the second *jmp* instruction is instrumented similarly to indirect branch instructions.

4.3 Configuration Space Layout Randomization (CSLR)

Inspired by Address Space Layout Randomization (ASLR) [42], we introduce Configuration Space Layout Randomization (CSLR) for FlexGE. The security of FlexGE is directly related to the entropy [4] of the CSLR implementation. Low entropy would allow an attacker to brute-force the entire search space and bypass CSLR if they can repeatedly attempt exploits. To counter this, FlexGE utilizes fine-grained randomization schemes to maximize CSLR entropy.

For DNN model-level fine-grained CSLR, FlexGE adopts a smaller partition size, called a randomization unit. These units have a fixed, configurable size (e.g., in a 20-layer DNN, a randomization unit size of 5 results in 4 partitions). The overall configuration space is determined by permutations between model partitions and back-end protection mechanisms. Consequently, CSLR entropy is proportional to the number of permutations of model partition sizes and available back-end protections. FlexGE is designed to support extensible back-end protection mechanisms, further enhancing system security.

5 Design Space Navigation

We propose a quantitative method (formal definition) given a configuration.

5.1 Quantitative Formalization

To systematically find the optimal model partition policy for a DNN model given a system configuration, we formalize the problem as an optimization problem. Formally, let S be a model partition solution that splits a DNN model into enclave and GPU-offloaded portions. Let P denote a configuration policy of S that specifies to what degree the model is put into an enclave. We define a security score and a performance score for a specific configuration, Security(P) and Performance(P), which quantify the security risk and performance cost of P, respectively. We define $Security_{max}$ as the security risk baseline of setting, which puts the whole DNN model in an enclave. $Security_{max}$ denotes the lower bound of the security risk (the strongest protection). We also define $Security_{min}$ as the security risk upper-bound of the setting, which puts all the layers out of the enclave and off-loads to GPU. $Security_{min}$ denotes the upper bound of the security risk (the weakest protection). Then, given the security requirement Δ, we formulate the optimal configuration $P^\star$ that satisfies:

$$P^\star = \underset{|Security(P)\text{-}Security_{max}|<\Delta}{argmin} Performance(P) \tag{4}$$

We define the security score, Security(P), using model stealing accuracy [37], which calculates how much test samples can be correctly classified by the attacker's surrogate model (We detail the model stealing procedure in Sect. 7). Achieving high accuracy is a primary goal of model stealing attacks [54]. Given the total test number N_{total}, and the sample number can be correctly classified $N_{correct}$. Security(P) is defined as:

$$Security(P) = (N_{correct}(P)/N_{total})\% \tag{5}$$

We define the performance score, Performance(P), using performance overhead. Suppose the inference latency of the DNN layers conducted in the enclave is $T\_Enclave$, and the total inference latency corresponding to the configuration P is T_P, then we define the Performance(P) as the ratio of inference latency in the TEE over the total inference latency of the DNN model:

$$Performance(P) = (T\_Enclave/T\_P)\% \tag{6}$$

Thus, a larger Performance(P) indicates fewer computations are offloaded on GPUs, leading to higher performance overhead.

5.2 Optimal Configuration Selection

For each of the five policies in Sect. 4.1, we iterate possible configurations to identify $P^\star$. In particular, for the policy that shields deep layers (P1), we use a lightweight, greedy clustering algorithm that assign layers into clusters. We begin the algorithm by placing the last layer into a cluster; we then proceed to perform repeated clustering one more layer operations until an assignment of layers produces the optimal $P^\star$. We shield different numbers of consecutive "deep" layers starting from the output layer with TEEs. Similarly, for (P2), which shields shallow layers, we put different amounts of consecutive layers starting from the DNN input layer. For ResNet models, we use the residual layers as the dividing boundaries. For VGG models and AlexNet models, we use convolution layers as boundaries. For shielding large-magnitude weights (P3), the number of protected weights is controlled by a configuration parameter mag_ratio. We gradually set range of mag_ratio from 0.1 to 0.9, until finding the optimal $P^\star$. Similarly, for putting filter with small value into enclave (P4), we set the filter_ratio. For shielding intermediate layers (P5), the number of shielded layers is also defined by a configuration parameter, inter_ratio. Similarly, we set the range of inter_ratio from 0.1 to 0.9. For P3, P4, and P5, setting mag_ratio, filter_ratio, and inter_ratio to 0 represents the $Security_{min}$ while setting the parameters to 1 is the $Security_{max}$.

6 Implementation

We implement FlexGE based on Darknet deep learning framework [39] and a tiny hypervisor: GEVisor [50]. We implement a build-time source transformations toolchain. We use the Vembyr PEG parser generator [36] to automate the development of a source-to-source translator. Vembyr provides a convenient extension interface that allows us to construct an abstract syntax tree (AST). The final compilation pass converts the AST into a concrete syntax tree (CST) to print out the C code. FlexGE's toolchain performs source transformations to (1) instantiate abstract gates, (2) instantiate data sharing code, (3) generate linker scripts, and (4) generate additional code in Gdev according to backend-provided GPU I/O protection recipes.

7 Evaluation

7.1 Evaluation Setting

Our experimental machine uses an Intel i7- 8700K 4.7 GHz CPU with Intel SGX (SDK v2.0), 6 cores, and 32 GB of main memory. We use a NVIDIA GeForce GTX TITAN Black GPU with 2,880 CUDA cores and 6,144 MB GDDR5 384-bit memory.

Datasets. We use four different datasets in our experiments, including CIFAR10 [21], CIFAR100 [21], STL10 [8], and UTKFace [53].

Models. The benchmark models include ResNet18 [14], VGG16_BN [44], AlexNet [22], ResNet34 and VGG19_BN. We use the public models [34] as initialization for all the experiments.

7.2   Case Study

In this section, we explore the vast performance/security design space enabled by FlexGE. In Fig. 5, we first partition the AlexNet model into four partitions and then plot the inference time with the CIFAR100 data set for each configuration. We variate different partitions with different protection backend configurations, while all of them apply policy P4. For example, having partition1 in an enclave with hardening, partition2 in an enclave with no hardening, partition3 in EPT_2M (EPT with 2M huge page optimization [50]) protection without hardening, and partition4 in EPT with hardening leads to a 30.8 s inference latency. Overall, we observe that FlexGE enables a very wide range of security configurations with significant performance variation. The configuration with all four partitions in an enclave with hardening performs worst, with 119 s of inference latency. It is worth mentioning that the red color does not attach with hardening because the GPU I/O encryption protection cannot come with software hardening. Existing approaches assume a one-size-fits-all all configuration are therefore suboptimal; in contrast, FlexGE enables users to easily navigate the security/performance trade-off inherent in their application.

CSLR Security Analysis. AlexNet employs an 8-layer CNN with five convolutional layers, two fully connected hidden layers, and one fully connected output layer. The randomization unit is 2 with 4 partitions. The configuration space in our study is 80. As a result, the entropy of CSLR is proportional to $80 * 4!$. It is very hard for an attacker to breach such a large randomization space.

Observation 1: *Even with a small DNN architecture, FlexGE can provide strong security protection with CSLR.*

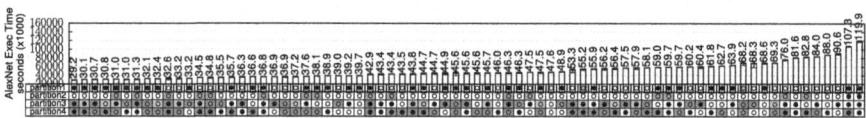

Fig. 5. AlexNet execution time for a range of configurations. Partitions are on the left. Software hardening can be enabled ● or disabled ○ for each partition. The white/blue/red/green color indicates the partitions are placed into. White: enclave, red: encryption, blue: EPT_2M, green: EPT. (Color figure online)

7.3   Quantitative Evaluation for Model Partition Policies

Model Stealing Attack. The attacker first analyzes the target defense scheme and then infers the architecture of the protected model based on the offloaded part, and the model output with existing techniques [6,7] to get an initialized model. Then, the attacker chooses a public model (with the same or an equivalent architecture), trains this model with queried data, and outputs the surrogate model (the recovered victim model). Lastly, the attacker transports the model weights in the offloaded part of the surrogate model to the corresponding parts of the initialized model.

Model Stealing Accuracy. We test the five partition policies and report the evaluation results over five models (AlexNet, ResNet18, ResNet34, and VGG16_BN VGG19_BN) in Table 2 (total 20 cases). As aforementioned, we also report the baseline settings ("$Security_{min}$" and "$Security_{max}$") for comparison. For each partition policy P in Sect. 4.1, we iterate possible configurations to identify $P^\star$. For ResNet models, we use the residual layers as the dividing boundaries. For VGG models and AlexNet models, we use convolution layers as boundaries. For each model and dataset, we mark the highest MS attack accuracy in red and the lowest accuracy in blue. The results show that TEESlice and FlexGE achieve the lowest MS attack accuracy. Due to the CSLR mechanism, FlexGE achieves the lowest MS attack accuracy in more cases than TEESlice. Moreover, it maintains the lowest MS attack accuracy across all policies.

Observation 2: All policies have the potential to achieve the lowest MS attack accuracy.

Table 2. MS attack accuracy for FlexGE with a randomized system configuration.

		$Security_{min}$	P1(DarkneTZ I FlexGE)	P2(Serdab I FlexGE)	P3(Magnitude I FlexGE)	P4(TEESlice I FlexGE)	P5(SOTER I FlexGE)	$Security_{max}$
AlexNet	C10	85.59%	70.95% I 34.64%	62.86% I 30.32%	60.79% I 29.88%	35.84% I 32.67%	70.76% I 33.98%	20.30%
AlexNet	C100	62.33%	38.99% I 19.22%	42.33% I 20.53%	47.90% I 22.65%	21.97% I 21.21%	50.37% I 24.99%	11.77%
AlexNet	S10	75.44%	73.55% I 35.22%	66.85% I 31.41%	68.49% I 31.89%	33.27% I 33.12%	37.98% I 18.71%	15.32%
AlexNet	UTK	91.03%	85.90% I 60.63%	80.03% I 59.44%	79.29% I 59.12%	63.80% I 60.13%	57.63% I 48.54%	46.44%
ResNet18	C10	92.38%	84.95% I 50.86%	89.86% I 52.13%	83.16% I 51.54%	56.62% I 49.66%	88.45% I 52.23%	17.87%
ResNet18	C100	80.24%	70.15% I 39.78%	75.02% I 41.22%	69.25% I 40.22%	41.88% I 38.95%	75.01% I 41.07%	13.44%
ResNet18	S10	86.53%	83.86% I 51.36%	83.31% I 51.16%	70.08% I 49.97%	55.22% I 48.33%	80.41% I 50.66%	21.36%
ResNet18	UTK	90.23%	83.54% I 53.19%	81.77% I 52.45%	61.37% I 49.88%	53.88% I 47.96%	75.66% I 51.85%	46.88%
ResNet34	C10	90.89%	84.33% I 59.22%	30.17% I 18.56%	20.85% I 15.33%	13.09% I 20.22%	90.06% I 62.39%	12.79%
ResNet34	C100	81.51%	71.32% I 50.88%	73.98% I 51.02%	75.48% I 51.39%	42.56% I 50.89%	79.67% I 52.77%	16.85%
ResNet34	S10	88.12%	83.69% I 54.98%	82.12% I 53.67%	65.97% I 45.33%	50.02% I 46.89%	79.64% I 52.17%	20.23%
ResNet34	UTK	86.91%	86.36% I 55.71%	76.78% I 50.49%	47.65% I 48.66%	47.94% I 47.25%	80.67% I 51.33%	47.06%
VGG16_BN	C10	91.53%	85.29% I 54.65%	90.35% I 56.17%	81.06% I 53.21%	56.01% I 55.02%	90.77% I 56.55%	14.33%
VGG16_BN	C100	71.84%	61.79% I 43.89%	70.48% I 48.13%	62.57% I 44.62%	46.91% I 46.66%	71.93% I 48.73%	10.65%
VGG16_BN	S10	90.22%	87.69% I 51.44%	87.66% I 51.43%	80.09% I 49.77%	52.89% I 50.21%	88.43% I 52.54%	19.13%
VGG16_BN	UTK	90.83%	85.45% I 53.88%	87.88% I 54.98%	55.69% I 53.91%	58.70% I 55.02%	90.21% I 55.28%	46.24%
VGG19_BN	C10	91.95%	89.34% I 55.66%	84.95% I 53.37%	79.66% I 48.65%	41.07% I 48.66%	82.01% I 48.93%	10.88%
VGG19_BN	C100	70.88%	62.97% I 44.65%	69.72% I 45.97%	61.38% I 45.11%	45.01% I 44.03%	46.72% I 45.67%	11.01%
VGG19_BN	S10	90.11%	88.44% I 55.09%	86.82% I 55.01%	82.79% I 54.33%	34.07% I 54.67%	55.44% I 54.89%	19.67%
VGG19_BN	UTK	90.14%	88.99% I 55.67%	87.02% I 55.34%	87.98% I 55.73%	58.32% I 55.62%	88.19% I 55.54%	43.87%

Optimal Configuration. For FlexGE, we use the system configuration: Enclave with SFI and ASLR software hardening, and GPU CUDA kernel encryption I/O protection and then measure the values of the Performance($P^\star$) as defined in Eq. 4 (the smallest value of performance overhead to achieve $Security_{max}$) for a given model partition policies P4. Table 3 reports the performance overhead to achieve $Security_{max}$ for each setting for MS. We highlight the results of FlexGE in orange, while marking the lowest performance ($P^\star$) of other works in blue and the highest values in red. Overall, Table 3 implies that the performance overhead to achieve $Security_{max}$ is distinct across different DNN models and datasets. For example, to protect AlexNet from MS by putting deep layers into enclave, DarknetTZ needs to take about 98.91% performance overhead, which also suggests it puts 98.91% of the protected model in TEE to achieve $Security_{max}$ for CIFAR10 (C10) and CIFAR100 (C100). However, for STL10

(S10) and UTKFace (UTK), it only needs to put 35% of the model in TEE to achieve $Security_{max}$. Compared to other defenses, FlexGE achieves the lowest performance cost in most cases. That is, FlexGE generally incurs less performance overhead while achieving the highest level of black-box defense. One interesting result we find is that, on average, *P2* (Serdab) has more performance cost to achieve $Security_{max}$ compared with *P1* (DarknetTZ), which illustrates that it is more secure to offload deep layers to GPU than offload shallow layers. The reason is that shallow layers are close to input data and respond more to low-level photographic information of the original inputs. In contrast, deep layers represent more abstract and specific feature information.

Observation 3: *Shallow layers are close to input data and are more easy to expose model privacy.*

Table 3. Different Performance($P^{\star}$) values of optimal configuration in front of MS. A lower value represents a lower performance overhead. The performance overhead for $Security_{min}$ and $Security_{max}$ baselines are 0% and 100%, respectively.

	AlexNet				ResNet18				ResNet34				VGG16_BN			
	c10	c100	s10	UTK	c10	c100	s10	UTK	c10	c100	s10	UTK	c10	c100	s10	UTK
DarkneTZ	98.91%	98.91%	35.28%	35.28%	99.07%	99.07%	100%	35.12%	99.23%	99.23%	100%	35.33%	100%	100%	81.55%	66.45%
Serdab	98.97%	98.97%	99.11%	99.12%	99.91%	99.91%	100%	35.56%	70.50%	100%	100%	50.05%	100%	100%	100%	100%
Magnitude	72.91%	80.58%	100%	68.21%	100%	80.25%	100%	58.48%	19.37%	100%	100%	5.01%	100%	80.34%	100%	100%
TEEslice	64.21%	55.09%	33.22%	31.10%	49.97%	49.97%	51.78%	50.44%	20.87%	41.59%	45.96%	8.27%	55.88%	50.99%	51.56%	52.34%
SOTER	99.98%	99.98%	100%	58.73%	100%	100%	100%	100%	100%	100%	81.23%	100%	100%	100%	100%	100%
FlexGE	49.23%	49.98%	31.17%	31.03%	47.83%	44.22%	41.37%	39.66%	19.56%	35.27%	31.23%	12.44%	48.34%	43.11%	49.69%	40.71%

7.4   Accuracy Loss

We compare the accuracy between FlexGE and TEESLICE as well as the baseline that puts the whole model into the enclave. Following TEESLICE, we set the queried data sizes as 1K, 2K, 4k 5K, 10K, 15K, 20K, 25K, 30K. We display MS accuracy on CIFAR100 and four models (AlexNet, ResNet18, ResNet34, and VGG16BN) in Fig. 6. In general, FlexGE does not lead to a considerable loss of accuracy corresponding to the baseline. We choose the optimal configuration for FlexGE with the same performance score (81.66%) with TEESLICE based on the evaluation methods in Sect. 7.2 and 7.3. Across most queried data, FlexGE achieves higher accuracy than TEESLICE, which depends on expensive supervised machine learning model training. This reliance can lead to overfitting, potentially resulting in suboptimal outcomes.

7.5   Performance

We run the TEESLICE's source code on the same Desktop PC with FlexGE. We choose the optimal configuration for FlexGE with the same security score with TEESLICE based on the results in Sect. 7.3. We ran all experiments ten times and got the average

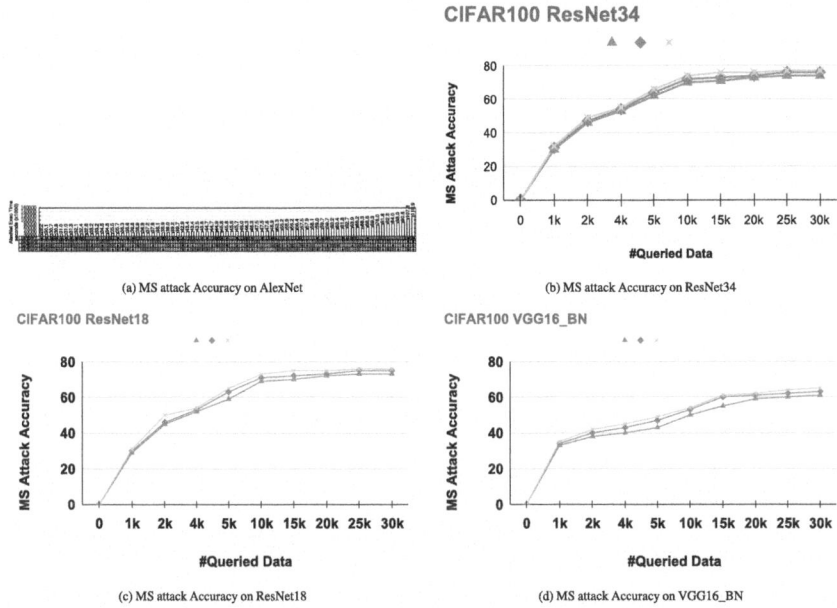

Fig. 6. Comparison of FlexGE, TEESlice, and the whole enclave protection against MS attacks with different sizes of queried data. $\triangle$ represents TEESLICE, $\diamond$ represents FlexGE, and $\times$ represents whole enclave.

inference time. The running time deviates is less than 10% from the average. We calculate the throughput (images per second) as it is a common criterion to evaluate the speed of machine learning systems.

Table 4 presents the throughput of FLexGE on three models (AlexNet, ResNet18, and VGG16BN), as well as two baselines: putting the whole model in the enclave and directly running on the GPU. The whole enclave is the throughput lower bound, and the direct GPU is the upper bound. From the results, we can see that the throughput of direct GPU (from 91.54 to 473.72) is much higher than that of the whole-enclave (from 1.52 to 7.63), demonstrating the efficiency of GPU. We choose the same security score (51.38%) for FlexGE and TEESLICE. The throughputs of TEESLICE ranges from 38.09 to 80.10, which is much faster than the whole-enclave baseline but slower than directly running on the GPU. In contrast, FlexGE's throughputs range from 38.77 to 82.01. Most of the results have performance speedups with reference to TEESLICE, which are, on average, 7.53% faster than TEESLICE. The performance speedup mainly comes from FLexGE's EPT I/O protection primitives and optimization mechanism, such as the super page optimization EPT_2 M, compared to the cryptographic I/O protection of TEESLICE.

To analyze the performance of FlexGE further, we also logged the latency of different parts during the inference phase. We break down the inference latency of FlexGE is divided into three parts: data transfer, partition in the enclave, and partition on GPU. Data Transfer refers to the time it takes to transfer data between SGX and GPU. Parti-

Table 4. The throughput comparison between shielding-whole-model, no-shield, TEESLICE, and FlexGE.

		AlexNet	ResNet18	VGG16_BN
	Whole Enclave	6.51	7.63	1.52
	Direct GPU	473.72	266.13	91.54
TEESLICE	CIFAR10	40.05	58.43	68.60
	CIFAR100	42.13	41.21	53.74
	STL10	80.10	61.04	66.23
	UTKFace	38.09	53.42	38.56
FlexGE	CIFAR10	44.88	62.78	71.25
	CIFAR100	48.51	40.85	56.62
	STL10	82.01	62.92	70.13
	UTKFace	42.44	52.89	38.77

tion in the enclave refers to the time it takes to compute the layers/filters inside SGX. Partition on GPU is the time to compute the layers on the GPU. Table 5(a) displays the percentage of each part over the total inference latency. From the table, we can see that partition in the enclave occupies 65.09% of the inference time due to the constrained computation resources inside SGX. Data Transfer occupies 32.18% of the inference time. In particular, the data transfer with EPT protection spends 9.44% of the inference time, the data transfer with EPT_2M has 8.32% of the inference time, and encryption protection occupies the most time of data transfer, which is 14.42%. Partition on GPU only occupies 2.73% of the time due to the strong computation ability of the GPU. Note that although FlexGE introduces the additional overhead of Data Transfer, FlexGE still accelerates the overall inference time to a large degree. The reason mainly derives from two aspects. On the one hand, the EPT-based I/O protection introduces a low-performance overhead to the data transfer. On the other hand, the partition in the enclave mainly comes from non-linear layers, and most of the linear layers go to GPU, which largely speeds up the performance.

Observation 4: *EPT based I/O protection introduces much lower performance than encryption based I/O protection.*

Table 5. (a) FlexGE inference time breakdown. (b) MS accuracy on BART

Data transfer			Partition(Enclave)	Partition(GPU)
EPT	EPT_2M	Encrypt		
9.44%	8.32%	14.42%	65.09%	2.73%

(a)

	SST-2	MRPC	RTE	Average
Whole-enclave	51.72%	69.67%	49.58%	57.15%
Direct-GPU	93.42%	86.57%	69.99%	82.26%
FlexGE	52.29%	70.34%	50.13%	58.65%

(b)

7.6   Overheads: Shared Data Allocations, Gate Latencies

In FlexGE, the weight matrix can be shared via shared memory between Enclave and the EPT backend or between VMs corresponding to the EPT backend and the EPT_2 M backend. To illustrate the benefits of the shared memory, we measure, for each of the mechanisms, the execution time of a DNN model that allocates 1 to 3 shared weight matrices (size 1K Byte) and returns immediately. The results are in Fig. 7a. Shared memory allocations between Enclave and EPT are about two times slower than typical VMs shared memory allocations.

Another source of FlexGE overhead is gate latency. To illustrate the raw performance of FlexGE's gates, we measure the gate latency of Enclave ECall gate, Enclave OCall gate, and EPT gate. The results are shown in Fig. 7b. The Enclave OCall gate is slightly faster than the Enclave ECall gate and 18x slower than the EPT gate. EPT latencies are similar to syscall latencies without KPTI, illustrating the practicability of the EPT backend.

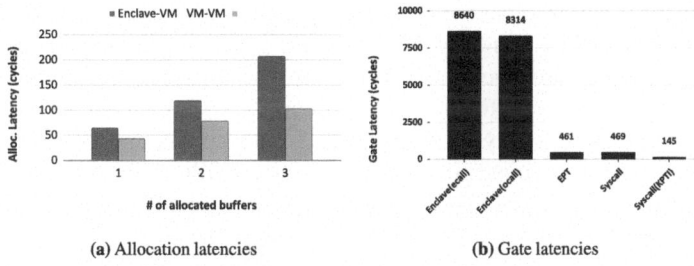

(a) Allocation latencies   (b) Gate latencies

Fig. 7. FlexGE latency microbenchmarks.

8   Discussion

Scalability to LargeLanguage Models (LLMs). Recently, LLMs (such as Chat-GPT [33] and LLaMA [11]) have been largely moved forward and widely used. LLMs bring new challenges to model privacy protection solutions because their sizes usually contain up to hundreds of billions of parameters that are much larger than traditional DNNs (only hundreds of millions of parameters [20]). However, the idea of FlexGE can also be applied to LLMs to protect the sensitive model privacy. To demonstrate the generalization of FlexGE, we evaluate FlexGE on a representative LLMs model, BART [24], and three datasets (SST-2, MRPC, and RTE) from the popular GLUE dataset [49]. We report MS accuracy in Table 5(b).

Other Metrices for Security and Performance Score. Currently we use the model stealing accuracy as the security score metric, however, fidelity [37] and Attack Success Rate (ASR) [37] can be another two metrics for model stealing. Fidelity is the percentage of test samples with identical predictions between the surrogate model and the victim model, including the samples that are misclassified by the victim model. ASR measures the transferability of adversarial samples. Moreover, for Membership Inference Attack (MIA) [52], we can also take gradient-based MIA accuracy, generalization gap [51], and confidence gap [52] as the metrics. For the performance score, we can use FLOPs to measure the utility cost of DNN models [16].

9   Related Work

Compartmentalization. Several compartmentalization frameworks [15,23,31,40] rely on code annotations for application porting. However, FlexGE targets data annotation by annotating the weight matrix offset of the DNN model. A few studies provide various degrees of porting automation through data flow analysis [5,27], and KSplit [17] applies compartmentalization and automation on driver isolation. Their principles can be applied to increase the degree of automation of FlexGE.

GPU TEEs. Recent work explored implementing trusted architectures directly inside GPUs to achieve isolation. Graviton [48] relies on architectural modification to the GPU to support TEE for GPU. Similarly, HIX [18] relies on hardware modification to the CPU, including SGX hardware and PCIe routing. Meanwhile, HETEE [56] supports large-scale confidential computing using PCIe Express-Fabric to distribute computation over server nodes that are physically isolated. StrongBox [9] and GEVisor [50] are software solutions based on existing hardware to build GPU TEE. StrongBox targets for ARM Endpoints based on TrustZone with an integrated GPU, while GEVISOR is designed to support trusted GPU execution with SGX enclaves.

Model Privacy Protection Methods. Existing model privacy protection approaches can be classified into four categories: two-party computation (2PC) based approaches, homomorphic encryption (HE) based approaches, trusted execution environment (TEE) based approaches, and obfuscation approaches. 2PC-based approaches [38,45], aim to protect the confidentiality of both user data in the client and DNN model on the cloud server. HE-based approaches [13] perform secure inference based on encrypted DNN model and encrypted data. TEE-based approaches as listed in Sect. 3 focus on securing DNN model computations in untrustworthy environments through TEE technologies. Obfuscation approaches can be categorized into approaches that protect the model's input [3] and approaches that safeguard its structure [25,55].

10   Conclusion

In this paper, we have proposed FlexGE, a secure model inference system for DNNs that supports arbitrary partitioning policies between TEE and GPU. It provides the flexibility for delegating partial computations to different backends with different protection primitives and various policies.

Acknowledgments. This work is supported in part by NSF grants CNS-2145744, ONR grant N00014-23-1-2157, and Wistron. Any findings and opinions expressed in this material are those of the authors and do not necessarily reflect the views of the funding agencies.

References

1. NVIDIA H100 tensor core GPU architecture (2022). https://resources.nvidia.com/en-us-tensor-core
2. Abadi, M., Budiu, M., Erlingsson, U., Ligatti, J.: Control-flow integrity principles, implementations, and applications. ACM Trans. Inf. Syst. Secur. (TISSEC) **13**(1), 1–40 (2009)
3. AprilPyone, M., Kiya, H.: Block-wise image transformation with secret key for adversarially robust defense. IEEE Trans. Inf. Forensics Secur. **16**, 2709–2723 (2021)
4. Barzasi, G.: The illusion of randomness: demystifying the entropy of ASLR on common operating systems (2022)
5. Bauer, M., Rossow, C.: Cali: compiler-assisted library isolation. In: Proceedings of the 2021 ACM Asia Conference on Computer and Communications Security, pp. 550–564 (2021)
6. Chen, J., et al.: Copy, right? A testing framework for copyright protection of deep learning models. In: 2022 IEEE Symposium on Security and Privacy (SP), pp. 824–841. IEEE (2022)
7. Chen, Y., Shen, C., Wang, C., Zhang, Y.: Teacher model fingerprinting attacks against transfer learning. In: 31st USENIX Security Symposium (USENIX Security 22), pp. 3593–3610 (2022)
8. Coates, A., Ng, A., Lee, H.: An analysis of single-layer networks in unsupervised feature learning. In: Proceedings of the Fourteenth International Conference on Artificial Intelligence and Statistics, pp. 215–223. JMLR Workshop and Conference Proceedings (2011)
9. Deng, Y., et al.: StrongBox: a GPU tee on arm endpoints. In: Proceedings of the 2022 ACM SIGSAC Conference on Computer and Communications Security, pp. 769–783 (2022)
10. Elgamal, T., Nahrstedt, K.: Serdab: an IoT framework for partitioning neural networks computation across multiple enclaves. In: 2020 20th IEEE/ACM International Symposium on Cluster, Cloud and Internet Computing (CCGRID), pp. 519–528. IEEE (2020)
11. Facebook AI: Introducing Llama: a foundational, 65-billion-parameter large language model (2023). https://ai.facebook.com/blog/large-language-model-llama-meta-ai/
12. Filippini, F., Lattuada, M., Jahani, A., Ciavotta, M., Ardagna, D., Amaldi, E.: Hierarchical scheduling in on-demand GPU-as-a-service systems. In: 2020 22nd International Symposium on Symbolic and Numeric Algorithms for Scientific Computing (SYNASC), pp. 125–132. IEEE (2020)
13. Gilad-Bachrach, R., Dowlin, N., Laine, K., Lauter, K., Naehrig, M., Wernsing, J.: CryptoNets: applying neural networks to encrypted data with high throughput and accuracy. In: International Conference on Machine Learning, pp. 201–210. PMLR (2016)
14. He, K., Zhang, X., Ren, S., Sun, J.: Deep residual learning for image recognition. In: Proceedings of the IEEE Conference on Computer Vision and Pattern Recognition, pp. 770–778 (2016)
15. Hedayati, M., et al.: Hodor: intra-process isolation for high-throughput data plane libraries. In: 2019 USENIX Annual Technical Conference (USENIX ATC 19), pp. 489–504 (2019)
16. Hou, J., Liu, H., Liu, Y., Wang, Y., Wan, P.J., Li, X.Y.: Model protection: real-time privacy-preserving inference service for model privacy at the edge. IEEE Trans. Dependable Secure Comput. **19**(6), 4270–4284 (2021)
17. Huang, Y., et al.: KSplit: automating device driver isolation. In: 16th USENIX Symposium on Operating Systems Design and Implementation (OSDI 22), pp. 613–631 (2022)
18. Jang, I., Tang, A., Kim, T., Sethumadhavan, S., Huh, J.: Heterogeneous isolated execution for commodity GPUs. In: 24th International Conference on Architectural Support for Programming Languages and Operating Systems (ASPLOS 2019), pp. 455–468. ACM, Providence, RI (2019). https://doi.org/10.1145/3297858.3304021
19. Kato, S., McThrow, M., Maltzahn, C., Brandt, S.: Gdev: first-class GPU resource management in the operating system. In: Presented as Part of the 2012 USENIX Annual Technical

Conference (USENIX ATC 12), pp. 401–412. USENIX, Boston, MA (2012). https://www.usenix.org/conference/atc12/technical-sessions/presentation/kato

20. Keras Contributors: Keras applications (2017). https://keras.io/api/applications/
21. Krizhevsky, A., Hinton, G., et al.: Learning multiple layers of features from tiny images (2009)
22. Krizhevsky, A., Sutskever, I., Hinton, G.E.: ImageNet classification with deep convolutional neural networks. In: Advances in Neural Information Processing Systems, vol. 25 (2012)
23. Lefeuvre, H., et al.: FlexOS: towards flexible OS isolation. In: Proceedings of the 27th ACM International Conference on Architectural Support for Programming Languages and Operating Systems, pp. 467–482 (2022)
24. Lewis, M.: BART: denoising sequence-to-sequence pre-training for natural language generation, translation, and comprehension. arXiv preprint arXiv:1910.13461 (2019)
25. Li, J., He, Z., Rakin, A.S., Fan, D., Chakrabarti, C.: NeurObfuscator: a full-stack obfuscation tool to mitigate neural architecture stealing. In: 2021 IEEE International Symposium on Hardware Oriented Security and Trust (HOST), pp. 248–258. IEEE (2021)
26. Lins, S., Pandl, K.D., Teigeler, H., Thiebes, S., Bayer, C., Sunyaev, A.: Artificial intelligence as a service: classification and research directions. Bus. Inf. Syst. Eng. **63**, 441–456 (2021)
27. Liu, S., Tan, G., Jaeger, T.: PtrSplit: supporting general pointers in automatic program partitioning. In: Proceedings of the 2017 ACM SIGSAC Conference on Computer and Communications Security, pp. 2359–2371 (2017)
28. Loukas, G.: Cyber-Physical Attacks: A Growing Invisible Threat. Butterworth-Heinemann (2015)
29. Mo, F., et al.: DarkneTZ: towards model privacy at the edge using trusted execution environments. In: Proceedings of the 18th International Conference on Mobile Systems, Applications, and Services, pp. 161–174 (2020)
30. Nair, V., Hinton, G.E.: Rectified linear units improve restricted Boltzmann machines. In: Proceedings of the 27th International Conference on Machine Learning (ICML-10), pp. 807–814 (2010)
31. Narayan, S., et al.: Retrofitting fine grain isolation in the Firefox renderer. In: 29th USENIX Security Symposium (USENIX Security 20), pp. 699–716 (2020)
32. Neiger, G., Santoni, A., Leung, F., Rodgers, D., Uhlig, R.: Intel virtualization technology: hardware support for efficient processor virtualization. Intel Technol. J. **10**(3) (2006)
33. OpenAI: ChatGPT (2023). https://chat.openai.com
34. Orekondy, T., Schiele, B., Fritz, M.: Knockoff nets: stealing functionality of black-box models. In: Proceedings of the IEEE/CVF Conference on Computer Vision and Pattern Recognition, pp. 4954–4963 (2019)
35. O'Shea, K., Nash, R.: An introduction to convolutional neural networks. arXiv preprint arXiv:1511.08458 (2015)
36. Rafkind, J.: Vembyr - multi-language peg parser generator written in Python, 2011 November. http://code.google.com/p/vembyr/
37. Rakin, A.S., Chowdhuryy, M.H.I., Yao, F., Fan, D.: DeepSteal: advanced model extractions leveraging efficient weight stealing in memories. In: 2022 IEEE Symposium on Security and Privacy (SP), pp. 1157–1174. IEEE (2022)
38. Rathee, D., et al.: CrypTFlow2: practical 2-party secure inference. In: Proceedings of the 2020 ACM SIGSAC Conference on Computer and Communications Security, pp. 325–342 (2020)
39. Redmon, J.: Darknet: open source neural networks in C (2013–2016). http://pjreddie.com/darknet/
40. Schrammel, D., et al.: Donky: domain keys–efficient in-process isolation for RISC-V and x86. In: 29th USENIX Security Symposium (USENIX Security 20), pp. 1677–1694 (2020)

41. Seo, J., et al.: SGX-shield: enabling address space layout randomization for SGX programs. In: NDSS (2017)
42. Shacham, H., Page, M., Pfaff, B., Goh, E.J., Modadugu, N., Boneh, D.: On the effectiveness of address-space randomization. In: Proceedings of the 11th ACM Conference on Computer and Communications Security, pp. 298–307 (2004)
43. Shen, T., et al.: SOTER: guarding black-box inference for general neural networks at the edge. In: 2022 USENIX Annual Technical Conference (USENIX ATC 22), pp. 723–738 (2022)
44. Simonyan, K., Zisserman, A.: Very deep convolutional networks for large-scale image recognition. arXiv preprint arXiv:1409.1556 (2014)
45. Srinivasan, W.Z., Akshayaram, P., Ada, P.R.: DELPHI: a cryptographic inference service for neural networks. In: Proceedings 29th USENIX Security Symposium, pp. 2505–2522 (2019)
46. Sun, Z., Sun, R., Liu, C., Chowdhury, A.R., Lu, L., Jha, S.: ShadowNet: a secure and efficient on-device model inference system for convolutional neural networks. In: 2023 IEEE Symposium on Security and Privacy (SP), pp. 1596–1612. IEEE (2023)
47. Tramer, F., Boneh, D.: SLALOM: fast, verifiable and private execution of neural networks in trusted hardware. arXiv preprint arXiv:1806.03287 (2018)
48. Volos, S., Vaswani, K., Bruno, R.: Graviton: trusted execution environments on GPUs. In: 13th USENIX Symposium on Operating Systems Design and Implementation (OSDI 2018), pp. 681–696. USENIX Association, Carlsbad, CA (2018). https://www.usenix.org/conference/osdi18/presentation/volos
49. Wang, A., Singh, A., Michael, J., Hill, F., Levy, O., Bowman, S.R.: Glue: a multi-task benchmark and analysis platform for natural language understanding. corr abs/1804.07461 (2018). arXiv preprint arXiv:1804.07461
50. Wu, X., Tian, D.J., Kim, C.H.: Building GPU tees using CPU secure enclaves with GEVisor. In: Proceedings of the 2023 ACM Symposium on Cloud Computing, pp. 249–264 (2023)
51. Yeom, S., Giacomelli, I., Fredrikson, M., Jha, S.: Privacy risk in machine learning: analyzing the connection to overfitting. In: 2018 IEEE 31st Computer Security Foundations Symposium (CSF), pp. 268–282. IEEE (2018)
52. Yuan, X., Zhang, L.: Membership inference attacks and defenses in neural network pruning. In: 31st USENIX Security Symposium (USENIX Security 22), pp. 4561–4578 (2022)
53. Zhang, Z., Song, Y., Qi, H.: Age progression/regression by conditional adversarial autoencoder. In: Proceedings of the IEEE Conference on Computer Vision and Pattern Recognition, pp. 5810–5818 (2017)
54. Zhang, Z., et al.: No privacy left outside: on the (in-) security of tee-shielded DNN partition for on-device ml. arXiv preprint arXiv:2310.07152 (2023)
55. Zhou, T., Ren, S., Xu, X.: ObfuNAS: a neural architecture search-based DNN obfuscation approach. In: Proceedings of the 41st IEEE/ACM International Conference on Computer-Aided Design, pp. 1–9 (2022)
56. Zhu, J., et al.: Enabling rack-scale confidential computing using heterogeneous trusted execution environment. In: 2020 IEEE Symposium on Security and Privacy (SP), pp. 1450–1465. IEEE (2020)

Poster: Exploring the Zero-Shot Potential of Large Language Models for Detecting Algorithmically Generated Domains

Tomás Pelayo-Benedet[1]([✉]) [iD], Ricardo J. Rodríguez[1] [iD],
and Carlos H. Gañán[2] [iD]

[1] Dpto. de Informática e Ingeniería de Sistemas, Universidad de Zaragoza, Zaragoza,
Spain
{tpelayo,rjrodriguez}@unizar.es
[2] Delft University of Technology, Delft, The Netherlands
C.HernandezGanan@tudelft.nl

Abstract. Domain generation algorithms enable resilient malware communication by generating pseudo-random domain names. While traditional detection relies on task-specific algorithms, the use of Large Language Models (LLMs) to identify Algorithmically Generated Domains (AGDs) remains largely unexplored. This work evaluates nine LLMs from four major vendors in a zero-shot environment, without fine-tuning. The results show that LLMs can distinguish AGDs from legitimate domains, but they often exhibit a bias, leading to high false positive rates and overconfident predictions. Adding linguistic features offers minimal accuracy gains while increasing complexity and errors. These findings highlight both the promise and limitations of LLMs for AGD detection, indicating the need for further research before practical implementation.

Keywords: Large Language Models · Algorithmically Generated Domains · DNS Traffic Analysis · Malware Detection

1 Introduction

Cybercrime has become a complex and persistent global threat, driven by the rapid evolution of attack techniques and technologies. As malicious actors develop increasingly sophisticated malware, defensive strategies must continually evolve to keep pace. In this context, the MITRE ATT&CK framework serves as a comprehensive, reality-based taxonomy of adversary behavior, categorizing both tactics (*why*) and techniques (*how*) employed by attackers at various stages of the intrusion lifecycle.

Within this framework, Command and Control (C2) plays an important role in maintaining communication between compromised systems and the attacker's infrastructure. A widely used technique to enable C2 is the use of Domain Generation Algorithms (DGAs) [5], first implemented in `Conficker` [9]. DGAs generate large volumes of pseudo-random domain names (*Algorithmically Generated*

© The Author(s), under exclusive license to Springer Nature Switzerland AG 2025
M. Egele et al. (Eds.): DIMVA 2025, LNCS 15748, pp. 86–92, 2025.
https://doi.org/10.1007/978-3-031-97623-0_5

Table 1. Overview of evaluated LLMs. "Context Window" and "Output Limit" indicate the token capacities as specified in the respective API documentation.

Model	Context Window	Output Limit	Release	Tier	API Tag
GPT-4o [6] †	128,000	16,384	Nov'24	Paid	gpt-4o-2024-11-20
GPT-4o-mini [6] ‡	128,000	16,384	Jul'24	Paid	gpt-4o-mini-2024-07-18
Claude Sonnet 3.5 [1] †	200,000	8,192	Oct'24	Paid	claude-3-5-sonnet-20241022
Claude Haiku 3.5 [1] ‡	200,000	8,192	Oct'24	Paid	claude-3-5-haiku-20241022
Gemini 1.5 Pro [2] †	2,097,152	8,192	Sep'24	Free	gemini-1.5-pro-002
Gemini 1.5 Flash [2] ‡	1,048,576	8,192	Sep'24	Free	gemini-1.5-flash-002
Gemini 1.5 Flash-8B [2] ‡	1,048,576	8,192	Oct'24	Free	gemini-1.5-flash-8b-001
Mistral Large [4] †	131,000	N/A	Nov'24	Free	mistral-large-2411
Mistral Small [4] ‡	32,000	N/A	Sep'24	Free	mistral-small-2409

† indicates larger models; ‡ indicates smaller models counterparts.

Domains, AGDs), predictable to both the malware and the attacker through a shared algorithm and a seed value. Infected systems repeatedly attempt to connect to these AGDs, improving resilience even if the domains are blocked.

While traditional AGD detection methods rely on personalized features or statistical learning, these approaches often struggle to generalize to new or obfuscated domains. This motivates the exploration of Large Language Models (LLMs), which can leverage pre-trained linguistic knowledge to identify patterns in domain names without requiring task-specific tuning. This work investigates whether LLMs, using only their pre-trained general knowledge in a zero-shot setting, can detect malicious AGDs. Specifically, we evaluate: (i) their ability to classify domains based solely on string analysis, and (ii) the impact of providing elaborate linguistic features on classification performance.

Our goal is to determine the feasibility of integrating LLMs as complementary tools into cybersecurity workflows to identify DGA-driven C2 activity.

2   Experimental Setup

To evaluate the effectiveness of LLMs in detecting AGD, we selected nine models from four major vendors, spanning both free and paid services. The models were selected based on their availability of public APIs and their proven natural language understanding and generation capabilities. A summary of the selected models is presented in Table 1.

To evaluate the models' performance, we built a dataset D composed of $25,000$ malicious domains generated by real-world DGAs and $25,000$ legitimate domains, with a total of $50,000$ domains. Malicious domains were obtained from DGArchive [7], selecting a diverse set of DGAs from 25 different malware families to capture diverse generation patterns. Legitimate domains were extracted from the Tranco list [10], which compiles widely visited and trusted domains. The legitimate domains were randomly selected to ensure an unbiased and representative sample of traffic to benign websites. While the Tranco list is generally considered a reliable source of legitimate domains, we acknowledge that it may occasionally include malicious entries [8].

Model evaluation was performed using two prompt variants, P_1 and P_2, designed following an iterative prompting approach [3]. The prompt P_1 represents a minimal formulation that introduces the classification task and specifies a structured output format: domain name, classification label, and confidence level. P_2 extends P_1 by incorporating explicit instructions for analyzing domain-specific lexical features and domain generation patterns [7]. The lexical features considered include: level of randomness, character frequency, digit-to-letter ratio, consonant-to-vowel ratio, pronounceability, presence of meaningful substrings, and similarity to known popular domains.

3   Evaluation Methodology and Discussion of Results

To evaluate the effectiveness of LLMs in detecting AGDs, we conducted a two-phase evaluation. First, we measured their baseline classification performance using the dataset D and the minimal prompt P_1. Then, we assessed whether performance improved by providing the models with domain-specific knowledge the lexical features of AGDs using the prompt P_2, while maintaining the same dataset. The evaluation results are detailed below.

LLM Performance. Table 2 presents the performance results for both prompting strategies (P_1 and P_2). The evaluated LLMs achieve accuracy, precision, recall, and F1-scores ranging from 77.3% and 97.4%, indicating strong initial performance. Particularly noteworthy are the Matthews correlation coefficients (MCC) values, which range from 55.4% to 79.7%, and Cohen's κ scores, which range from 0.41 to 0.68, suggesting moderate agreement between model predictions and ground truth labels across all models.

The results reveal three key findings: (i) **incorporating additional lexical context on top of the AGDs using P_2 produces a minimal performance improvement**; (ii) **the larger models consistently outperform their smaller counterparts, indicating a positive correlation between model size and classification quality**; and (iii) **the high FPR observed across all models represent a significant challenge for real-world deployment**.

Figure 1 shows the classification accuracy for malicious (red) and benign (blue) domains with both prompts. A systematic and consistent bias toward malicious classification is evident across all LLMs. The observed accuracy gap (up to 16% in some models) between malicious and benign predictions highlights a critical imbalance. This hypersensitivity to malicious domains increases the risk of misclassifying legitimate domains, posing operational risks such as service outages and false alarms, even with the best-performing model (i.e., Mistral Large).

Unclassified Domains. Sometimes, LLMs produced output that did not conform to the expected format or omitted essential information. As shown in Fig. 2, we

Table 2. Performance metrics of the LLMs evaluated using the D dataset under two prompt strategies, P_1 (minimal) and P_2 (lexically based). Metrics include classification accuracy (Acc), precision (Prec), recall (Rec), F1-score (F1), false positive rate (FPR), true positive rate (TPR), Matthews correlation coefficient (MCC), and Cohen's kappa (κ).

Model	P	Acc	Prec	Rec	F1	FPR	TPR	MCC	κ
GPT-4o	P_1	86.80	83.60	91.40	87.30	17.90	91.40	73.80	0.603
	P_2	87.00	84.60	90.50	87.40	16.50	90.50	74.20	0.603
GPT-4o-mini	P_1	77.30	73.00	86.40	79.20	31.90	86.40	55.40	0.415
	P_2	78.50	74.60	86.50	80.10	29.40	86.50	57.80	0.435
Claude 3.5 Sonnet	P_1	89.30	83.80	**97.40**	**90.10**	18.80	**97.40**	**79.70**	**0.682**
	P_2	**89.40**	84.20	96.80	**90.10**	18.20	96.80	79.50	0.678
Claude 3.5 Haiku	P_1	85.60	84.00	87.90	85.90	16.80	87.90	71.20	0.563
	P_2	85.20	84.70	86.00	85.40	15.60	86.00	70.50	0.548
Gemini 1.5 Pro	P_1	87.70	83.80	93.50	88.40	18.10	93.50	76.00	0.632
	P_2	87.60	84.20	92.60	88.20	17.40	92.60	75.60	0.625
Gemini 1.5 Flash	P_1	84.80	83.50	86.90	85.10	17.20	86.90	69.70	0.544
	P_2	84.90	83.60	86.80	85.20	17.10	86.80	69.80	0.545
Gemini 1.5 Flash-8B	P_1	81.70	78.20	87.90	82.80	24.50	87.90	63.90	0.494
	P_2	82.70	79.80	87.60	83.50	22.10	87.60	65.80	0.510
Mistral Large	P_1	88.70	**87.30**	90.60	88.90	**13.20**	90.60	77.40	0.639
	P_2	88.50	87.10	90.50	88.80	13.40	90.50	77.10	0.636
Mistral Small	P_1	85.10	82.60	89.00	85.70	18.80	89.00	70.40	0.560
	P_2	85.50	83.70	88.10	85.80	17.10	88.10	71.10	0.562

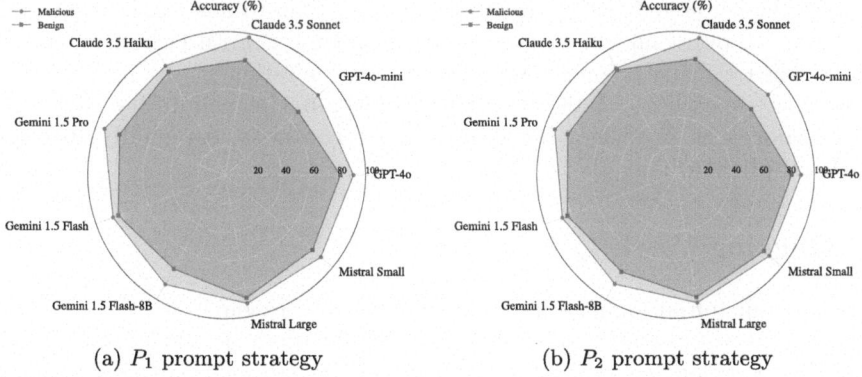

(a) P_1 prompt strategy   (b) P_2 prompt strategy

Fig. 1. Accuracy of each LLM in classifying malicious (red) versus benign (blue) domains using prompts P_1 (left) and P_2 (right). (Color figure online)

identified three recurring error patterns in the unclassified domains: (i) violations of the required output structure, (ii) transcription errors in domain names, and (iii) complete omission of domain entries. These problems were more prevalent with the more complex prompt P_2, suggesting that while additional lexical guidance may facilitate reasoning, it may also increase the cognitive load of the LLMs, resulting in poor output formatting. Simpler instructions like P_1 resulted in more reliable structural conformance.

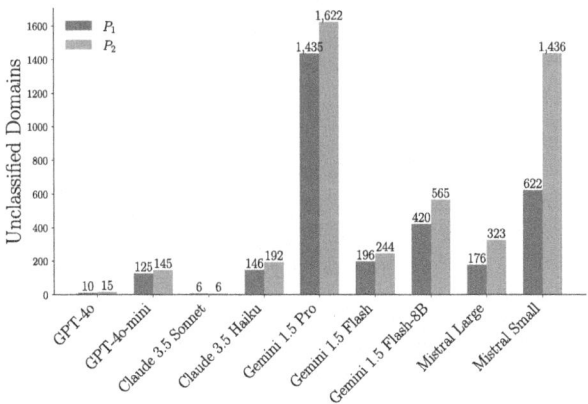

Fig. 2. Number of unclassified domains per LLM on the first assessment attempt.

Confidence in Predictions. The analysis of model confidence, shown in Fig. 3, reveals a worrying trend: all LLMs report consistently high confidence scores (medians above 85%) for both malicious and benign classifications. A greater confidence bias is observed toward malicious predictions. While models tended to assign higher confidence to correct predictions than to misclassifications, the prevalence of overconfident false predictions and the general lack of nuance in the confidence distribution reveal a significant limitation in the models' self-assessment capabilities. **This calls into question the reliability of confidence scores as practical indicators of correctness, especially in real-world applications.**

4   Ongoing Work

Based on our initial findings, we are expanding our research in two key directions. First, we investigate the multiclass classification capabilities of LLMs to determine whether they can distinguish between different malware families based on their distinctive DGA patterns. Second, we evaluate LLMs' performance in benign, real-world domains that share structural features with DGAs to test their robustness to more realistic and ambiguous scenarios.

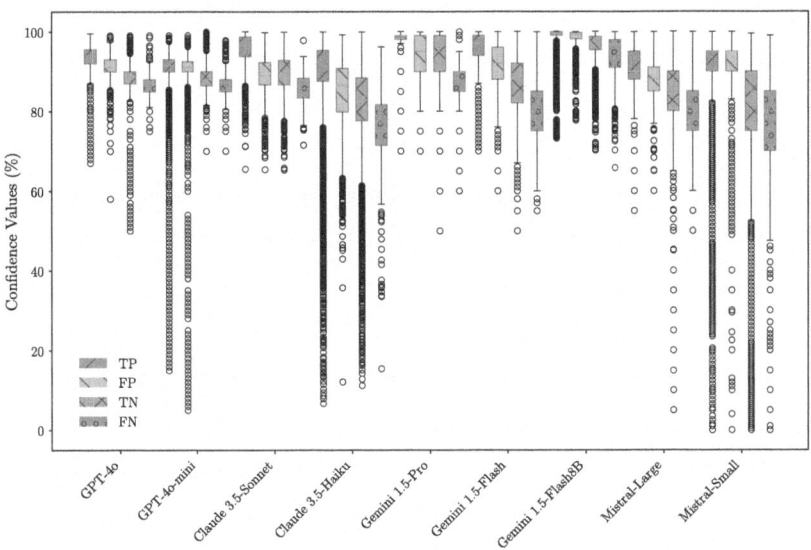

Fig. 3. Distribution of classification confidence values according to prediction types (true/false positives/negatives) for each LLM evaluated in this work.

Acknowledgments. This research was supported in part by grant PID2023-1514 67OA-I00 (CRAPER), funded by MICIU/AEI/10.13039/501100011033 and by ERDF/ EU, by grant TED2021-131115A-I00 (MIMFA), funded by MICIU/AEI/10.13039/ 501100011033 and by the European Union NextGenerationEU/PRTR, and the University of Zaragoza, by grant *Proyecto Estratégico Ciberseguridad EINA UNIZAR*, funded by the Spanish National Cybersecurity Institute (INCIBE) and the European Union NextGenerationEU/PRTR, by grant *Programa de Proyectos Estratégicos de Grupos de Investigación* (DisCo research group, ref. T21-23R), funded by the University, Industry and Innovation Department of the Aragonese Government, and by the RAPID project (Grant No. CS.007) financed by the Dutch Research Council (NWO). We extend our gratitude to the DGArchive team for providing the current dataset in advance, allowing us to begin experimentation sooner. We used OpenAI's ChatGPT to improve the grammar, clarity, and coherence of the paper.

References

1. Anthropic: Sonnet 3.5 and Haiku 3.5 (2024). https://www.anthropic.com. Accessed 1 Dec 2024
2. Google: Gemini Pro 1.5, Gemini Flash 1.5 and Gemini Flash 8b 1.5. https:// deepmind.google/technologies/gemini/. Accessed 1 Dec 2024
3. Marvin, G., Hellen, N., Jjingo, D., Nakatumba-Nabende, J.: Prompt engineering in large language models. In: Jacob, I.J., Piramuthu, S., Falkowski-Gilski, P. (eds.) Data Intelligence and Cognitive Informatics, pp. 387–402. Springer, Singapore (2024)

4. MistralAI: Mistral AI Large and Mistral AI small (2024). https://mistral.ai/. Accessed 1 Dec 2024
5. MITRE: Dynamic Resolution: Domain Generation Algorithms. https://attack. mitre.org/techniques/T1568/002/. Accessed 23 Aug 2023
6. OpenAI: GPT-4o and GPT-4o-mini (2024). https://openai.com. Accessed 1 Dec 2024
7. Plohmann, D., Yakdan, K., Klatt, M., Bader, J., Gerhards-Padilla, E.: A comprehensive measurement study of domain generating malware. In: 25th USENIX Security Symposium, pp. 263–278. USENIX Association, August 2016
8. Pochat, V.L., Goethem, T.V., Tajalizadehkhoob, S., Korczynski, M., Joosen, W.: Tranco: a research-oriented top sites ranking hardened against manipulation. In: 26th Annual Network and Distributed System Security Symposium, San Diego, California, USA, 24–27 February 2019. The Internet Society (2019)
9. Porras, P.A., Saïdi, H., Yegneswaran, V.: A foray into Conficker's logic and rendezvous points. LEET **9**, 7 (2009)
10. Tranco: Tranco List (2024). https://tranco-list.eu/. Accessed 15 Aug 2024

Poster: Using Machine Learning to Infer Network Structure from Security Metadata

Asfa Khalid$^{(\boxtimes)}$ ⓘ, Seán Óg Murphy ⓘ, Cormac J. Sreenan ⓘ, and Utz Roedig ⓘ

School of Computer Science and Information Technology (CSIT), University College Cork, Cork, Ireland
123111938@umail.ucc.ie, {seanogmurphy,cormac.sreenan,u.roedig}@ucc.ie
https://ucc.ie/nascresearch/

Abstract. In distributed cloud-edge environments, data-driven decision-making is essential for enhancing operational efficiency and maintaining a competitive advantage. Achieving this requires strong guarantees of data integrity and authenticity, as any compromise can lead to inaccurate insights, loss of trust, and financial damage. To address the cybersecurity risks posed by data transmission across complex, heterogeneous networks, Data Confidence Fabrics have been introduced. These enhance data security by generating metadata at each stage of transmission and storing it using distributed ledgers, which ensures the immutability and verifiability of this metadata. However, despite these benefits, the public accessibility of ledgers introduces significant privacy concerns. While previous research has focused on hostname obfuscation to protect network structure, timestamps often remain exposed, creating an exploitable vulnerability. We demonstrate that one can use K-means clustering on exposed timestamp patterns to reconstruct the obfuscated network structure, even when hostnames are fully obfuscated. Our findings reveal a critical gap in existing metadata protection mechanisms and highlight the need for defense against timestamp-based inference attacks.

Keywords: Distributed Systems · Data Confidence Fabrics · Machine Learning · K-means Clustering · Timestamp Analysis · Metadata Privacy · Network Structure Inference

1 Introduction

Research has shown that metadata, even when seemingly innocuous, can pose serious privacy threats. Studies across social networks, ISPs, and financial systems demonstrate how elements like timestamps, activity logs, and network structures can be exploited to identify users, infer behaviors, and reveal organizational patterns. For networks, techniques such as DNS encryption [3], network topology disguising [4], and web data hiding [10] aim to shield sensitive structures and activities.

Data Confidence Fabrics (DCFs) have been introduced to enhance data security for cloud-edge systems by generating metadata (annotations) at each stage of transmission, recording details about network nodes [1], and storing this information in distributed ledgers. This is illustrated in Fig. 1, which depicts the Linux Foundation's Alvarium Data Confidence Fabric (DCF). Since ledgers ensure immutability, metadata remains secure and verifiable. However, despite these advantages, the public accessibility of metadata on ledgers presents significant privacy concerns: attackers can analyze metadata to infer network structure. By analyzing network metadata, attackers can uncover structural details, identify critical nodes, and assess vulnerabilities, enabling them to plan precise attacks. In particular, stored metadata includes a timestamp field and host identifiers, allowing adversaries to infer the relationships between network nodes. Therefore, safeguarding the confidentiality of annotations has practical significance.

To address the challenge of protecting network structure information, encryption-based obfuscation offers stronger security by preventing meaningful pattern extraction. Prior work by us and others used encryption to effectively concealing hostname information [6,7], but timestamp data remains a critical vulnerability. We show in this short paper that adversaries can use timestamp clustering techniques, such as K-means clustering, to identify communication patterns, reconstruct network layers, and infer host relationships.

This research explores how the timing of network communications can reveal relationships between hosts using machine learning techniques such as K-means clustering, even when hostnames are encrypted. Using a dataset of network annotations with encrypted hostnames, we analyzed time intervals between nodes and applied K-means clustering to group hosts with similar timing patterns. Our method accurately maps encrypted hostnames to their real counterparts, effectively reconstructing much of the original network structure. Experimental results demonstrate that timing-based clustering remains highly effective for inferring network structure, even without access to hostname information. These findings highlight a significant vulnerability in current metadata protection strategies and underscore the need for greater consideration of timestamp privacy in the design of secure distributed systems.

2   Methodology

Our approach uses K-means clustering on timestamp metadata to successfully infer the original network hosts Fig. 2. By carefully finding the exact K-means value and accurately clustering hosts based on their timestamps, we reconstructed most of the network structure.

K-means clustering [11], widely applied in pattern analysis, has been adapted for spatio-temporal and large-scale datasets. Rather than enhancing clustering accuracy, we highlight how K-means can be used maliciously to extract network insights from timestamp metadata. This perspective reframes clustering not just as an analytical tool but also a potential vector for privacy attacks, reinforcing the need for metadata obfuscation within DCFs.

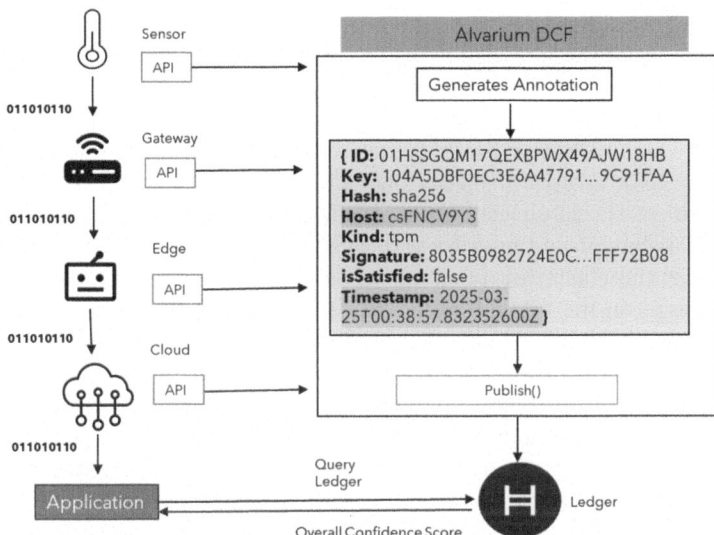

Fig. 1. Alvarium DCF publishes the generated annotations to the ledger, revealing host and timestamp information that attackers could exploit to infer the network structure.

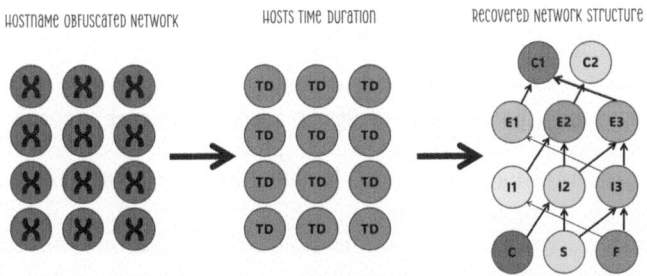

Fig. 2. Inferring obfuscated network based on timestamp clustering

The first step in our approach involves organizing the network into distinct layers by analyzing annotation keys in combination with their corresponding timestamps. Each annotation key, generated through a hashing process, is unique to a specific data point at a particular node. This uniqueness allows us to consistently trace and identify the location of each host within the network. By examining when and where these keys appear, we can assign each host to a specific layer within the system. In this context, a layer represents a logical level within the network, used to categorize hosts based on the timing and sequence of their communications. As a result, a hierarchical model of the network emerges, clearly outlining whether a host operates in the first, second, or subsequent layer based on its interaction patterns.

The second step involves preparing the annotation dataset for clustering analysis. We restructure the data to focus on temporal differences between hosts, allowing us to analyze the timing relationships among them. To identify the optimal number of clusters (denoted as K), we employ an iterative K-means clustering approach. Starting with a small value for K, we incrementally increase it, assessing the quality and cohesion of the resulting clusters at each step. The goal is to find a value of K that leads to meaningful and interpretable groupings of hosts rather than arbitrary separations. Through this process, we identify that K = 9 produces the most accurate and consistent clustering, aligning well with known host behavior and network structure.

Following the clustering process, we assign each host a unique, structured identifier based on its original layer and its cluster number, which groups hosts by the time gaps in their communication patterns. The merged value containing host's original layer with cluster number form a unique identifier that reflects both temporal and structural characteristics. For example, a host in Layer 1 assigned to Cluster 1 would be labeled as "H1-1." Similarly, a host in Layer 4 grouped into Cluster 9 becomes "H4-9." These new identifiers replace the original encrypted hostnames in the dataset.

With hosts now labeled according to these structured identifiers, we proceed to reconstruct the network topology. This involves mapping relationships between hosts based on their cluster groupings.

3   Evaluation

To evaluate the selection of the K value and the effectiveness of our clustering approach in grouping original hosts to recover network structure based on them, we use two methods. First, we compare the results of different cluster evaluation metrics (Silhouette Score, Davies-Bouldin Index, Dunn Index, and the Elbow Method) to find the optimal K value. Second, we use the original hostnames as a reference dataset and compare them with our clustering results. We base our analysis on a dataset of 4,000 annotated timestamped communication events, where each original host within an annotation is encrypted. These annotations were generated using the Alvarium SDK [2], simulating network behavior, including timing durations and host communication patterns.

Evaluation of Optimal K for Clustering. To determine the optimal number of clusters for grouping similar communication patterns, we used several clustering metrics across K values ranging from 5 to 15. Specifically, we used the Silhouette Score [8], Davies-Bouldin Index (DBI) [12], Dunn Index [9], and the Elbow Method [5], each offering a different perspective on clustering quality. These metrics assess factors such as cohesion within clusters, separation between clusters, and the balance between compactness and spread. By analyzing these scores collectively, we aimed to identify the most meaningful and stable value of K for our dataset.

Based on this multi-metric analysis (Fig. 3), both $K = 9$ and $K = 10$ emerge as optimal choices. The Dunn Index suggests $K = 10$ as a strong alternative, with slightly better inter-cluster separation. However, $K = 9$ is the most consistently

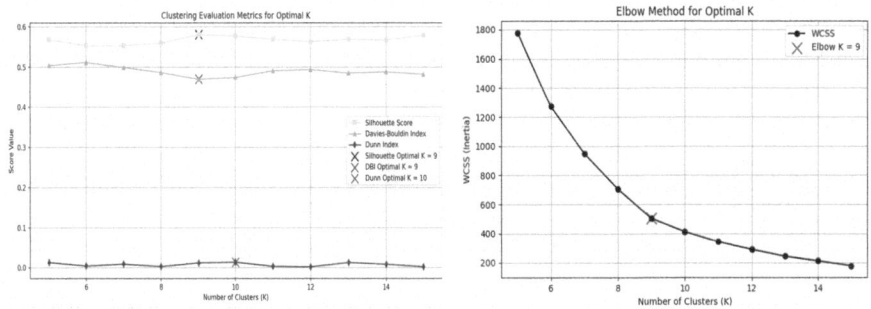

Fig. 3. Scoring result of cluster evaluation metrics for optimal K

optimal choice, as reflected by the Silhouette Score, DBI, and Elbow Method, all of which indicate well-separated and compact clusters.

Evaluation of Clustering Accuracy Through Bidirectional Validation.
To further evaluate the accuracy of our clustering approach, we employed a bidirectional validation method that compares the clustered hostnames against the real hostnames. This evaluation involved two complementary perspectives. The first, referred to as forward mapping, examined how instances of each real hostname were distributed across the resulting clusters. The second, reverse mapping, assessed how well each cluster corresponded to a specific original hostname.

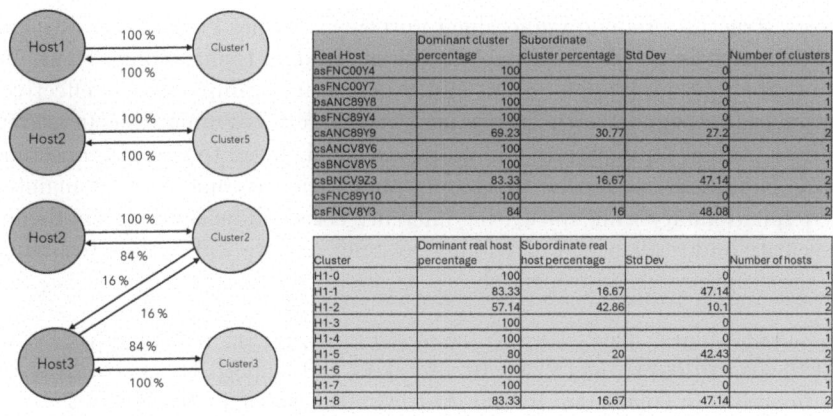

Real Host	Dominant cluster percentage	Subordinate cluster percentage	Std Dev	Number of clusters
asFNC00Y4	100		0	1
asFNC00Y7	100		0	1
bsANC89Y8	100		0	1
bsFNC89Y4	100		0	1
csANC89Y9	69.23	30.77	27.2	2
csANCV8Y6	100		0	1
csBNCV8Y5	100		0	1
csBNCV9Z3	83.33	16.67	47.14	2
csFNC89Y10	100		0	1
csFNCV8Y3	84	16	48.08	2

Cluster	Dominant real host percentage	Subordinate real host percentage	Std Dev	Number of hosts
H1-0	100		0	1
H1-1	83.33	16.67	47.14	2
H1-2	57.14	42.86	10.1	2
H1-3	100		0	1
H1-4	100		0	1
H1-5	80	20	42.43	2
H1-6	100		0	1
H1-7	100		0	1
H1-8	83.33	16.67	47.14	2

Fig. 4. Bidirectional validation of clustering accuracy.

We conducted a series of experiments using different values of K and analyzed the resulting clustering performance. The most accurate results were observed at $K = 9$, as depicted in Fig. 4. Showing strong alignment between the real and assigned cluster hostnames. In the forward mapping, we observed that most clustered hosts mapped exclusively to a single real host, achieving 100% accuracy with zero standard deviation. This outcome indicates excellent cluster purity. In other cases, although some hostnames were split across multiple clusters, the split was often limited and containing high standard deviation combined with a low number of clusters per host suggests that each host is primarily split across a few clusters in a dominant-subordinate pattern (e.g., 84%–16%). This pattern still suggests effective clustering, as the majority of instances from a single host were fully linked to a single cluster. The reverse mapping reinforced these findings. Several real hostnames were found to align perfectly with individual clusters, showing strong one-to-one correspondence. Furthermore, clusters with high standard deviation but few hostnames typically indicated that those clusters were dominated by one host, again pointing to well-separated groupings.

While a small number of misclassifications did occur, these were minimal relative to the overall dataset size. Furthermore, the high standard deviation in the affected clusters suggests that most errors were concentrated within a small subset of hosts. Overall, the results confirm that by relying solely on timing data and without any access to real hostnames, we can infer functional relationships and communication structures within the system.

Clustering performs well on medium-sized networks, but for large networks it poses computational and memory challenges, meaning that attacks would necessitate more targeted approaches.

4   Conclusion

Network structures remain vulnerable to privacy breaches even when hostnames are obfuscated, as attackers can exploit timestamp data to infer connections. Our study shows that applying K-means clustering to timestamp data can effectively reconstruct these hidden structures, thus exposing critical limitations in existing metadata protection strategies and underscoring the need to address timestamp privacy. Developing advanced timestamp obfuscation techniques that maintain system functionality while mitigating inference risks will be key to strengthening the security and resilience of distributed systems and we will explore techniques that alter timestamps appropriately.

Acknowledgments. This work was conducted as part of the CLEVER project, EU Grant Number 101097560 and EI No: IR-2022-0065, and supported in part by a research grant from Science Foundation Ireland (SFI), Grant Number 13/RC/2077 P2.

References

1. Dell Technologies: Data confidence drives better decisions. https://www.delltechnologies.com/asset/en-us/solutions/business-solutions/industry-market/data-confidence-drives-better-decisions.pdf. Accessed 1 May 2025

2. GitHub - Project-Alvarium: Repository containing work related to the forthcoming Java SDK implementation. https://github.com/project-alvarium/alvarium-sdk-java. Accessed 1 May 2025
3. Herrmann, D., Maaß, M., Federrath, H.: Evaluating the security of a DNS query obfuscation scheme for private web surfing. In: Cuppens-Boulahia, N., Cuppens, F., Jajodia, S., Abou El Kalam, A., Sans, T. (eds.) SEC 2014. IAICT, vol. 428, pp. 205–219. Springer, Heidelberg (2014). https://doi.org/10.1007/978-3-642-55415-5_17
4. Hou, T., Wang, T., Lu, Z., Liu, Y.: Combating adversarial network topology inference by proactive topology obfuscation. IEEE/ACM Trans. Network. **29**(6), 2779–2792 (2021)
5. Humaira, H., Rasyidah, R.: Determining the appropriate cluster number using elbow method for K-means algorithm. In: Proceedings of 2nd Workshop on Multidisciplinary and Applications, pp. 1–8 (2020)
6. Khalid, A., Óg Murphy, S., Sreenan, C.J., Roedig, U.: Obfuscating network structure from blockchain analysis. In: Barolli, L. (ed.) Advanced Information Networking and Applications, pp. 95–104. Springer, Cham (2025)
7. Khandkar, V.S., Hanawal, M.K.: Masking host identity on internet: encrypted TLS/SSL handshake. arXiv preprint arXiv:2101.04556 (2021)
8. Lovmar, L., Ahlford, A., Jonsson, M., Syvänen, A.C.: Silhouette scores for assessment of SNP genotype clusters. BMC Genomics **6**, 1–6 (2005)
9. Luna-Romera, J.M., García-Gutiérrez, J., Martínez-Ballesteros, M., Riquelme Santos, J.C.: An approach to validity indices for clustering techniques in big data. Progress Artif. Intell. **7**, 81–94 (2018)
10. Masood, R., Vatsalan, D., Ikram, M., Kaafar, M.A.: Incognito: a method for obfuscating web data. In: Proceedings of the 2018 World Wide Web Conference, pp. 267–276 (2018)
11. Wu, J.: Cluster analysis and K-means clustering: an introduction. In: Advances in K-Means Clustering: A Data Mining Thinking, pp. 1–16. Springer (2012)
12. Xiao, J., Lu, J., Li, X.: Davies Bouldin index based hierarchical initialization K-means. Intell. Data Anal. **21**(6), 1327–1338 (2017)

Android and Patches

More Than You Signed Up For: Exposing Gaps in the Validation of Android's App Signing

Norah Ridley, Enrico Branca, and Natalia Stakhanova[✉]

University of Saskatchewan, Saskatoon, Canada
{norah.ridley,enrico.branca}@usask.ca, natalia@cs.usask.ca

Abstract. Android's ubiquitous flexibility has helped it to become one of the most widely used mobile operating systems in the world. However, the convenient and extensive access to phone resources adopted by Android has revealed inefficiencies of its existing protections. Among these protections is application (app) signing. This process is intended to maintain the integrity of the app after it is released, and it provides users with confidence in the app's authenticity. We analyze the functionality of Android signature verification and identify the logical gaps in the process that can be used to hide a malicious payload. We demonstrate the security implications of these gaps.

1 Introduction

Android has become one of the most widely used mobile operating systems (OS) in the world. Central to Android's ecosystem is its app signing mechanism, a security feature designed to ensure the integrity and authenticity of mobile applications (apps). By requiring developers to digitally sign their apps, this process helps to ensure that the app has not been tampered with since its release and to verify the identity of developers. Despite its importance, this mechanism has flaws and has faced scrutiny over the years. Numerous examples have highlighted the limitations of app signing.

In this paper, we examine the app signing process in recent Android versions. We demonstrate how its design flaws and inconsistent implementation can be exploited by attackers, rendering mobile attacks virtually undetectable.

Over the years, Android has strengthened its app integrity protection by gradually introducing more restrictive signing schemes. The v1 signature scheme was the Android's original mechanism for signing applications. Based on the Java Archive (JAR) file signing format, it used to aggregate multiple files into a single file for easier distribution. In the v1 scheme, a digital signature was created for each file within the Android Package Kit (APK) format, which offered integrity protection of individual files but not of the entire APK. The limited protection provided by the v1 signature scheme resulted in the infamous Janus vulnerability that allowed attackers to inject malicious code into APK files without altering

M. Egele et al. (Eds.): DIMVA 2025, LNCS 15748, pp. 103–123, 2025.
https://doi.org/10.1007/978-3-031-97623-0_7

their original cryptographic signatures [13]. Similarly, the Master Key vulnerability enabled modifications to previously signed applications while preserving the validity of their original signatures [12]. Due to these issues, Android introduced more robust v2 and v3 signature schemes that implement several mechanisms to provide stronger guarantees for app integrity.

In this work, *we show fundamental issues in the design and validation of the APK signing block.*

First, both schemes leverage a separate APK signing block to store app signatures. This block contains a sequence of signatures in the form of ID-value pairs prefixed and suffixed with the data size. The key observation that makes our attacks possible is that the sequence of signatures is never verified. During app signature verification, known ID-value pairs are evaluated while unknown pairs are ignored. As a result, an APK signature block provides space for embedding an attack payload.

Second, unlike v1, which only verified individual files within the APK, v2 and v3 validate the integrity of the entire APK file to prevent unauthorized modifications to the APK after signing. The verification process is simplified by using a single hash of the entire APK. To be precise, the hash is calculated only for APK content (ZIP entries), the ZIP Central Directory, and the End of Central Directory (EoCD). Although the integrity of the signed data inside the APK signing block is protected, the signing block as a container is left without integrity verification, making it possible to modify the signing block after signing.

Third, the APK signing block is intentionally variable in size, allowing it to store digital signatures as well as any additional metadata. This design inherently offers significant space for embedding attack payloads.

In this paper, *we leverage these observations and show how the discovered issues in the v2 and v3 signature schemes make it possible to craft an Android app that can deliver and execute a malicious payload without raising any flags during the official app verification process.*

Our proposed attack leverages the internal structure of the signing block, which provides a variable space to embed an attack payload. We demonstrate that this approach can successfully deliver a wide range of attack payloads, including text and code, to Android devices through the APK signing block of a seemingly benign app that does not request any dangerous permissions. In addition, we show that these seemingly benign apps containing the payloads can be successfully distributed through traditional app distribution platforms. We upload the proof-of-concept app to the APKPure[1] and Aptoide[2] markets and bypass their security checks.

Responsible Disclosure. We followed the responsible disclosure process and informed Google of the discovered issue in November 2024. Upon receiving our report, the Android security team assigned it a critical severity. In January 2025, Google determined that the identified issue was not a security vulnerability, which nevertheless required remediation. As a result of our disclosure,

---

[1] https://apkpure.com/.

[2] https://en.aptoide.com/.

Google informed us that the Google Play's developer console will be updated to reject pre-signed APKs that contain unknown or unsupported IDs.

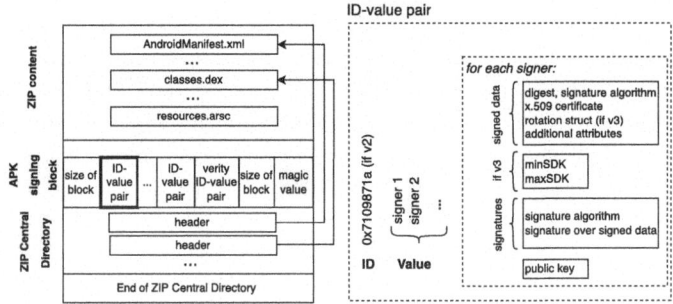

Fig. 1. APK structure

2   Background

The Android application file is a self-contained ZIP container that includes the compiled code, assets, and resources. Android apps can be installed on the device or published on the app distribution platforms using the APK or Android App Bundle (AAB) formats. The Play Store, Google's official store, is one option for distributing apps. Alternatively, a developer can upload an app to third-party markets, many of which are popular in some countries (e.g., China, India).

Regardless of the format, Android requires all apps to be digitally signed before they are installed on a device or updated by the developer. For apps using APKs, the developer has to manually sign the app using the app signing key. The app signing key refers to a pair of cryptographic keys: a public key and a private key. The public key is embedded within the X.509 certificate along with the app's developer and key information. The certificate is then included in the APK file. Apps distributed in Android App Bundle format are signed with an upload key first and then uploaded to Google Play Store where, after verification of the upload certificate, they are signed with the app signing key. While many markets support apps in both formats (e.g., Samsung Galaxy Store), Google Play Store began restricting APK format in 2021. This change only allows developers to update apps in APK format.

App Signing. Application signing serves two main purposes: developer identification and integrity verification. Essentially, this key mechanism verifies that an app delivered to a user's phone is unmodified. Currently, Android supports four APK signature schemes [6]: v1, v2, v3, and v4.

The v1 signature scheme, the original signing scheme, was introduced in Android 1.6 and is based on the JAR signing mechanism. It verifies the integrity of individual files within the APK, ensuring that none of the signed files has been tampered with. The v1 scheme is used to verify older apps designed for Android 7.0. To address the limitations of v1, Android introduced *the v2 signature scheme* in Android 7.0 [4]. Unlike v1, v2 validates the entire APK, including ZIP metadata. This feature ensures that any modification to the APK invalidates the signature. *The v3 signature scheme*, introduced in Android 9.0, builds on v2 by adding support for key rotation. This allows developers to specify new signing keys for future app updates, enabling them to replace compromised or outdated keys. The most recent addition, *the v4 signature scheme*, was introduced in Android 11. The v4 scheme requires a complementary v2 or v3 signature. By default, the v4 signature is stored in a separate file [5]. While v4 is used by Google Play, it is not mandatory for APKs.

Once an app is installed on an Android device, the package manager performs signature verification and records metadata about the app including its signing data and the signing certificate.

APK Structure. An overview of the APK structure is shown in Fig. 1. From a signing perspective, the APK is divided into four sections: the contents of ZIP entries (APK content), the APK Signing Block, the ZIP Central Directory (which contains offsets to ZIP entries), and the End of Central Directory (EoCD). The APK Signing Block, introduced with the v2 signing scheme, contains v2 and v3 signatures while maintaining backward compatibility with the v1 scheme. The APK signing block starts with an eight-byte *size of block* field that specifies the total size of the APK signing block. The size field is followed by a sequence of ID-value pairs, then a second eight-byte *size of block* field, which is expected to store the same value as the first field. The final element in the signing block is the magic value "APK Sig Block 42" that is used to identify the signing block during parsing and also to mark the end of the signing block.

The ID-value pair sequence contains information about the app's signers, with each pair representing a specific signing scheme identified by its unique ID. For every signer, this includes the *signed data*, *signature*, and the *public key*. The signed data part contains APK digests (digests of APK content, the ZIP Central Directory, and EoCD), signature algorithms, signer's x.509 certificate, and signer's public key. In addition to signing schemes identified by their unique IDs, Android defines several other IDs recognized by the verification process. One such ID corresponds to the *verity ID-value pair*. The primary purpose of this pair is to maintain the alignment of the APK signing block. It does not contain meaningful signing information and consists of padding to align the block size as a multiple of 4,096 bytes although this alignment is not enforced. During signing, `apksigner` positions the verity ID-value as the last entry in the ID-value pair sequence. The verity pair is followed by the size field and the magic value.

Integrity verification of an APK is based on the digests of the file content. The digest is computed over APK content, the ZIP Central Directory, and EoCD by splitting them into chunks and calculating the digest for each of them separately. The resulting digests are combined in a Merkle tree fashion into one or more APK digests that are stored in the signed data part of the block. Signature algorithm

IDs used in the generation of digests are stored in both the signed data part and in the signature part of the signing block.

3   Related Work

Android Malware and Attacks. Over the years, several studies have demonstrated that app repackaging is an effective method for distributing malware onto users' devices [9,17,21]. Poeplau et al. [14] showed how malware can use dynamic code-loading techniques to circumvent offline vetting mechanisms (e.g., Google's Bouncer). Their analysis of apps from the Google Play Store demonstrated that although benign apps have legitimate reasons for using code-loading techniques, developers often improperly implement them. Zhang et al. [20] showed that Android's data cleanup mechanism is ineffective, leaving data residue (including private information) after an app is uninstalled. Du et al. [7] identified significant vulnerabilities in Android's shared storage system. To illustrate the severity of their findings, the authors crafted an end-to-end attack that impersonated both users during a voice chat session. Ruggia et al. [15] mounted a state inference-based phishing attack by exploiting Linux's *inotify* component. Their proposed attack assumed that the attacker knew the installation path of their target app, and thus, had the potential to affect all apps regardless of the Android version. While these attacks have an impact on signed apps, they do not specifically target app integrity protections. In contrast, our work focuses on exploiting the limitations of these protection mechanisms to introduce arbitrary content into the signing block.

Android App Signing Issues. Numerous studies have considered app signing issues in real-world apps. Vidas and Christin [16] gathered 41,057 apps from 194 alternative markets and 35,423 apps from the Google Play Store. Their subsequent analysis of the dataset showed the heavy reuse of signing certificates. For example, 48% of the certificates from Google Play were reused. Fahl et al. [8] assessed 989,935 free apps from Google Play and found that app signing practices were generally not transparent or secure. The dataset collected by Lindorfer et al. [11] contained over one million apps from between 2010 and 2014 with approximately 40% being identified as malware. They noted that 2.26% of the apps were signed with a publicly available test key, making them potentially vulnerable to attacks. They also observed the Master Key vulnerability being exploited in 1,152 samples (0.11%), all from 2013 and 2014, and only in malware. The study by Wang et al. [18] showed that in spite of the introduction of v2 and v3 signing schemes, over 93% of the apps only used the v1 scheme. Furthermore, over 65,000 apps were signed with publicly known keys, which could allow attackers to arbitrarily modify such apps without breaking their original signatures. These studies do not examine the underlying reason for app signing issues. In this work, we consider these problems by introducing a new form of attack that exploits the failures of the verification process.

4   Attack

4.1   Threat Model

In our model, we assume that the attacker's goal is to successfully install malware on the target user's device. Specifically, the malicious app carrying the payload must: ① remain undetected during the verification and review process that apps undergo before being distributed to users, and ② successfully install on the device and execute the payload.

The app can be an original app developed and signed by the attacker that appears benign but conceals the malicious payload in the signing block. Similar to other attacks on Android, our approach assumes the victim has been deceived into installing a malicious app [7, 10, 15, 19, 20]. Once installed on the device, the app must be run locally by the user to trigger the payload, which will be executed within its application sandbox. As such, the effectiveness of the payload is limited by the privileges and data access that the app itself is granted. To reduce the chances of raising suspicion, the app's permission requirements are kept minimal. In fact, no permissions are needed to activate the embedded payload. The app in this attack must have a signing block, i.e., it is signed with the v2 and/or v3 schemes. Furthermore, in the case of arbitrary code execution, it must execute without causing the app to crash. The success of the attack depends on a combination of these factors, and if any of the required conditions are not satisfied, then the vulnerability cannot be triggered.

4.2   Android Specifications Loopholes

We analyze the official Android specifications on app signing and signature block structure for the v2 and v3 signature schemes to identify potential weaknesses. Next, we examine the implementation of app signing and signature verification to confirm these vulnerabilities and to assess how they might enable malicious manipulations. For our implementation analysis, we focus on the process of manually signing apps using `apksigner` [2]. Google offers several methods for developers to sign their apps: they can sign their apps manually using Android Studio or `apksigner`, or they can opt for Google to manage app signing through the Google Play Store. Google's official tool, `apksigner`, is included in the Android SDK Build Tools package [1]. According to Google, `apksigner` can sign and verify signatures for apps intended to run on all versions of the Android platform [2]. Our analysis of the signing and verification process reveals several loopholes:

① *The APK signing block is not signed.* According to the official Android documentation, the v2 and v3 signature schemes protect the integrity of the contents of the ZIP entries, the Central Directory, and the EoCD. However, the APK signing block remains largely unprotected with only the "signed data" section inside the v2 and v3 ID-value pairs being protected by digest(s). This limited protection allows an attacker to modify the APK signing block after the app was signed without breaking the app's integrity.

② *Verification ignores unknown ID-value pairs.* The signing block contains a sequence of app signatures in the form of ID-value pairs, prefixed and suffixed with fields that indicate the block size. The verification process ensures that the size fields in the APK signing block match and that the block conforms to the specified size. However, only known ID-value pairs are evaluated and unknown pairs are ignored. This oversight allows an attacker to embed arbitrary content into the APK signing block without compromising the required structure of the signing block.

③ *The APK signing block has variable size.* It is intended to accommodate not only digital signatures but also any additional metadata or information required during the signing process. The block size varies based on the number of signature schemes used and the size of the cryptographic keys and signatures. This flexible design provides significant space for embedding attack payloads.

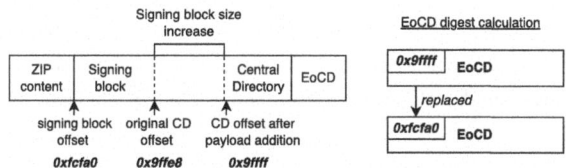

Fig. 2. The Central Directory offset treatment

④ *The Central Directory offset is never verified.* Since the APK signing block precedes the Central Directory section, any change in size of the APK signing block will also change the Central Directory's offset. When this occurs, the Central Directory offset that is stored in the EoCD must also be updated for the APK file to be parsed. If this offset in the EoCD is updated, the digest calculated over the modified EoCD will no longer match the original digest. In theory, this design should limit modifications that alter the APK signing block's size. However, to support legitimate modifications such as the addition of a new signature, during integrity verification Android treats the offset of the APK signing block as the offset of the ZIP Central Directory (Fig. 2).

During verification, the ZIP components are extracted using the offset that correctly reflect the position of the Central Directory section after the signing block's size modification. This offset is then disregarded, and the offset of the APK signing block is instead used during digest calculation. In other words, the actual offset of the Central Directory is never verified, and the size of the signing block is altered without causing integrity verification to fail. As a result, an attacker can inject an unlimited payload into the APK signing block without invalidating the APK's signature.

4.3   Attack Strategy

The design and implementation issues with the APK signing block can lead to various security vulnerabilities. In this work, we leverage the APK signing block as a vehicle for our attack, focusing specifically on the v2 and v3 signing schemes.

We demonstrate how a malicious Android app can deliver and execute a payload without triggering any flags during the official app verification process.

Given an Android app signed with the v2 and/or v3 signing schemes, we can craft a malicious payload and embed it inside the app's signing block. We parse the APK and extract a copy of its signing block. We then locate the bytes representing the sequence of ID-value pairs in the signing block and inject our malicious payload at a target location. Afterwards, we overwrite the APK's original signing block with our modified version. *These modifications do not invalidate the app's original signature.*

In most cases, inserting the payload into the signing block increases its size. Since the signing block is designed to be extensible, we exploit this feature to conceal the payload in the signing block. To ensure that the modified APK can still be parsed accurately by Android tools, we make several adjustments:

① We replace the values in the two *size of block* fields with the new signing block size. ② We rewrite the Central Directory and the End of Central Directory (EoCD). ③ We update the Central Directory offset in the EoCD.

Payload. For our experiments, we test four different types of the payload.

– *EICAR Antivirus Test File.* Developed by the European Institute for Computer Antivirus Research[3] and the Computer Antivirus Research Organization, this file is used to test the response of antivirus programs without using a real malicious payload. This test allows us to assess whether antivirus programs can detect malicious content concealed in the APK signing block.
– *Text.* The text payload is a sequence of ASCII characters stored in a text file. For our experiments, we use a 7KB text file that contains 1,000 words of *lorem ipsum.* The text payload serves as an initial exploration on the types of content that can be embedded in the signing block.
– *Image.* Our image payload is a 829KB JPEG image file. Since image files can be used to conceal malicious content, this payload helps us to investigate the limits (if any) on the type of content that can be inserted into the APK signing block.
– *Code.* For our code payload, we implement a simple JavaScript code snippet that adds two integer numbers. Despite its brevity, this type of payload demonstrates that syntactically valid code can be concealed in the APK signing block and then executed after app installation.

Payload Packaging. Android does not limit the size of the APK signing block, and although Android suggests that the size of the signing block should be a multiple of 4,096, it is not enforced. Consequently, there is a variable-size space for embedding an attack payload. The structure of the APK Signing Block is defined as a sequence of ID-value pairs that is prefixed and suffixed with fields containing the size of the signing block. This format potentially allows attackers

---

[3] https://www.eicar.org/.

to disguise a malicious payload as a legitimate ID-value pair. In our experiments, we explore the following approaches:

– *v2 and v3 ID-value pair.* We explore the use of IDs that correspond to a v2 signing scheme and a v3 signing scheme. For both v2 and v3 signature blocks, we create a new pair with the payload packaged as a value in a pair.
– *Verity ID.* We use an existing verity pair by inserting the payload into the value of the existing pair. To ensure that the size of the verity pair remains consistent, we adjust the remaining padding bytes by removing extra padding (if payload is smaller than the original padding) or increase the size of the pair (if the length of the payload exceeds the length of the verity pair).
– *Arbitrary ID.* We create a new ID-value pair with an arbitrary ID (0x716f43c0) and the payload embedded as value.
– *Raw byte form.* Our last method of adding the payload does not rely on the ID-value pair organization of the APK signing block. We directly insert the payload into the bytes of the APK signing block.

Original APK signing block	size of block	ID-value pair	ID-value pair	size of block	magic value
Between ID-value pairs	size of block	ID-value pair	**injected ID-value pair**	ID-value pair	size of block	magic value
First ID-value pair	size of block	**injected ID-value pair**	ID-value pair	ID-value pair	size of block	magic value
Last ID-value pair	size of block	ID-value pair	ID-value pair	**injected ID-value pair**	size of block	magic value

Fig. 3. Payload injection locations

Payload Injection Location. We define three locations in the APK signing block where the packaged payload can be inserted as shown in Fig. 3.

– *First ID-value pair.* The packaged payload is inserted in the APK signing block after the first *size of block* field and before the existing sequence of ID-value pairs. As a result, when the payload is packaged as an ID-value pair, this injected pair becomes the first pair in the sequence. If the payload is in the raw byte form, the payload is inserted directly after the first eight bytes of the APK signing block (the defined length of size field).
– *Between ID-value pairs.* To maintain consistency across our tests, this case is treated as the midpoint of the ID-value pairs sequence. The insertion index is calculated by dividing the length of the sequence by two, and we use integer division to obtain a whole number index without the fractional part.
– *Last ID-value pair.* The payload is inserted in the APK signing block after the last pair in the sequence, which is directly before the second *size of block* field and the magic value. This location is the last place in the signing block that the payload can be inserted without breaking the signing block's structure.

5   Attack Implementation

To validate our proposed attack strategies, we implemented a custom app and tested the strategies on physical phones with the two latest versions of Android.

5.1   Testing App

We developed a custom Android app ($minSDK = 33$ and $targetSDK = 34$) in Android Studio Ladybug 2024.2.1, which was the latest version at the time of writing. Our app was created using the Kotlin programming language. The app's primary function is to process embedded payloads hidden within the APK signing block. Specifically, the app locates its file path on the device, parses itself, and extracts its signing block. It retrieves the sequence of ID-value pairs and parses each pair to identify the injected payload. The app does not require user interaction and is designed as a single-page application with two user interface elements: a TextView for displaying the embedded text payload and an empty ImageView for displaying the embedded image payload. The app handles the payloads as follows:

- *EICAR.* We do not define a trigger for this payload form. Instead, we only check that the app can be installed and opened without crashing.
- *Text.* The text payload is converted from an array of bytes to a string, which is then displayed using a TextView element.
- *Image.* We decode the byte array that represents the image payload to a Bitmap and then set the Bitmap as the content of the ImageView element.
- *Code.* We use the RhinoScriptEngine[4] from the JSR223[5] application programming interface (API), a scripting interface for Java Virtual Machine (JVM) languages, to execute the code payload (represented as a string) in our app. To ensure that the payload extracted from the signing block is syntactically correct JavaScript, we remove all zero bytes from the byte array that represents the payload and convert the filtered array to a string. The string is passed to the script engine and executed, and the result is returned and set as the value of the TextView element.

After building an aligned release version of the app using Android Studio, we signed the app using `apksigner` (revision 35.0.0 of Android SDK Build Tools). For signing we used a 384-bit elliptic curve cryptography (ECC) key. We refer to this app as *base app*. Since our experiments examined both the v2 and v3 schemes, we created two versions of this base testing app: one app signed with the v2 scheme and the other app signed with the v3 scheme. These two base apps serve as a baseline in our experiments.

---

[4] https://github.com/APISENSE/rhino-android.
[5] https://www.openhab.org/docs/configuration/jsr223.html.

Algorithm 1: Inserting the payload into the APK signing block

---

    /* locate APK signing block   */

1 read the bytes of the APK file;

2 get the starting offset of the Central Directory;

3 position = Central Directory offset - 16 bytes;

4 magicBytes = read the bytes from $position$ to ($position$ + 16);

5 **if** $magicBytes \neq$ *"APK Sig Block 42"* **then**

6 | return;   // signing block not found

7 **end**

    /* extract APK signing block bytes   */

8 position = Central Directory offset - 24 bytes;

9 sizeOfBlockField2 = read the bytes from $position$ to ($position$ + 8 bytes);

10 position = $-sizeOfBlockField2$ + 8 bytes;

11 sizeofBlockField1 = read the bytes from $position$ to ($position$ + 8 bytes);

12 **if** $sizeOfBlockField1 \neq sizeOfBlockField2$ **then**

13 | return;   // size fields must be equal

14 **end**

15 position = $position$ - 8 bytes;

16 originalSigningBlockOffset = $position$;

17 signingBlock = read bytes from $position$ to ($position$ + $sizeOfBlockField2$);

    /* add payload to a copy of the APK signing block   */

18 parse $signingBlock$;

19 get ID-value pair sequence;

20 package payload;

21 **if** $insertion\ location\ is\ first$ **then**

22 | newSigningBlock = insert payload at the front of the pair sequence bytes;

23 **end**

24 **if** $insertion\ location\ is\ between$ **then**

25 | newSigningBlock = insert payload at midpoint of the pair sequence bytes;

26 **end**

27 **if** $insertion\ location\ is\ last$ **then**

28 | newSigningBlock = insert payload at the end of the pair sequence bytes;

29 **end**

    /* save Central Directory and EoCD offsets and bytes   */

30 newSigningBlockSize = sizeOf($newSigningBlock$);

31 rewrite $newSigningBlock$ two $size\ of$ fields with $newSigningBlockSize$;

32 cdAndEocd = extract Central Directory and EoCD bytes from APK;

33 eocdOffset = save offset of EOCD;

34 cdOffset = save offset of CD;

35 sizeDifference = sizeOf($newSigningBlock$) - sizeOf(old signing block);

    /* replace original APK signing block with the modified copy   */

36 open APK in binary read and write mode;

37 position = $originalSigningBlockOffset$;

38 write $newSigningBlock$ to APK file at $position$;

39 write $cdAndEocd$ to APK file at ($position$ + $newSigningBlockSize$);

40 resize bytes of APK file to $position$;

41 newCdOffset = $cdOffset$ + $sizeDifference$;

42 position = $eocdOffset$ + $sizeDifference$ + 16;

43 write $newCdOffset$ to APK file at $position$;

5.2   Implementation of Attack Scenarios

To demonstrate each of the attack strategies, we modified two original apps creating 120 apps with a payload. For each strategy, we extract the signing block from the original APK file, inject a payload and replace the original signing block with the modified version following Algorithm 1. We implemented this approach using the Python3 programming language.

To parse APK file and locate the signing block, we follow the approach typically used for parsing any ZIP archive. We start at the End of Central Directory (EoCD) to locate the starting offset of the Central Directory. We use the extracted offset to jump directly to the start of the Central Directory. We then backtrack to locate the magic "APK Sig Block 42" bytes. Since Android uses the signing block to support the v2 and v3 signing schemes—and indirectly the v4 scheme since it requires a complementary v2 or v3 signature—these magic bytes (and the signing block) are not present only when the APK is signed with the v1 scheme. If the signing block is indeed located before the Central Directory, these magic bytes are found in the 16 bytes directly before the starting offset of the Central Directory. Moving backwards from this offset, we parse the second *size of block* field that provides the starting offset of the APK signing block. We save this offset to use during the last steps of the APK modification.

We extract a copy of the signing block and locate the bytes between the two *size of block* fields. These bytes represent the sequence of ID-value pairs within the signing block, and all of our insertions occur in this section. Locating the offsets for payload insertion as the *First pair* and *Last pair* is straightforward.

Finding the correct offset for the *Between pairs* insertion is more complicated, as it requires clearly delineating the ID-value pairs in the sequence. According to the Android specifications, the first 12 bytes of each pair represent the ID and length of the pair. We follow this approach to delineate the pairs and compute the offset that represents the midpoint between the pairs in the sequence.

After finding the insertion location, we package the payload as follows:

v2 and v3 ID. Upon their build, the original v2 and v3 base apps have two ID-value pairs, i.e., one ID-value pair containing original signer information and the verity ID-value pair (ID = 0x42726577). For both apps, we create a new pair corresponding to v2 (ID = 0x7109871a) and v3 (ID = 0xf05368c0) signing schemes, respectively. We package the payload as a value in these new pairs. As a result, the modified app's signing block contains 3 pairs.

Verity ID. Verity ID-value pairs present a special case. As opposed to other cases where original pairs are not altered, we modify the value of the original verity ID-value pair (ID = 0x42726577) already present in the original base apps to include the payload.

Consequently, if the payload injection location is first, the modified verity pair is moved to be the first pair in the sequence. If the payload location is last, the modified verity pair remains in its original position in the sequence.

Finally, if the payload location is between two valid pairs, we duplicate an existing first ID-value pair containing signing information. Since the signing information in the second pair is identical to the original first pair, it does not affect the signing profile of the app. We then insert the modified verity pair containing the payload between two legitimate ID-value pairs.

Arbitrary ID. This case is similar to the treatment of v2 and v3 ID pairs. We create a new pair with a non-existent (ID = 0x716f43c0) and inject the payload as a value in this new pair.

Raw Bytes. We follow the same approach for inserting this form of the payload in the signing block as we do for the ID-value pair packaging forms. Unlike these forms, however, we do not prefix the payload with the 12 bytes that specify pair ID and length.

Adjusting the Size. Android does not limit the size of the signing block. Depending on its size, injecting the payload may increase the size of the APK signing block. If the payload (packaged as a verity pair) is smaller than the verity pair padding, it is padded with zero bytes to maintain the original verity pair size. However, if the payload is larger than the original verity pair length, the length is adjusted to reflect the new size of the pair. In cases where the signing block increases, we rewrite the values stored in the signing block's two *size of block* fields to avoid breaking the structure of the signing block.

At this point in the process, we have not modified the original APK. For our changes to take effect in the APK, we must overwrite the original signing block and adjust the APK file's metadata to ensure that the APK can still be parsed after our modification. We extract the bytes representing the original Central Directory and EoCD and their corresponding offsets. We will use this information to realign the APK's structure after modification.

To assemble a modified APK file, we jump to the saved starting offset of the APK signing block and overwrite the original signing block with our modified copy of the signing block. To ensure that the APK's data aligns with its structure, we update the offset of the Central Directory in EoCD so the signing block is reachable when parsed. Finally, we append the APK's Central Directory and EoCD bytes directly after the end of the signing block.

As the final step in this process, we manually verify that the payload was correctly packaged and inserted in the right location. We use Python script to print out all of the pairs in the signing block to confirm that the payload was inserted at the correct index in the sequence and that the other pairs were not affected by the insertion.

5.3   Verification

We verify the modified apps using `apksigner`. We use the *--print-certs* option, which allows us to verify that the original APK's signing certificate remains in place and is correctly readable by `apksigner`. Although our embedded payloads are not supposed to be inside the signing block, most of them are innocuous. We test the detection tools' abilities to identify the embedded content in APK signing block on the example of the EICAR payload since it is intended to text antivirus software. We use VirusTotal's API v3 to scan the modified APKs.

5.4   Testing

For our evaluation of attack strategies, we use the following devices: Google Pixel 8 (Android 14, API 34) and Google Pixel 7 Pro (Android 13, API 33). Our testing apps are designed for $minSDK = 33$ and $targetSDK = 34$, which means the apps are compatible with devices running Android 13 (API level 33) or higher and are optimized to work with Android 14 (API level 34). This is the primary justification for the selection of evaluation devices.

Each modified app is installed on an Android device using `Android Debug Bridge (adb)`, a command-line tool that enables us to issue actions on devices [3]. If the installation is successful, we further verify that the payload is triggered successfully. Since the main functionality of the app is to handle and display the payload (or the result of executing the payload), we check that the payload is triggered by opening the app and manually verifying the content displayed by the app.

6   Experimental Results

6.1   Verification

We verified all 120 apps with an embedded payload. We used `apksigner` to confirm that apps' original signatures were verified successfully on the two versions

Table 1. Experimental results

Signature scheme	Payload packaging	Payload	apksigner verification			Payload triggering					
						Android 13			Android 14		
			First	Between	Last	First	Between	Last	First	Between	Last
v2	v2 ID	EICAR	✗	✓	✓	○	●	●	○	●	●
		Text	✗	✓	✓	○	●	●	○	●	●
		Image	✗	✓	✓	○	●	●	○	●	●
		Code	✗	✓	✓	○	●	●	○	●	●
	v3 ID	EICAR	✗	✗	✗	○	○	○	○	○	○
		Text	✗	✗	✗	○	○	○	○	○	○
		Image	✗	✗	✗	○	○	○	○	○	○
		Code	✗	✗	✗	○	○	○	○	○	○
	Arbitrary ID	EICAR	✓	✓	✓	●	●	●	●	●	●
		Text	✓	✓	✓	●	●	●	●	●	●
		Image	✓	✓	✓	●	●	●	●	●	●
		Code	✓	✓	✓	●	●	●	●	●	●
	Verity ID	EICAR	✓	✓	✓	●	●	●	●	●	●
		Text	✓	✓	✓	●	●	●	●	●	●
		Image	✓	✓	✓	●	●	●	●	●	●
		Code	✓	✓	✓	●	●	●	●	●	●
	Raw byte form	EICAR	✗	✓	✓	○	●	●	○	●	●
		Text	✗	✓	✓	○	◐	◐	○	◐	◐
		Image	✗	✓	✓	○	◐	◐	○	◐	◐
		Code	✗	✓	✓	○	◐	◐	○	◐	◐
v3	v2 ID	EICAR	✓	✓	✓	●	●	●	●	●	●
		Text	✓	✓	✓	●	●	●	●	●	●
		Image	✓	✓	✓	●	●	●	●	●	●
		Code	✓	✓	✓	●	●	●	●	●	●
	v3 ID	EICAR	✗	✓	✓	○	●	●	○	●	●
		Text	✗	✓	✓	○	●	●	○	●	●
		Image	✗	✓	✓	○	●	●	○	●	●
		Code	✗	✓	✓	○	●	●	○	●	●
	Arbitrary ID	EICAR	✓	✓	✓	●	●	●	●	●	●
		Text	✓	✓	✓	●	●	●	●	●	●
		Image	✓	✓	✓	●	●	●	●	●	●
		Code	✓	✓	✓	●	●	●	●	●	●
	Verity ID	EICAR	✓	✓	✓	●	●	●	●	●	●
		Text	✓	✓	✓	●	●	●	●	●	●
		Image	✓	✓	✓	●	●	●	●	●	●
		Code	✓	✓	✓	●	●	●	●	●	●
	Raw byte form	EICAR	✗	✓	✓	○	●	●	○	●	●
		Text	✗	✓	✓	○	◐	◐	○	◐	◐
		Image	✗	✓	✓	○	◐	◐	○	◐	◐
		Code	✗	✓	✓	○	◐	◐	○	◐	◐

✓- APK is verified by `apksigner`, × - APK is not verified by `apksigner`
● - Installs and triggers the payload, ◐ - Installs but crashes, ○ - Does not install

of the Android platform supported by our test APK (API 33 and 34). Table 1 presents the results of verifying the test apps. The majority of our test apps were successfully verified by `apksigner`. In these cases, the embedded payload did not affect the app's signing profile.

Arbitrary ID. Regardless of the payload injection location, all apps in which the payload was packaged as an ID-value pair using an arbitrary ID were successfully verified. This outcome is not surprising, as Android's policy is to ignore ID-value pairs with unknown IDs.

Verity Pairs. Similarly, test apps with the payload embedded in the verity pair were all verified by `apksigner`. Our manual inspection of `apksigner`'s source code revealed that the contents of the verity block are not checked during verification. As a result, the embedded payload remains undetected.

v2 and v3 Pairs. The results of the tests where we packaged the payload as a pair with a v2 ID or v3 ID varied. Apps with EICAR, text, or code payloads inserted as the first pair in the sequence failed verification due to a malformed list of signers. Similarly, the image payload failed verification, but this was due to a *java.lang.IllegalArgumentException: Negative length* exception. The root cause of these failures is consistent: the format of the pair did not adhere to Android's defined specifications, leading to unexpected behaviour during parsing by `apksigner`. The difference between these errors lies in their handling: the malformed signer error was anticipated and appropriately caught by `apksigner`, while the conditions triggering the Java exception were not.

When the payloads were inserted into the APK signing block between two valid ID-value pairs or as the last pair in the sequence (v2 payload in originally v2 apps, v3 payload in originally v3 apps), `apksigner` verification was successful. This is due to the verification logic of `apksigner`, which verifies the first known ID-value pair that it expects according to the app's *maxSDK*. Consequently, when the pair with the payload was inserted behind a valid ID-value pair (i.e., between and last insertion locations), `apksigner` verified the first expected pair and did not proceed to the pair that contained the payload resulting in successful app verification.

If the first encountered pair is an older signing scheme (e.g., v2) compared to what is expected, `apksigner` proceeds until it finds a pair corresponding an expected scheme. In all cases, where the payload, packaged as a v3 pair, was inserted into an APK originally signed with the v2 scheme, all tests failed. For APKs with a *maxSDK* $\geq$ *API 33* (Android 13), `apksigner` expected the v3.1 and v3 signature schemes. As a result, `apksigner` proceeded to a v3 ID-value pair to verify. Since the format of the v3 pair's value (which contained the payload) did not match the expected structure, verification failed. A similar issue occurs when the payload, packaged as a v2 pair, is inserted as the first pair into an APK already signed with the v2 scheme. In this case, since no v3 signing pair is present, `apksigner` proceeds to verify the first v2 pair it encounters. However, this pair contains the embedded payload, which does not conform to the expected format and leads to verification failure. Likewise, a v3 APK with the payload embedded in the first v3 pair also fails verification for the same reason.

Raw Byte Form. The test apps with the payload added to the APK signing block in its raw byte form at the start of the sequence failed verification. Our investigation revealed that `apksigner` mistakenly interpreted the first 12 bytes of the raw payload as the ID and length fields of a valid ID-value pair. This misinterpretation caused `apksigner` to treat a larger chunk of bytes—extending beyond the boundaries of our embedded payload—as the value field. As a result, the legitimate ID-value pairs following our payload and containing the signing information were incorrectly interpreted. This effectively rendered the APKs' signing data inaccessible to `apksigner`, leading to verification failure.

Overall, our results show that there are several viable payload packaging and insertion location combinations that an attacker can reliably exploit.

VirusTotal Analysis. To establish a baseline, we analyzed the EICAR text file[6] using VirusTotal. Out of 76 detection engines included in the VirusTotal platform, 66 flagged this file as malicious. We then analyzed the 30 modified APK files with the EICAR payload embedded in their signing block. For 29 of the modified APK files, the detection rate dropped significantly to 5/76. The same five engines consistently detected the EICAR payload (Alibaba, Ikarus, Fortinet, Google, and SymantecMobileInsight) while the majority of engines categorized the files as undetected. A small number of engines failed or timed out during analysis. The one exception was the case in which the EICAR payload was packaged as a v3 ID-value pair and inserted between other pairs in the testing app that was signed with the v3 scheme. Only four engines (Alibaba, Ikarus, Fortinet, and SymantecMobileInsight) detected the EICAR payload; the Google engine reported a failure.

6.2 Testing

After verification, we confirmed that ① the modified apps can be installed on the physical devices, and ② the embedded payloads can be successfully accessed or triggered.

The results of our analysis are shown in Table 1. Overall, the installation successes and failures align with the verification results with a few exceptions. *Injecting the payload as part of a verity ID-value pair or a pair with an arbitrary ID consistently results in successful verification, installation, and reliable triggering of the payload.*

Packaging the payload as a v2 or v3 pair is less consistent. The v2 APKs with the payload packaged as a v3 pair were not verified and, as a result, did not install on any of the devices. Similarly, the v2 and v3 APKs with the payload packaged as the first v2 and v3 pairs, respectively, did not install. The rest of

---

[6] https://www.eicar.org/download-anti-malware-testfile/.

the APKs were verified and successfully installed on the devices, allowing access to the embedded payload.

The apps whose signing blocks contained the payloads in the raw byte form behaved differently if the payload was inserted between or after valid pairs. Although the apps installed on the device successfully, they crashed when they were opened. Further manual investigation showed that this behaviour was due to an out-of-bounds error that caused an incorrect interpretation of the payload bytes as a length-prefixed field. When the text, image, and code payloads (raw bytes) were inserted between pairs or as the last pair in the sequence, their initial bytes were interpreted as the ID and length fields of a pair. In most cases, these initial bytes did not correspond to the actual payload length (as they were not intended for this purpose). As a result, parsing exceeded the number of available bytes, causing an out-of-bounds exception. The EICAR payload did not cause this behaviour due to its size. In either case, crafting a payload to match the expected length can potentially resolve these errors and allow for more consistent parsing of payloads.

6.3   App Distribution Through App Market

The final step in our analysis is to verify whether our crafted app with an embedded payload can be successfully distributed through an app distribution platform. For this experiment, we designed an app that allows users to increment or decrement a counter. Since app markets do not accept applications without clear functionality, we could not use our base app. We signed this app with a v2 signature and embedded our code payload as an ID-value pair with the arbitrary ID that we used in our previous experiments. We inserted the packaged payload between the v2 pair containing the legitimate signing information and the verity pair. We then uploaded this app to APKPure and Aptoide, which are alternative app markets. Since the Google Play Store requires developers to upload their apps in the AAB format, we do not include it in our analysis.

Table 2. Results of malicious app analysis (case study)

Strategy	Verified by apksigner	VirusTotal analysis		Installation status	
		Detection rate	Google engine detection	Android 13	Android 14
Baseline	✓	27/75	Yes	●	●
Direct injection	✓	5/76	Yes	●	●
Compress	✓	6/76	Yes	●	●
Encode	✓	0/76	No	●	●
Reverse	✓	0/76	No	●	●
Split ($n = 2$)	✓	3/76	Yes	●	●
Split ($n = 4$)	✓	0/76	No	●	●

✓- APK is verified by apksigner,● - APK installs on the device

APKPure and Aptoide verified the app without issues and made it publicly available to users. We downloaded the app's file from each market to inspect its signing block. In both cases, the code payload remained intact, confirming that neither market had stripped our signature during the upload process. The APK files from APKPure and Aptoide were successfully installed on both devices, and once opened, each app displayed the results of the code execution.

7 Case Study

Until now, our evaluation of attack strategies was confined to innocuous payloads. As a proof-of-concept, we modified the test app with the v3 signing scheme to include a malicious payload in the form of an APK file obtained from the MalwareBazaar repository.[7] This executable was a Trojan malware that targets Android mobile banking apps by disguising itself as a legitimate app.[8] We packaged the malware as an ID-value pair with the same arbitrary ID that we used in our experiments. Similar to our previous experiments, we inserted the malicious payload between the original v3 ID-value pair and the verity pair of our app (Table 2).

To establish a baseline, we analyzed the sample using VirusTotal. Out of 75 engines, 27 (including Google's detection engine) classified it as malicious. In contrast, the detection rate for our malicious app was significantly lower. Only 5 out of 76 engines, including Google, identified it as malware.

Since the Google engine detected our app as malicious, it is unlikely that such an app would be allowed to be distributed through the Google Play platform. To evade Google detection and obscure the payload, we explored four strategies:

- *Compress.* We compressed the malicious payload to the ZIP format before its insertion into the signing block. The detection rate increased compared to the previous test as 6 out of 76 engines detected its malicious content.
- *Encode.* We encoded the payload as the Base64 format before its insertion. This strategy was successful, and none of the engines detected the malware.
- *Reverse.* We reversed the order of the payload's byte sequence. As with the encoding strategy, none of the 76 engines detected malicious payload in the manipulated APK.
- *Split.* We divided the payload's byte sequence into n equal subsequences and packaged each as a separate arbitrary ID-value pair. For our first test, we split the payload into 2 subsequences creating 2 pairs. The pair containing the first half of the payload was inserted between the v3 signing block and the verity pair while the pair containing the second half was inserted as the last pair in the sequence. Only 3 engines, including Google, detected the malicious payload. When we divided the payload into 4 subsequences and distributed them throughout the signing block, none of the engines, including Google, detected the malicious content.

For all apps generated in this case study, we used **apksigner** to verify the signatures of the modified malicious APKs. As **apksigner** is not designed to detect malicious content, it is unsurprising that all the APKs produced during

---

[7] https://bazaar.abuse.ch/.

[8] SHA256: b8ea74902684dcced62a5ca2c1d6932659decfefcbdb2615bfe5899e05eb1451.

our tests were verified successfully. We also attempted to install each of the modified APKs on the two Google devices. Like the EICAR payload experiments, we did not define a trigger and only checked that the APK could be installed and would not crash when opened. All apps successfully installed and could be opened without crashing.

8 Discussion

Recommendations for the Android Development Team:

– *APK signing block content verification.* The lack of signing block verification is a key factor in enabling our attacks. This issue could be addressed by implementing a digest over the entire APK (as the v4 scheme does). However, because the v4 signature file is a separate file that is stored alongside the APK, this approach poses challenges for APK distribution. Another approach would be to implement a digest over the APK signing block, which may ultimately limit the extensibility of the signing mechanism. Additionally, the more conservative approach of only implementing a digest over the ID-value pair sequence still leaves room for payload injection. While each of these three proposed approaches have their drawbacks, they can reduce the likelihood of signing block exploitation when combined with the recommendations presented below.
– *Address the mismatch between signing and validation functionalities.* Updates made to the signing block content during the signing process should be reflected in corresponding checks during the verification process. For example, the integrity and expected format of the verity pair should be verified.
– *Comprehensive verification.* All content within the APK signing block should be verified, and unverified content should be rejected. This would eliminate scenarios where ID-value pairs with unrecognized IDs are ignored and allowed to remain in the block.
– *Unambiguous official specifications.* The official Android documentation assumes the presence of only one v2 or v3 signing pair, leading to the verification of the first encountered pair. This oversight effectively allows an attacker to exploit additional pairs for malicious purposes.
– *ZIP offset verification.* The digest calculation should include the actual Central Directory offset. Using this offset would require a significant redesign of the verification process especially with respect to the addition of new signing material. Nevertheless, it is a necessary step towards preventing modifications to the APK signing block after the app is signed.

Recommendations for App Distribution Market Operators:

– *Strengthen app vetting.* Market operators could further enhance their vetting process for apps submitted to their markets, particularly focusing on the APK signing block. The vetting process could ensure compliance with official specification, employ deeper analysis of embedded content, and leverage advanced malware detection techniques.

Limitations. Our work has two major limitations. *First,* we assume that the APK signing block's internal structure and position within the APK file adheres to Android documentation. Our method for parsing the APK file to extract the signing block follows the Android-defined approach. However, we are aware that some third-party app markets (e.g., F-Droid) use different APK signing block parsing methods. In such scenarios, our embedded payload may be detectable, and consequently, our attacks may not succeed in those markets. *Second,* we do not include APKs with mixed signature schemes in our analysis (i.e., v1 and v2, v1 and v3, v2 and v3). We do not address the v1 signing scheme due to the differences in signing implementation between the v1 and v2 schemes. In other words, modifying an APK signed with the v1 scheme changes its content and makes the manipulation detectable by `apksigner`.

9   Conclusion

App signing is a critical protection mechanism that ensures the integrity and authenticity of mobile apps. Our approach exploits logical weaknesses in the design and implementation of the verification process to embed arbitrary content in the APK signing block. This attack is feasible because validation tools operate under the assumption that all components present are created according to Google's guidelines for generating and signing APKs. This assumption creates significant blind spots in Android tools, leaving ample space within the APK file for embedding attack payloads.

References

1. Android: Android SDK platform tools (2023). https://developer.android.com/tools#tools-build
2. Android: apksigner (2023). https://developer.android.com/tools/apksigner
3. Android: Android debug bridge (ADB) (2024). https://developer.android.com/tools/adb
4. Android: APK signature scheme v2 (2024). https://source.android.com/docs/security/features/apksigning/v2
5. Android: APK signature scheme v4 (2024). https://source.android.com/docs/security/features/apksigning/v4
6. Android: Application signing (2024). https://source.android.com/docs/security/features/apksigning
7. Du, S., Zhu, P., Hua, J., Qian, Z., Zhang, Z., Chen, X., Zhong, S.: An empirical analysis of hazardous uses of Android shared storage. IEEE Trans. Dependable Secure Comput. **18**(1), 340–355 (2021)
8. Fahl, S., Dechand, S., Perl, H., Fischer, F., Smrcek, J., Smith, M.: Hey, NSA: stay away from my market! future proofing app markets against powerful attackers. In: 2014 ACM CCS, pp. 1143–1155. ACM (2014)
9. Jung, J.H., Kim, J.Y., Lee, H.C., Yi, J.H.: Repackaging attack on Android banking applications and its countermeasures. Wireless Pers. Commun. **73**, 1421–1437 (2013)

10. Kar, A., Stakhanova, N.: Exploiting Android browser. In: CANS, pp. 162–185. Springer (2023)
11. Lindorfer, M., Neugschwandtner, M., Weichselbaum, L., Fratantonio, Y., Veen, V.V.D., Platzer, C.: ANDRUBIS – 1,000,000 apps later: a view on current Android malware behaviors. In: BADGERS, pp. 3–17 (2014)
12. NVD: CVE-2013-4787 (2013). https://nvd.nist.gov/vuln/detail/CVE-2013-4787
13. NVD: CVE-2017-13156 (2019). https://nvd.nist.gov/vuln/detail/CVE-2017-13156
14. Poeplau, S., Fratantonio, Y., Bianchi, A., Kruegel, C., Vigna, G.: Execute this! Analyzing unsafe and malicious dynamic code loading in Android applications. In: NDSS, pp. 1–16 (2014)
15. Ruggia, A., Possemato, A., Merlo, A., Nisi, D., Aonzo, S.: Android, notify me when it is time to go phishing. In: IEEE EuroS&P, pp. 1–17 (2023)
16. Vidas, T., Christin, N.: Sweetening Android lemon markets: measuring and combating malware in application marketplaces. In: ACM CODASPY, pp. 197–208. ACM (2013)
17. Vidas, T., Votipka, D., Christin, N.: All your droid are belong to us: a survey of current Android attacks. In: WOOT, pp. 1–10. USENIX Association (2011)
18. Wang, H., Liu, H., Xiao, X., Meng, G., Guo, Y.: Characterizing android app signing issues. In: IEEE/ACM ASE, pp. 280–292. IEEE Press (2019)
19. Xu, H., Yao, M., Zhang, R., Dawoud, M.M., Park, J., Saltaformaggio, B.: DVa: extracting victims and abuse vectors from Android accessibility malware. In: USENIX, pp. 701–718. USENIX Association (2024)
20. Zhang, X., Ying, K., Aafer, Y., Qiu, Z., Du, W.: Life after app uninstallation: are the data still alive? Data residue attacks on android. In: NDSS, pp. 1–15 (2016)
21. Zhou, Y., Jiang, X.: Dissecting android malware: Characterization and evolution. In: IEEE S&P, pp. 95–109 (2012)

An Empirical Study of Multi-language Security Patches in Open Source Software

Shiyu Sun[1][✉], Yunlong Xing[1], Grant Zou[2], Xinda Wang[3], and Kun Sun[1]

[1] George Mason University, Fairfax, VA, USA
ssun20@gmu.edu
[2] University of Virginia, Charlottesville, VA, USA
[3] University of Texas at Dallas, Richardson, TX, USA

Abstract. Vulnerabilities in software repositories written in multiple programming languages present a major challenge to modern software quality assurance, especially those resulting from interactions between different languages. Existing static and dynamic program analysis tools are generally constrained to single-language analysis, while current deep-learning models lack the capability to process cross-language interactions effectively. To gain deeper insights into vulnerability patterns and patching behaviors in multi-language code, we conduct a measurement study on commits associated with multi-language security patches. We first collect a large-scale dataset of multi-language security patches from the MITRE corporation. We then analyze trends in language combinations, assess their proneness to vulnerabilities, and compare the severity of these vulnerabilities to those in single-language patches. Additionally, we classify patch patterns based on the types of language interactions to support automated program repair. To encourage further research, we release our dataset to the community, fostering deeper investigation into multi-language security patch development and enhancement.

Keywords: Multi-Languages Open Source Software · Security Patches

1 Introduction

Multi-language development is increasingly prevalent and essential in modern software deployment, as applications often integrate multiple technologies to meet performance, scalability, and functionality demands [6]. Among the millions of commits involving modifications across multiple languages [5], approximately 6–18% [8,31] are intended to fix cross-language vulnerabilities, remedying critical risks caused by interactions between different languages.

Several efforts have been made to analyze multi-language software. Li et al. [17] studied multi-language software and revealed a statistical connection between language selection, vulnerability proneness, and interfacing mechanisms of multilingual projects. However, it cannot represent the association at the commit level since not all commits in the multilingual projects involve multiple

© The Author(s), under exclusive license to Springer Nature Switzerland AG 2025
M. Egele et al. (Eds.): DIMVA 2025, LNCS 15748, pp. 124–146, 2025.
https://doi.org/10.1007/978-3-031-97623-0_8

programming languages. Moreover, other papers have worked on detecting multi-language vulnerabilities [12,22,23,32,33]. However, they either only focus on one language combination (e.g., Python-C or Java-C) or neglect the fixes for the bug. Additionally, they fail to explain why and how vulnerabilities are triggered or fixed across multiple languages. Furthermore, large language models (LLMs) can analyze code and have different domain knowledge to understand multiple programming languages, but they struggle to capture the full context when the input is large. Therefore, we can conclude that there is no existing labeled multi-language security patch dataset and the existing tools cannot be transferred to analyze different language combinations.

In this paper, we fill this gap by constructing a multi-language security patch dataset to reveal why different programming languages have been committed together and how they work together to fix vulnerabilities. Since different languages can interact with other programming languages across files and in the same file, we consider two types of commits: (1) commits that modify multiple files written in different programming languages, and (2) commits that modify a single file containing inline code from another language. First, we perform a statistical analysis of the patches, examining their language combinations, patch complexity, vulnerability categories, and severity levels. Second, we manually review each security patch to understand how vulnerabilities are fixed based on interfacing mechanisms. This analysis provides insight into the critical context needed for vulnerability remediation and its relationship with different types of vulnerabilities. Finally, we propose a taxonomy summarizing common fixing patterns, which can help advance automated program repair research.

According to the statistics from IEEE Spectrum [11] and the available data from MITRE, we focus on 16 popular programming languages, including C, C#, C++, Go, HTML, Java, JavaScript, Kotlin, Perl, PHP, Python, Ruby, Rust, Shell, SQL, and TypeScript. We collect 2,798 multi-language security patches, consisting of 1,253 multi-language security patches across multiple files and 1,545 in a single file. We summarize 11 fixing patterns, including eight for multi-file fixing and three patterns for single-file fixing.

Among our findings, we identify that multi-language security patches are more complex than single-language security patches in terms of lines of modification, branch structure changes, and dependency updates. Multi-language security patches are also more prone to injection-related vulnerabilities (e.g., OS command injection and SQL injection), web-related vulnerabilities (e.g., cross-site scripting (XSS) and cross-site request forgery (CSRF)), and authentication vulnerabilities. The severity of multi-language vulnerabilities is similar to that of single-language vulnerabilities. From our manual analysis of patches, we further observe that multi-language security patch patterns are associated with their interfacing mechanisms. Single-language security patch patterns can combine to form part of a multi-language security patch pattern (e.g., one language sanitizes the input while another validates it). Additionally, non-fixing changes are often entangled with multi-language security fixes, making code review more difficult.

The findings of our analysis provide insights that suggest paths forward for the security community to improve vulnerability management.

Our work provides several contributions. First, we collect the first dataset of multi-language security patches across 16 languages. Second, we analyze the characteristics of multi-language security patches. Moreover, we summarize the fixing patterns of multi-language security patches, revealing the reasons why different languages have been committed together and highlighting the need for automated tools to analyze multi-language programs. We release the dataset at https://figshare.com/Multi-language_Security_Patch_Dataset.

2   Background

2.1   Language Interoperability and Interfacing Mechanisms

Multilingual software leverages the distinct capabilities of various programming languages. To enable these languages to work together smoothly, they require effective inter-language interaction or interoperability mechanisms to ensure cohesive functionality [24]. We categorize four common interaction mechanisms with code-level evidence: client-server, foreign function interface (FFI), database queries, and subprocess execution.

Client-Server. The interaction occurs in three ways: (1) from the client to the server, (2) from the server to the client, and (3) bidirectionally between both. Client-side code is typically implemented using HTML for structure, JavaScript or TypeScript for dynamic behavior, whereas server-side code leverages languages and frameworks such as JavaScript/Node.js, Python, Java, PHP, C#, Rust, Go, or Ruby. These components interact bidirectionally, with the client sending requests to the server and the server processing and returning data or actions. There are several commonly used methods, often via HTTP requests.

Foreign Function Interfaces (FFI). An FFI is a mechanism that allows code written in one programming language to call functions or use services written in another language. FFI is crucial in scenarios where developers leverage existing libraries, optimize performance, or integrate systems written in multiple languages. For example, a Python application might use FFI to call a high-performance C library for numerical computations, or a Rust program might expose functions via FFI to be used in a Node.js application.

Database Queries. Python, Java, JavaScript, PHP, Ruby, Go, Rust, and C#, have libraries or frameworks that enable them to access databases and execute SQL commands. Some of these libraries provide direct SQL execution (*e.g.*, psycopg2 in Python), while others offer object-relational mapping (ORM) (*e.g.*, Entity Framework in C# or ActiveRecord in Ruby) to simplify database interactions by using an object-oriented approach. In this scenario, SQL statements are often composed as strings in the function call.

Subprocess Execution. Similar to database queries, many programming languages offer libraries or built-in methods to execute shell commands, allowing developers to run system commands directly from their code, including Python, JavaScript, Java, C/C++, Ruby, PHP, Go, Perl, and Rust.

2.2 Multi-language Security Patch

A software security patch is a set of changes between two versions of source code to address specific vulnerabilities. Listing 1.1 shows a security patch that sanitizes user input using `htmlentities()` to prevent XSS attacks.

```
1  diff --git a/web/edit/web/index.php b/web/edit/web/index.php
2  --- a/web/edit/web/index.php
3  +++ b/web/edit/web/index.php
4  +  $user_plain=htmlentities($_GET['user'])
5  diff --git a/.../edit_server.html b/web/templates/pages/edit_server.html
6  --- a/web/templates/pages/edit_server.html
7  +++ b/web/templates/pages/edit_server.html
8  - <span id="generate-csr"> / <a class="generate" target="_blank" href="/generate/ssl/?
       domain=<?=\$v_hostname?>"><?=_('Generate CSR');?></a></span>
9  + <span id="generate-csr"> / <a class="generate" target="_blank" href="/generate/ssl/?
       domain=<?=htmlentities(trim($v_hostname,'"'));?>"><?=_('Generate CSR');?></a></
       span>
```

Listing 1.1. Security Patch for CVE-2022-0986 on PHP and HTML files.

We refer to security patches that contain more than one language as multi-language security patches. Language interaction can happen across multiple files or within a single file, depending on the language, environment, and tools used. In this paper, we consider both situations and use *multi-file multi-language* security patches and *single-file multi-language* security patches to refer to them.

Multi-file Multi-language Patch. Listing 1.1 shows a security patch fixing the vulnerability CVE-2022-0986 by modifying the PHP file and the HTML file. The `index.php` file escapes the HTML entities of the `user`, and `edit_server.html` escapes the HTML entities of the `v_hostname`. These two files work together to resolve the XSS vulnerability.

Single-file Multi-language Patch. Languages can interact in a single-file setup, which is referred to as embedding or interfacing between languages. Embedding is a program written in one language (like C) that uses another language (like a shell script) to handle specific tasks. For example, a C program might execute shell commands for file manipulation using `system()`. Interfacing happens when different languages work together through defined boundaries or protocols, such as using APIs, libraries, or inter-process communication, which is similar to multi-file interaction. As shown in Listing 1.2, the patch fixes the SQL injection by adding a Python sanity check to the field of the vulnerable SQL query.

```
1  diff --git a/gamespy/gs_database.py b/gamespy/gs_database.py
2  --- a/gamespy/gs_database.py
3  +++ b/gamespy/gs_database.py
4  @@ -367,12 +367,12 @@ def update_profile(self, profileid, field):
5  -     with Transaction(self.conn) as tx:
6  -         q = "UPDATE users SET \"%s\" = ? WHERE profileid = ?"
7  -         tx.nonquery(q % field[0], (field[1], profileid))
8  +     if field[0] in ["firstname", "lastname"]:
9  +         with Transaction(self.conn) as tx:
10 +             q = "UPDATE users SET \"%s\" = ? WHERE profileid = ?"
11 +             tx.nonquery(q % field[0], (field[1], profileid))
```

Listing 1.2. Security Patch for CVE-2020-36631 on SQL in Python file.

3   Patch Collection

To explore multi-language vulnerabilities and their fixes, we collect security patches from open-source software. According to the statistics from IEEE Spectrum [11] and the available data from MITRE [2], we focus on 16 popular programming languages, including C, C#, C++, Go, HTML, Java, JavaScript, Kotlin, Perl, PHP, Python, Ruby, Rust, Shell, SQL, and TypeScript. We summarize the language type, usage, and language interaction in Table 1.

Table 1. Language Selection.

Languages	Type	Level	Application	Language Interaction
C	Compiled	Low	Operating system	Assembly and high-level languages
C#	Compiled	High	Game	.NET languages and native code
C++	Compiled	Low	System and game	C and other system languages
Go	Compiled	High	Cloud and system	C and statically typed languages
HTML	Markup	-	Web page structure	Web languages
Java	Compiled	High	Mobile apps	JVM and native interfaces
JavaScript	Interpreted	High	Web	Web and server-side languages
Kotlin	Compiled	High	Mobile and web	Java and other JVM languages
Perl	Interpreted	High	System and web	C/C++ and web technologies
PHP	Interpreted	High	Web	HTML, JavaScript, and SQL
Python	Interpreted	High	Web and data science	C/C++ and web languages
Ruby	Interpreted	High	Web and scripting	Scripting languages
Rust	Compiled	Low	System and web	C and other system languages
Shell	Scripting	-	System and automation	Shell and command line tools
SQL	Domain-specific	-	Database management	Database languages
TypeScript	Interpreted	High	Web	JavaScript and other frameworks

Our data collection process begins by mining commits associated with CVE records indexed by MITRE [2]. First, we retrieve vulnerabilities that have been assigned CVE IDs. Next, we parse the vulnerability reports to extract patch hyperlinks and save the corresponding commits. We then download all security patches. To keep only the security patches written in the specified languages, we analyze the modified file extensions and select those that match the chosen languages, e.g., .c for C. If there is more than one type of file extension from different languages, we classify it as a multi-file, multi-language security patch. If only one programming language is involved, we include it for further analysis. For multi-file multi-language security patches, we retain all related files and manually analyze their interactions. For single-file multi-language security patches, we first review the official documentation to identify the possible languages within the file. Then, we develop a Python script to check for the presence of embedded language keywords in the host language syntax, e.g., <?php> as an embedded language inside a Java program. We match the presence of languages that are not listed in their documentation and flag their existence. After that, we manually check the matched single-file patches to identify the single-file multi-language security patches and analyze their interactions.

4   Data Characterization

Based on the method in Sect. 3, we collect a multi-language security patch dataset consisting of publicly indexed vulnerability fixes. We conduct a set of experimental studies to investigate the quantitative characteristics of the dataset.

4.1   Dataset Composition

We collect 14,453 security patches containing at least one of the selected languages (from the 1990s to September 24, 2024). Among them, 4,567 security patches modify more than one file type, and 1,253 of those modify at least two of the selected languages (which we consider as *multi-file multi-language* security patches), accounting for 27.44%. 9,886 security patches only modify one file type, and 1,545 of them embed other languages (which we call *single-file multi-language* security patches), making up 15.63%. In total, we collect 2,798 multi-language patches.

4.2   Language Combinations

Among 1,253 multi-file multi-language security patches, there are 111 combinations. Of these, 62 combinations include two languages, 34 combinations include three languages, 11 combinations include four languages, three combinations include five languages, and one combination includes six languages (*i.e.,* the patch for CVE-2016-10096 modifies Shell, C, HTML, PHP, SQL, and JavaScript files in the same commit). We use a Python script to count the existence of different languages. Table 2 shows the top-20 combinations. The table reveals that JavaScript, HTML, and Python stand out as the most popular languages for interacting with other languages. This popularity can be attributed to three key reasons. First, Python's versatility allows it to be applied across various domains and seamlessly integrated with other languages. Second, JavaScript's ability to function in both front-end and back-end development increases its likelihood of collaborating with other programming languages. Finally, HTML, as the backbone of most front-end websites, is inherently supported by back-end technologies, creating more opportunities for its use alongside other languages.

Table 2. The Language Combination of Multi-file Multi-language Security Patch.

Language Combination	Count	Language Combination	Count
JavaScript & PHP	178	C++ & PHP	34
C++ & Python	163	Java & JavaScript	27
JavaScript & TypeScript	89	C++ & JavaScript	27
HTML & JavaScript	64	HTML & Java	25
C & Python	55	JavaScript & Ruby	23
HTML & Python	54	C & Perl	20
C & Shell	43	HTML & PHP	18
C & C++	41	HTML & JavaScript & Python	17
JavaScript & Python	36	SQL & JavaScript & PHP	15
HTML & Ruby	35	SQL & PHP	15

Table 3. The Language Combination of Single-file Multi-language Security Patch.

Host Language	Embedded Language	Host Language	Embedded Language
JavaScript	SQL, HTML, Shell	C#	SQL, HTML, JavaScript
Kotlin	Java	C++	Shell, SQL, PHP
Perl	Shell, SQL, HTML	Go	SQL, HTML, Shell, C
Python	HTML, Shell, SQL	HTML	SQL, Java, JavaScript, PHP
Ruby	SQL, Shell, C	Java	SQL, PHP, HTML, Shell
Rust	C, Shell		

Among 1,545 single-file multi-language security patches, there are 50 combinations. We list languages that embed with one of the selected 16 programming languages in Table 3. The results show that Shell, HTML, and SQL are the most popular languages for direct use in other languages. They often serve as embedded languages because they excel in tasks that complement general-purpose languages like Python, Java, or C++. Shell scripting automates system tasks and manages files, HTML structures web pages, and SQL handles database queries. When embedded in frameworks like Django (Python) or Rails (Ruby), HTML defines web page templates, enabling dynamic content generation. Similarly, embedding SQL allows direct database interaction for efficient data management. This combination leverages the strengths of both host and embedded languages—host languages handle application logic, while embedded languages optimize specific tasks like database operations or system management.

4.3 Security Patch Complexity

Security patch complexity reveals the difficulty in creating, applying, and verifying a security patch. We assess the complexity of security patches from three perspectives: the modification complexity, the branch structural complexity, and the dependency complexity. We directly count the lines of code that have changed, identifying additions and deletions marked with '+' and '−' for modification complexity. Branch structural complexity is evaluated by counting updated conditions, loops, or branches introduced by the patch. Dependency complexity changes are measured by noting any added or removed libraries or frameworks. We develop a Python script to count the added and deleted lines, identify branch statements by matching keywords and grammar, *e.g.,* `for a in A:` in Python, and identify library usage with keywords and grammar matching, *e.g.,* `#include<A>` in C.

As shown in Fig. 1, we uncover that multi-language security patches (in red) are more complex compared to single-language ones (in blue). The complexity of the multi-language security patches mainly comes from three aspects. First, the scope and nature of the vulnerability in multi-language security patches are more complex, affecting multiple components or systems and necessitating solutions that span several languages. Second, the software architecture of multi-language software is more sophisticated. In a multi-language environment, patches might need to address cross-language interactions, which can introduce additional modifications and branch structure complexity. For example, if a web application uses JavaScript for the front-end and Python for the back-end, a security vulnerability that affects both might require coordinated patches across these technologies, considering how data is handled and passed between them. Third, languages like Python can interact with multiple languages and often have more libraries/APIs

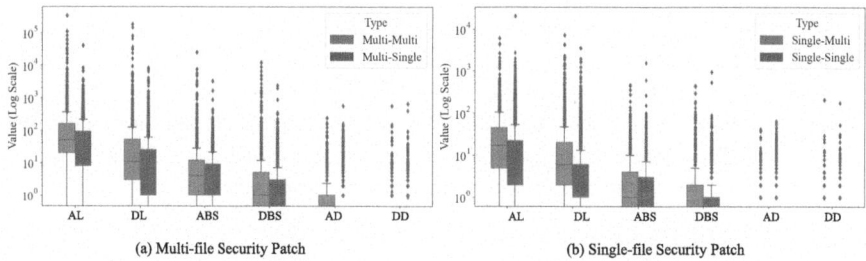

(a) Multi-file Security Patch   (b) Single-file Security Patch

Fig. 1. Patch Complexity. AL (added lines) and DL (deleted lines) represent the modification complexity. ABS (added branch structure) and DBS (deleted branch structure) represent branch structural complexity. AD (added dependency) and DD (deleted dependency) represent dependency complexity. (Color figure online)

to facilitate this interaction. The availability and reliability of these libraries can affect the dependency complexity of implementing security measures.

4.4 Vulnerability Categories Proneness

To measure the vulnerability categories' proneness of multi-language security patches, we use CWE (Common Weakness Enumeration) [4] identifiers. We first collect the CWE identifiers associated with each patch. Next, we count and visualize the number of patches that fall into each CWE category. We get 7,605 security patches associated with 329 CWE-IDs, including 1,776 multi-language patches with 185 CWE-IDs and 5,829 single-language patches with 311 CWE-IDs. Multi-language and single-language security patches share 167 CWE-IDs. We plot the top-10 CWE distribution for multi-file and single-file patches in Fig. 2.

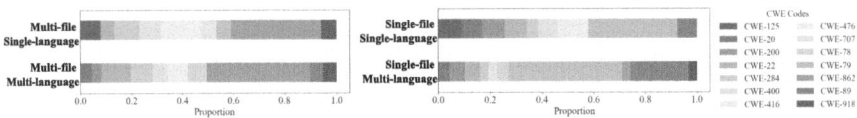

Fig. 2. CWE Distribution of Security Patches

Multi-file Single-Language. The common vulnerabilities are XSS, information leakage, DoS, path traversal, heap-based buffer overflow, SSRF, and use-after-free. The distribution indicates a focus on memory management and DoS issues.

Multi-file Multi-language. The common vulnerabilities are XSS, improper input validation, path traversal, CSRF, SQL injection, SSRF, and CSRF. The distribution shows a mix of both client-side and server-side vulnerabilities, indicating a broad surface area due to the complex interactions between multiple file types and programming languages.

Single-File Single-Language. The common vulnerabilities are XSS, out-of-bounds read, path traversal, NULL pointer dereference, use after free, and improper access control. The distribution emphasis on direct memory manipulation and access vulnerabilities, common in environments where a single language is used without the added complexity of multiple languages.

Single-File Multi-language. The common vulnerabilities are XSS, SQL injection, path traversal, OS command injection, improper neutralization, SSRF, and missing authorization. This category heavily features vulnerabilities that involve embedded languages in a single programming language.

The figure reveals that both multi-language and single-language environments exhibit a higher prevalence of XSS and CSRF vulnerabilities, indicating challenges in user data handling and session management. Compared with single-language environments that tend to show more severe memory and resource handling issues like buffer overflows and use-after-free, multi-language environments are more prone to injection-related vulnerabilities, including command injection and SQL injection. Therefore, we conclude that the different combinations of programming languages lead to more concentrated types of vulnerabilities. This language combination tendency is also influenced by the type of application and the architecture used in multi-language combinations. For example, client-server architectures are common in the open-source ecosystem, and web technology stacks often include a database in the backend, shaping the types of SQL injection vulnerabilities.

4.5   Vulnerability Severity

We choose the score calculated by Common Vulnerability Scoring System 3.1 [3] to measure the severity of the vulnerability. Among the patched vulnerabilities, there are 2,200 multi-file and 3,508 single-file security patches associated with the CVSS V3.1 score, including 673 and 668 multi-language patches, respectively. We plot the histogram of the scores and compare them with single-language patched vulnerabilities. As shown in Fig. 3, we can tell that the average score does not vary a lot between single-language and multiple-language patched vulnerabilities.

5   Data Analysis

We manually analyze the patches to answer two questions: (1) Why are various languages committed together? and (2) How do these languages work together to fix the vulnerability? We then showcase the fixing patterns for each language combination of multiple-file security patches and single-file security patches.

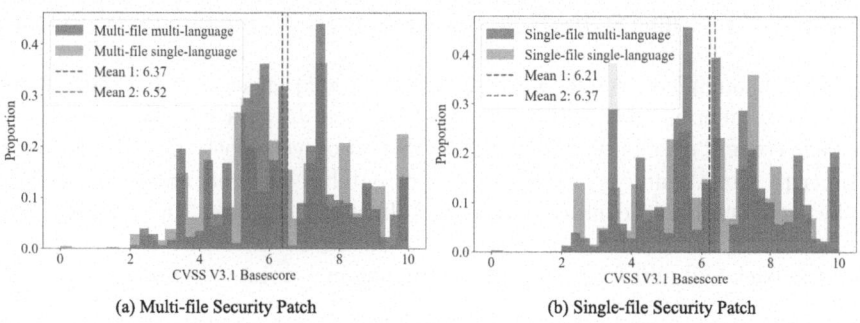

(a) Multi-file Security Patch   (b) Single-file Security Patch

Fig. 3. CVSS Distribution of Security Patches.

5.1   Patterns in Multiple Files

Among all multi-file multi-language security patches, we identify eight patterns. Six of these involve interoperability, which includes: (1) string sanitization to ensure input strings are properly sanitized from the user and server sides; (2) attribute validation to check and validate attributes to ensure they meet security requirements; (3) attribute or token addition, *e.g.*, the backend adds a token to support frontend validation and prevent CSRF attacks; (4) HTTP method update; (5) function and usage update; and (6) testing. Additionally, we observe two other patterns: (7) consistent changes across languages and (8) code and documentation sync that fixes the issue in one language while updating documentation in another language. The patterns cover 93% of examples, and we showcase each pattern along with the common language combinations and associated vulnerability types.

Programming Language Interoperability Patterns. Among all the multi-file multi-language security patches, the modified programming languages interact with each other with cross-language data/call flow.

String/Input Sanitization. Input sanitation is essential to prevent security vulnerabilities arising from untrusted user data. If input is not properly sanitized, attackers can exploit it through techniques such as SQL injection, XSS, and path traversal. In multi-language software systems, data shared across different languages is referred to as *global data* (*e.g.*, global variables), which is crucial for enabling interaction between different languages or system components. In web applications, global data can act as a bridge between server-generated content (*e.g.*, PHP) and client-side scripts (*e.g.*, JavaScript). In lower-level programming, global data is often used to share data between different languages or components. For instance, global data in C might be accessed directly in an assembly function. For multi-language systems where programs in different languages (*e.g.*, Python, Java, C++) communicate, environment variables can serve as global data accessible across languages. In more complex, distributed systems, databases or key-value stores (*e.g.*, Redis, Memcached) are often used to maintain a global state accessible to multiple applications or services, possibly written in different languages. As shown in Listing 1.3, `shortdesc` is global data accessible across different files in this system (PHP and JavaScript), likely as part of the page context or a globally defined JavaScript object. Both the PHP code and JavaScript directly access `shortdesc`, so this patch secures `shortdesc` usage both in the server-rendered content and in any client-side updates. In summary, this patch adds sanitization to the `shortdesc` by ensuring it is consistently escaped in both the back-end and front-end, protecting against unintended script injection.

Attribute Validation. Adding a sanity check to validate security attributes, such as size and value, is a common way to fix vulnerabilities. This pattern applies across all languages and frameworks. A sanity check ensures that an attribute (or input) follows expected rules before being processed, preventing vulnerabilities such as XSS, SQL injection, and out-of-bounds access. As shown in Listing 1.4, the patch for CVE-2024-47227 is applied to enforce strict validation

```
1 diff --git a/includes/Hooks/ActionsHooks.php b/includes/Hooks/ActionsHooks.php
2 --- a/includes/Hooks/ActionsHooks.php
3 +++ b/includes/Hooks/ActionsHooks.php
4 @@ -32,7 +32,7 @@ public function onInfoAction( $context, &$pageInfo ) {
5                 $context->msg( 'shortdescription-info-label' ),
6 -               $shortdesc
7 +               htmlspecialchars( $shortdesc )
8 diff --git a/modules/ext.shortDescription.js b/modules/ext.shortDescription.js
9 --- a/modules/ext.shortDescription.js
10 +++ b/modules/ext.shortDescription.js
11 @@ -7,7 +7,7 @@ function main() {
12         tagline.classList.add( 'ext-shortdesc', 'shortdescription' );
13 -       tagline.innerHTML = shortdesc;
14 +       tagline.innerHTML = mw.html.escape( shortdesc );
```

Listing 1.3. Input Sanitization (CVE-2022-21710).

of the `order_name` parameter. The update ensures that only predefined values
("`name`" and "`quota`") are allowed, preventing unexpected or malicious inputs. A
similar validation check is added to the template logic to prevent incorrect values
from affecting UI behavior. By doing so, the patch improves code consistency,
prevents invalid input from being processed, and reduces risks.

```
1 diff --git a/controllers/sql/user.py b/controllers/sql/user.py
2 --- a/controllers/sql/user.py
3 +++ b/controllers/sql/user.py
4 @@ -23,7 +23,13 @@ def GET(self, domain, cur_page=1, disabled_only=False):
5 +    # Currently only sorting by 'name' and 'quota' are supported.
6 +    if order_name not in ["name", "quota"]:
7 +        order_name = "name"
8 diff --git a/templates/default/sql/list.html b/templates/default/sql/list.html
9 --- a/templates/default/sql/list.html
10 +++ b/templates/default/sql/list.html
11 @@ -230,10 +230,19 @@ <h2>
12         {% else %}
13 +        {% set url_suffix = "" %}
14 +        {% if order_name in ["name", "quota"] %}
15 +            {% set url_suffix = "?order_name=" + order_name %}
16 +            {% if order_by_desc %}
17 +                {% set url_suffix = url_suffix + "&order_by=desc" %}
18 +            {% endif %}
19 +        {% endif %}
```

Listing 1.4. Attribute Validation (CVE-2024-47227).

Attribute or Token Addition/Update. This is where attributes or tokens have been
added or updated, commonly used for web applications to enhance protection
against common attacks, particularly cross-site request forgery (CSRF). This
pattern involves injecting security tokens or attributes into backend logic and
frontend templates to ensure that user actions are properly authenticated. It is
frequently seen in applications built with PHP, Python, Java, JavaScript, and
HTML. Listing 1.5 shows the fix of CVE-2016-9456 by generating a CSRF token
in the backend logic and then adding it as a hidden input field in the frontend.

HTTP Method Update. This pattern replaces GET requests with POST requests
when handling sensitive data, user authentication, or modifying application

```
1 diff --git a/lib/max/Admin/TrackerAppend.php b/lib/max/Admin/TrackerAppend.php
2 --- a/lib/max/Admin/TrackerAppend.php
3 +++ b/lib/max/Admin/TrackerAppend.php
4 @@ -46,6 +46,7 @@ function __construct()
5 +           $this->csrf_token     = phpAds_SessionGetToken();
6            $this->advertiser_id = MAX_getValue('clientid', 0);
7 diff --git a/lib/max/themes/TrackerAppend.html b/lib/max/themes/TrackerAppend.html
8 --- a/lib/max/themes/TrackerAppend.html
9 +++ b/lib/max/themes/TrackerAppend.html
10 @@ -26,6 +26,7 @@
11 +      <input type="hidden" name="token" value="{csrf_token}" />
12        <input type="hidden" name="clientid" value="{advertiser_id}" />
```

Listing 1.5. Attribute Addition (CVE-2016-9456).

state. It is widely used in web applications built with frameworks like Python
(Django/Flask), JavaScript (AJAX), PHP, and others. As demonstrated in Listing 1.6, the patch addresses CVE-2016-10766 by switching from GET to POST
when retrieving user information, preventing sensitive user identifiers from being
exposed in the URL and instead placing them securely in the request body. The
test case has been updated to confirm that the API call uses POST rather than
GET in JavaScript, ensuring that user data is sent correctly in the request body.
This approach mitigates risks such as data exposure in URLs, CSRF, and cache-
based data leakage. Furthermore, it is a crucial security measure across various
frameworks, including Django, JavaScript, PHP, and Java-based environments,
helping to safeguard user data throughout different application layers.

```
1 diff --git a/lms/djangoapps/instructor/api.py b/lms/djangoapps/instructor/api.py
2 --- a/lms/djangoapps/instructor/api.py
3 +++ b/lms/djangoapps/instructor/api.py
4 @@ -874,10 +841,10 @@ def modify_access(request, course_id):
5 -    user = get_student_from_identifier(request.GET.get('unique_student_identifier'))
6 +    user = get_student_from_identifier(request.POST.get('unique_student_identifier'))
7      except User.DoesNotExist:
8          response_payload = {
9 -            'unique_student_identifier': request.GET.get('unique_student_identifier'),
10 +            'unique_student_identifier': request.POST.get('unique_student_identifier'),
11            'userDoesNotExist': True,
12 diff --git a/spec/staff_debug_actions_spec.js b/spec/staff_debug_actions_spec.js
13 --- a/spec/staff_debug_actions_spec.js
14 +++ b/spec/staff_debug_actions_spec.js
15 @@ -91,7 +91,7 @@ define([
16                  StaffDebug.reset(locationName, location);
17 -                expect($.ajax.calls.mostRecent().args[0].type).toEqual('GET');
18 +                expect($.ajax.calls.mostRecent().args[0].type).toEqual('POST');
```

Listing 1.6. HTTP Method From GET to POST (CVE-2016-10766).

Function and Usage Update. Vulnerabilities can be patched by modifying function calls, updating method names, or replacing deprecated functions to enforce
access control and validation. It is commonly used in applications that allow
dynamic function execution, API calls, or remote procedure calls, which may
otherwise be exploited for unauthorized access, privilege escalation, or injection
attacks. This pattern appears across different software architectures, *e.g.*, web
applications (Python (Frappe/Django/Flask) + JavaScript (AJAX) + REST

APIs), and embedded systems (C/C++ + Python). FFI is the most common interfacing mechanism for this pattern, which is a way to call functions, use variables, or access features written in different languages. When the function definition is updated, *e.g.*, adding new parameters, the caller needs to update the function call. Therefore, two files should be patched together to complete a fix. As shown in Listing 1.7, the backend modifies the `execute_cmd` function to ensure that method validation and whitelisting are correctly applied, preventing unauthorized execution, and the frontend JavaScript code is updated to use the new function `run_doc_method`, ensuring that frontend requests comply with the updated security rules and do not trigger insecure function calls.

```
1  diff --git a/frappe/handler.py b/frappe/handler.py
2  --- a/frappe/handler.py
3  +++ b/frappe/handler.py
4  @@ -64,8 +69,9 @@ def execute_cmd(cmd, from_async=False):
5  -    is_whitelisted(method)
6  -    is_valid_http_method(method)
7  +    if method != run_doc_method:
8  +        is_whitelisted(method)
9  +        is_valid_http_method(method)
10 @@ -75,31 +81,10 @@ def is_valid_http_method(method):
11 +@frappe.whitelist()
12 +def run_doc_method(method, docs=None, dt=None, dn=None, arg=None, args=None):
13 +# for backwards compatibility
14 +runserverobj = run_doc_method
15 diff --git a/frappe/public/js/frappe/request.js b/frappe/public/js/frappe/request.js
16 --- a/frappe/public/js/frappe/request.js
17 +++ b/frappe/public/js/frappe/request.js
18 @@ -55,7 +55,7 @@ frappe.call = function(opts) {
19            $.extend(args, {
20 -            cmd: "runserverobj",
21 +            cmd: "run_doc_method",
```

Listing 1.7. Updated Function Usage (CVE-2022-23057).

Testing. When a function is written in one language but needs to be compatible with or used in applications written in another language, testing from the second language ensures that the function behaves as expected when integrated. This is often seen in APIs or libraries designed to be accessible in multiple languages (*e.g.*, a C++ library tested with Python to ensure compatibility in Python applications). Moreover, when a function is written in a low-level language like C or C++ for performance reasons, it may still be easier to test it using a higher-level language like Python, which is more flexible and readable. This approach allows developers to test the function's logic, edge cases, and integration without sacrificing the performance benefits of the low-level implementation. As shown in Listing 1.8, C++ code adds checks when processing the matrix, and Python code tests the newly added code. This is a common pattern among Python and C/C++, C and Shell, C and Perl, and C and PHP combinations.

Other Patterns. We also discover two more patterns that involve more than one type of file being updated while they do not interact with each other.

```
1  diff --git a/tensorflow/matrix_solve_op.cc b/tensorflow/matrix_solve_op.cc
2  +    for (int dim = 0; dim < ndims - 2; dim++) {
3  +        OP_REQUIRES_ASYNC(
4  +            context, input.dim_size(dim) == rhs.dim_size(dim),
5  +            errors::InvalidArgument(
6  +                ""All input tensors must have the same outer dimensions.""), done);
7  +    }
8  diff --git a/tensorflow/matrix_solve_op_test.py b/tensorflow/matrix_solve_op_test.py
9  +    matrix = np.random.normal(size=(2, 6, 2, 2))
10 +    rhs = np.random.normal(size=(2, 3, 2, 2))
11 +    with self.assertRaises((ValueError, errors_impl.InvalidArgumentError)):
12 +        self.evaluate(linalg_ops.matrix_solve(matrix, rhs))
```

Listing 1.8. Testing.

Consistent Changes Across Languages. Developers may migrate the software from one language to another to adapt the new features, *e.g.*, Linux from C to Rust. In the middle of the migration, to ensure functionality and compatibility, the code repository may contain similar code in the two programming languages. Besides, to be compatible with different platforms or meet different expectations, a piece of software may be developed in different languages, *e.g.*, Mozilla Firefox, C++/Rust for Desktop, Java/Kotlin for Android. In such scenarios, although multiple different files have been modified, the semantics of the modified files stay the same to guarantee all versions are patched. Listing 1.9 shows C and C++ perform the same change to avoid buffer overflow.

```
1  diff --git a/dcraw/dcraw.c b/dcraw/dcraw.c
2  -    if(tiff_bps <= 8)
3  -        gamma_curve(1.0/imgdata.params.coolscan_nef_gamma,0.,1,255);
4  +    if(!image)
5  +        throw LIBRAW_EXCEPTION_IO_CORRUPT;
6  +    int bypp = tiff_bps <= 8 ? 1 : 2;
7  diff --git a/internal/dcraw_common.cpp b/internal/dcraw_common.cpp
8  -    if(tiff_bps <= 8)
9  -        gamma_curve(1.0/imgdata.params.coolscan_nef_gamma,0.,1,255);
10 +    if(!image)
11 +        throw LIBRAW_EXCEPTION_IO_CORRUPT;
12 +    int bypp = tiff_bps <= 8 ? 1 : 2;
```

Listing 1.9. Same Change (CVE-2018-5812).

Code and Documentation. Another common pattern in non-interactive multi-language security patches is that one language is responsible for fixing the vulnerability, while the other updates the version information or related documentation. As shown in Listing 1.10, the vulnerability is addressed through Java code changes, whereas the HTML update is limited to modifying the bug description. In this pattern, the language responsible for fixing the vulnerability can vary, but HTML is commonly used for documentation updates.

```
 1  diff --git a/changelog.html b/changelog.html
 2  --- a/changelog.html
 3  +++ b/changelog.html
 4  @@ -61,6 +61,10 @@
 5  +    <li class=bug>
 6  +      may refuse set <a href="https://www.owasp.org/index.php/SecureFlag">secure</a>.
 7  +      Deal with it gracefully.
 8  +      (<a href="https://issues.jenkins-ci.org/browse/JENKINS-25019">issue 25019</a>)
 9  diff --git a/core/src/main/java/WebAppMain.java b/core/src/main/java/WebAppMain.java
10  --- a/core/src/main/java/WebAppMain.java
11  +++ b/core/src/main/java/WebAppMain.java
12  @@ -56,6 +56,7 @@
13  +          markCookieAsHttpOnly(context);
14  +          private void markCookieAsHttpOnly(ServletContext context) {
15  +          try {
16  +              LOGGER.log(Level.WARNING, "Failed to set HTTP-only cookie flag", e);
17  +          }
18  +      }
```

Listing 1.10. Documentaion Update (CVE-2014-9635).

5.2   Patterns in Single File

Among security patches, when multiple languages interact within a single file, embedding is a popular way to handle specific tasks. As shown in Table 3, SQL, Shell, and HTML are the three most popular programming languages that are embedded.

SQL Query. Many programs, *e.g.*, mobile apps and web apps, interact with databases, so embedding SQL allows developers to combine SQL's data-handling strengths with the computational capabilities of general-purpose languages like Python, Java, or C. The interaction workflow spans four phases: (1) Connection: Establish a connection to the database using credentials and connection strings; (2) Query Execution: SQL commands are sent as strings or method calls from the language to the database; (3) Result Handling: The database processes the query and sends results back, which are parsed and used in the program; (4) Error Handling: SQL errors (*e.g.*, syntax issues, connection failures) are handled by the host language's mechanisms. The fix patterns are associated with each workflow phase. Listing 1.11 shows a patch that modifies the `machine.py` file to improve how SQL queries are handled. Specifically, it replaces string interpolation with parameterized queries to enhance security and maintainability. The updated code uses a parameterized query and passes the `cc_number` as a parameter. This ensures that the database driver handles escaping, preventing SQL injection.

Subprocess Execution. Embedding shell scripts allows developers to combine the logic and features of high-level programming languages with the system-level capabilities of shell commands. Most programming languages provide a way to execute shell commands or scripts using system calls. These system calls create a child process to run the shell script and capture its output. As shown in Listing 1.12, this patch modifies the `apkleaks.py` script to improve the handling of command-line arguments when decompiling APK files. It addresses potential issues like command injection and ensures proper escaping of arguments passed

```
1  diff --git a/machine.py b/machine.py
2  --- a/machine.py
3  +++ b/machine.py
4  @@ -151,8 +151,7 @@ def is_card_pin_at_session(request):
5   def get_card(request, cc_number):
6  -    q = "select * from cards where cc_number = '%s'" % cc_number.replace('-', '')
7  -    row = request.db.execute(q).fetchone()
8  +    row = request.db.execute("select * from cards where cc_number = ?", (cc_number.
        replace('-', ''),)).fetchone()
9  @@ -172,7 +171,7 @@ def update_failed_attempts(request, failed_attempts):
10  def block_card(request):
11     card = request.session['card']
12 -    request.db.execute("update cards set status='blocked' where id=%s" % card['id'])
13 +    request.db.execute("update cards set status='blocked' where id=?", (card['id'],))
14     request.db.commit()
```

Listing 1.11. SQL Query (CVE-2015-10069).

to the `os.system` call. Besides, subprocess libraries offer more control over shell execution, such as capturing output, handling errors, and managing inputs. As shown in Listing 1.13, this patch modifies the `gluon/messageboxhandler.py` file to replace the use of `os.system` with `subprocess.run` for sending system notifications using the `notify-send` command. From these two examples, we can tell that the shell script can be constructed as a string, or the tokens that form the shell statement will be used as parameters for library methods.

```
1  diff --git a/apkleaks/apkleaks.py b/apkleaks/apkleaks.py
2  --- a/apkleaks/apkleaks.py
3  +++ b/apkleaks/apkleaks.py
4  @@ -2,6 +2,7 @@
5  + from pipes import quote
6  @@ -84,8 +85,9 @@ def decompile(self):
7  -        dec = "%s %s -d %s --deobf" % (self.jadx, dex, self.tempdir)
8  -        os.system(dec)
9  +        args = [self.jadx, dex, "-d", self.tempdir, "--deobf"]
10 +        comm = "%s" % (" ".join(quote(arg) for arg in args))
11 +        os.system(comm)
```

Listing 1.12. Subprocess (CVE-2021-21386).

```
1  diff --git a/gluon/messageboxhandler.py b/gluon/messageboxhandler.py
2  --- a/gluon/messageboxhandler.py
3  +++ b/gluon/messageboxhandler.py
4  @@ -1,6 +1,6 @@
5  -import os
6  +import subprocess
7  @@ -36,4 +36,4 @@ def __init__(self):
8      def emit(self, record):
9          if tkinter:
10             msg = self.format(record)
11 -            os.system("notify-send '%s'" % msg)
12 +            subprocess.run(["notify-send", msg], check=False, timeout=2)
```

Listing 1.13. Subprocess (CVE-2023-45158).

Embedded in HTML. Embedding HTML introduces security risks such as XSS, HTML injection, Clickjacking, and CSRF. These vulnerabilities arise when user input is not properly escaped, sanitized, or validated, allowing attackers to inject malicious scripts, manipulate HTML structures, or execute unauthorized actions. To mitigate these risks, developers often use escaping functions, enforce Content Security Policies (CSP) to restrict inline scripts, sanitize user input using libraries like DOMPurify (JavaScript) or Bleach (Python), and implement CSRF tokens in forms. Additionally, avoiding direct execution of embedded HTML and using templating engines ensures structured and secure rendering, reducing the likelihood of security breaches while maintaining flexibility in content presentation. As shown in Listing 1.14, the patch shows HTML embedded in PHP using Laravel Blade templates. The code includes HTML elements (a `<form>` tag) while also using Laravel's Blade directives like `@csrf` and `route('admin.logout')` , which generate PHP code dynamically. This update ensures that logout requests are sent as POST requests with a CSRF token, preventing CSRF attacks where attackers might trick users into logging out unknowingly.

```
1  diff --git a/views/layouts/main.blade.php b/views/layouts/main.blade.php
2  --- a/views/layouts/main.blade.php
3  +++ b/views/layouts/main.blade.php
4  @@ -71,6 +71,11 @@
5     @include('twill::partials.footer')   </section>   </div>
6  +   <form class="visually-hidden" method="POST" action="{{ route('admin.logout') }}"
          data-logout-form> @csrf </form>
```

Listing 1.14. Embedded HTML (CVE-2021-3932).

6   Discussion

We outline various usage scenarios for our work. Additionally, we discuss its limitations and propose directions for future work.

6.1   Usage Scenarios

We release the dataset, including patch patterns, and will expand it with multi-language non-security patches. This dataset will serve as a benchmark, training data for multi-language program analysis tools, and automatic program repairs.

Benchmark. To the best of our knowledge, the dataset we released is the largest multi-language security patch dataset, encompassing the most languages. It serves as a robust benchmark for evaluating the learning or transfer capabilities of target techniques, as it includes interactions across 16 programming languages and 185 vulnerability types.

Multi-language Program Analysis Tool. Previous program analysis tools, often focused on single languages, may miss vulnerabilities in embedded or interacting languages. Our dataset addresses this gap by offering language combinations and test cases to evaluate such tools. By analyzing existing vulnerability statistics, researchers can prioritize studying language pairs more prone to vulnerabilities, *e.g.*, JavaScript and PHP, to enhance tool effectiveness.

Auto Program Repair. The primary goal of releasing the multi-language security patch dataset is to advance automatic program repair. The dataset includes both vulnerable (pre-patch) and fixed (post-patch) code, enabling the detection of vulnerabilities and guiding their remediation. The pre-patch versions can be used to design static or dynamic analysis tools or train deep learning models to identify vulnerability patterns across languages. Meanwhile, the post-patch versions provide remediation examples, which can train language models to transform vulnerable code into secure code. Notably, our dataset covers single-file multi-language security patches, meaning it can address cases where embedded languages appear as strings within a host language.

6.2   Limitation and Future Work

Our dataset is sourced from MITRE, but the analysis and indexing of CVEs by CVE Numbering Authorities (CNAs) [1] may introduce biases, as the 421 diverse CNAs do not catalog every reported vulnerability. As a result, our dataset does not include all multi-language security patches found in open-source repositories. Additionally, not all vulnerability records include fixes, nor are all fixes publicly released, so our dataset lacks patches for these cases. Furthermore, some vulnerability reports omit CWE and CVSS scores, potentially limiting the statistical analysis of vulnerability types.

Due to differing advisory policies, some vendors release security patches without public disclosure. Research [31] shows that 6–10% of GitHub commits are security patches, prompting our future focus on identifying and analyzing silent multi-language patches. Furthermore, as our MITRE-based dataset lacks certain interaction types, we plan to generate synthetic examples to incorporate unseen language combinations and interaction patterns.

7   Related Work

Patch Dataset. Security commits offer valuable insights into existing vulnerabilities and their corresponding fixes, making them essential for creating datasets used in security commit detection and automated program repair. However, all [9,13,14,28,29,31,34] of the existing datasets are limited to patches written in one programming language. Besides, most of the existing datasets are limited to specific projects [29,35] or contain only a small number of security commits linked to CVEs [9,14]. While some researchers include commits indexed by NVD as well as silent fixes [13,28,30,31], their focus has been primarily on C/C++, Java, and Python, overlooking widespread multi-language repositories. Although

Li et al. [17] have released a commit dataset of multilingual projects, the samples in this dataset are not in multi-languages nor indexed by the MITRE, and the interfacing mechanisms included are limited. Additionally, the quality of the data cannot be assured, as only less than 5% of the samples were manually verified. Furthermore, their method of using keyword matching to identify vulnerable commits does not guarantee the accuracy of the identified vulnerabilities.

Multi-language Software and Patch Analysis. Building software using multiple programming languages has been a normal practice for a long time. Researchers have measured the language selection [19,20,25], challenges [26], and bad practices [7]. However, it remains uncertain if multilingual code construction has significant security implications or leads to real security consequences. Li et al. [17,18] found statistically significant associations between the proneness of multilingual code to vulnerabilities, which is correlated with the language interfacing mechanism, not that of individual languages. They also introduce the first taxonomy of language interfacing. However, their analysis is limited to the project level. In multi-language software, not all commits involve multiple languages. As a result, their conclusions apply to multi-language software as a whole, not to multi-language patches. The former cannot fully represent the latter. Even within the same repository, interaction mechanisms can vary. For the same language combination, both vulnerability patterns and interface mechanisms may differ. Moreover, they define only four types of interfacing and overlook interaction details that occur within single-file multi-language cases.

Motivated by the security impact of the multi-language trend, researchers propose approaches to analyze the multi-language patches. However, these approaches either focus on limited language combinations or specific application types. Buro et al. [10] show formal properties of interest of multi-language abstractions. Li et al. [33] and Yang et al. [21] propose to locate and detect multilingual bugs in Python and C/C++ repositories, *i.e.*, TensorFlow and NumPy. Lei et al. [16] propose intermediate representations to detect vulnerabilities across C/++ and Java. Figueiredo et al. [15] detects multi-language vulnerabilities for web applications. Negrini [27] introduces a generic framework for multi-language analysis for Go, Java, and C++ in smart contract development. Yang et al. [32] characterize Python-C and Java-C patches into 5 aspects, including their symptoms, locations, manifestations, root causes, and fixes.

The scarcity of data on language combinations, insufficient explanations of language interoperability, and the lack of summarized fixing patterns have motivated us to create a comprehensive dataset of multi-language security patches from diverse projects. Furthermore, we seek to extract fixing patterns to encourage advancements in automated program repair.

8   Conclusion

In this paper, we investigate why commits in multi-language repositories involve updates across multiple languages and how these languages collaborate to address vulnerabilities. We analyze 2,798 multi-language security patches indexed by MITRE, comprising 1,253 multi-file and 1,545 single-file patches. Our findings reveal that multi-language patches are more complex than single-language ones, with greater modifications, intricate branch structures, and higher dependency complexity. We also find that multi-language security patches are more likely to involve injection-related and web-related vulnerabilities, mainly due to the general system design and choice of languages used in multi-language software systems. We summarize 11 fixing patterns and observe their correlation with interfacing mechanisms. To promote further research, we release our dataset, including fixing patterns, to support the development of automated multi-language program repairs.

Acknowledgments. We appreciate the helpful comments from our shepherd and the reviewers. This work was partially supported by the US Office of Naval Research grant N00014-23-1-2122.

References

1. CNAs (2024). https://www.cve.org/ProgramOrganization/CNAs
2. CVE (2024). https://cve.mitre.org/index.html
3. CVSS_v3 (2024). https://nvd.nist.gov/vuln-metrics/cvss/v3-calculator
4. CWE (2024). https://cwe.mitre.org/
5. GitHub commits language count (2025). https://api.github.com/search/commits?q=language:languagecombination
6. GitHub repositories language count (2025). https://api.github.com/search/repositories?q=language:languagecombination
7. Abidi, M., Grichi, M., Khomh, F., Guéhéneuc, Y.G.: Code smells for multi-language systems. In: Proceedings of the 24th European Conference on Pattern Languages of Programs, pp. 1–13 (2019)
8. Bandara, V., et al.: Fix that fix commit: a real-world remediation analysis of Javascript projects. In: 2020 IEEE 20th International Working Conference on Source Code Analysis and Manipulation (SCAM). IEEE (2020)
9. Bhandari, G., Naseer, A., Moonen, L.: CVEfixes: automated collection of vulnerabilities and their fixes from open-source software. In: Proceedings of the 17th International Conference on Predictive Models and Data Analytics in Software Engineering, pp. 30–39 (2021)
10. Buro, S., Crole, R., Mastroeni, I.: On multi-language abstraction: towards a static analysis of multi-language programs. Formal Methods Syst. Des. (2023)
11. Cass, S.: The top programming languages 2024, August 2024. https://spectrum.ieee.org/top-programming-languages-2024
12. Chakraborty, P., Alfadel, M., Nagappan, M.: BLAZE: cross-language and cross-project bug localization via dynamic chunking and hard example learning. arXiv preprint arXiv:2407.17631 (2024)
13. Chen, Y., Ding, Z., Chen, X., Wagner, D.: DiverseVul: a new vulnerable source code dataset for deep learning based vulnerability detection. arXiv preprint arXiv:2304.00409 (2023)
14. Fan, J., Li, Y., Wang, S., Nguyen, T.N.: AC/C++ code vulnerability dataset with code changes and CVE summaries. In: Proceedings of the 17th International Conference on Mining Software Repositories, pp. 508–512 (2020)

15. Figueiredo, A., Lide, T., Correia, M.: Multi-language web vulnerability detection. In: 2020 IEEE International Symposium on Software Reliability Engineering Workshops (ISSREW), pp. 153–154. IEEE (2020)
16. Lei, T., Xue, J., Wang, Y., Liu, Z.: IRC-CLVul: cross-programming-language vulnerability detection with intermediate representations and combined features. Electronics **12**(14), 3067 (2023)
17. Li, W., Li, L., Cai, H.: On the vulnerability proneness of multilingual code. In: Proceedings of the 30th ACM Joint European Software Engineering Conference and Symposium on the Foundations of Software Engineering, pp. 847–859 (2022)
18. Li, W., Li, L., Cai, H.: PolyFax: a toolkit for characterizing multi-language software. In: Proceedings of the 30th ACM Joint European Software Engineering Conference and Symposium on the Foundations of Software Engineering, pp. 1662–1666 (2022)
19. Li, W., Marino, A., Yang, H., Meng, N., Li, L., Cai, H.: How are multilingual systems constructed: characterizing language use and selection in open-source multilingual software. ACM Trans. Softw. Eng. Methodol. **33**(3), 1–46 (2024)
20. Li, W., Meng, N., Li, L., Cai, H.: Understanding language selection in multi-language software projects on GitHub. In: 2021 IEEE/ACM 43rd International Conference on Software Engineering: Companion Proceedings (ICSE-Companion), pp. 256–257. IEEE (2021)
21. Li, W., Ming, J., Luo, X., Cai, H.: PolyCruise: a cross-language dynamic information flow analysis. In: 31st USENIX Security Symposium (USENIX Security 22), pp. 2513–2530 (2022)
22. Li, Z., Ji, J., Liang, P., Mo, R., Liu, H.: An exploratory study on just-in-time multi-programming-language bug prediction. Inf. Softw. Technol. **175**, 107524 (2024)
23. Li, Z., Wang, W., Wang, S., Liang, P., Mo, R.: Understanding resolution of multi-language bugs: an empirical study on apache projects. In: 2023 ACM/IEEE International Symposium on Empirical Software Engineering and Measurement (ESEM), pp. 1–11. IEEE (2023)
24. Malone, T.: Interoperability in programming languages. Sch. Horiz. Univ. Minn. Morris Undergraduate J. **1**(2), 3 (2014)
25. Mayer, P., Bauer, A.: An empirical analysis of the utilization of multiple programming languages in open source projects. In: Proceedings of the 19th International Conference on Evaluation and Assessment in Software Engineering, pp. 1–10 (2015)
26. Mushtaq, Z., Rasool, G.: Multilingual source code analysis: state of the art and challenges. In: 2015 International Conference on Open Source Systems & Technologies (ICOSST), pp. 170–175. IEEE (2015)
27. Negrini, L.: A generic framework for multilanguage analysis (2023)
28. Nikitopoulos, G., Dritsa, K., Louridas, P., Mitropoulos, D.: CrossVul: a cross-language vulnerability dataset with commit data. In: Proceedings of the 29th ACM Joint Meeting on European Software Engineering Conference and Symposium on the Foundations of Software Engineering, pp. 1565–1569 (2021)
29. Ponta, S.E., Plate, H., Sabetta, A., Bezzi, M., Dangremont, C.: A manually-curated dataset of fixes to vulnerabilities of open-source software. In: 2019 IEEE/ACM 16th International Conference on Mining Software Repositories (MSR). IEEE (2019)
30. Sun, S., Wang, S., Wang, X., Xing, Y., Zhang, E., Sun, K.: Exploring security commits in python. In: 2023 IEEE International Conference on Software Maintenance and Evolution (ICSME), pp. 171–181. IEEE (2023)
31. Wang, X., Wang, S., Feng, P., Sun, K., Jajodia, S.: PatchDB: a large-scale security patch dataset. In: 2021 51st Annual IEEE/IFIP International Conference on Dependable Systems and Networks (DSN), pp. 149–160. IEEE (2021)

32. Yang, H., Cai, H.: Dissecting real-world cross-language bugs (2025)
33. Yang, H., Nong, Y., Zhang, T., Luo, X., Cai, H.: Learning to detect and localize multilingual bugs. Proc. ACM Softw. Eng. (FSE) (2024)
34. Zheng, Y., et al.: D2A: a dataset built for AI-based vulnerability detection methods using differential analysis. In: 2021 IEEE/ACM 43rd International Conference on Software Engineering: Software Engineering in Practice (ICSE-SEIP) (2021)
35. Zhou, Y., Siow, J.K., Wang, C., Liu, S., Liu, Y.: SPI: automated identification of security patches via commits. ACM Trans. Softw. Eng. Methodol. (TOSEM) **31**(1), 1–27 (2021)

Red Light for Security: Uncovering Auto Feature Check and Access Control Gaps in AAOS

Jumana$^{(\boxtimes)}$, Parjanya Vyas, and Yousra Aafer

University of Waterloo, Waterloo, Canada
{jjumana,parjanya.vyas,yousra.aafer}@uwaterloo.ca

Abstract. The Android Automotive Operating System (AAOS) is a specialized version of the Android OS designed for in-vehicle infotainment and system control. Prominent automakers such as Honda, General Motors (GM), Volvo, and Ford have already adopted it in their latest vehicles. Despite its popularity, the security of AAOS integration has hardly been evaluated, particularly at the framework layer, where auto feature and access control anomalies are likely to arise. To bridge the gap, we perform the first security evaluation of automotive entry points in AAOS. Our study is enabled by *AutoAcRaptor*, an automated pipeline that leverages static analysis to identify automotive entry points, generate their access control and auto feature specifications, and analyze them for potential security risks. Our evaluation of *AutoAcRaptor* on two AOSP and eight automaker AAOS images demonstrates that it is able to identify 23 auto feature and access control anomalies, on average per ROM. We report ten cases to the corresponding automakers. At the time of writing, five have been acknowledged while the rest are pending verification.

Keywords: Android Automotive · Access Control · Security

1 Introduction

The Android Automotive OS (AAOS) has rapidly gained prominence in the automotive industry as a fully integrated, auto-specific OS. Leading automakers, including Volvo, Polestar, and Stellantis, have already integrated AAOS into their vehicles, with others, such as Ford and Porsche, expected to follow [32].

Unlike Android Auto, which simply projects smartphone content on infotainment screens, AAOS is a base Android platform that extends Android[1] to support automotive functions such as entertainment, navigation, and system control. The extension, which spans various layers of the Android software stack,

---

Jumana and P. Vyas—The two lead authors contributed equally to this research.

[1] We refer to the unmodified version of the OS that does not support automotive functions as "Android" and "Traditional Android" interchangeably.

M. Egele et al. (Eds.): DIMVA 2025, LNCS 15748, pp. 147–166, 2025.
https://doi.org/10.1007/978-3-031-97623-0_9

has been gradually carried out over 18 months before AAOS was put into production. This design choice is concerning from a security perspective, given that Android has *evolved* to support AAOS rather than having it weighted-in from the beginning. Further complicating the situation is AAOS openness model, which offers automakers the flexibility to tailor the AAOS codebases to their needs, all without a central entity that oversees the customization process.

Although prior works [13, 22, 25–29] have studied some aspects of AAOS security, they are limited to investigating its external attack surface [27, 29] and specific enabling technologies [22, 26, 27]. *Evolution-related* vulnerabilities, which are likely to be introduced due to the AAOS design model and underregulated customization, remain understudied. For each new AAOS version and/or vehicle model, Google and automakers alter the latest AAOS codebase to embed new AAOS-specific and/or vendor-specific functional requirements. When such an alteration is not carried out properly, vulnerabilities are inevitable. Several prior works have investigated the impact of vendor customization in traditional Android setting (i.e., in mobile smartphones and tablets). However, to the best of our knowledge, none has looked at the security impact of customization in auto devices. To bridge the gap, we perform an empirical investigation of AAOS customization in a diverse set of AAOS ROMs. Our study aims to answer whether AAOS customization leads to potential anomalies, by analyzing ten AAOS images from four automakers. We focus our investigation on the framework layer, which implements AAOS APIs.

To facilitate the study, we develop a new automated analysis pipeline, *AutoAcRaptor*, which locates AAOS-related entry points and evaluates their security aspects. *AutoAcRaptor* performs static analysis of traditional and car-specific Android system services to identify entry points related to AAOS, that is, app-accessible APIs that are newly added or adapted for AAOS implementation. To ensure accurate identification, our tool relies on static patterns and naming indicators.

Evaluating the security aspects of AAOS entry points faces the following two main challenges: (1) due to the lack of functional specifications, it is unclear which Android functionality should be blocked or adjusted (e.g., disabling video playing when the car is in drive mode), and similarly (2) due to the lack of formal security documentation, it is unclear whether new APIs should implement access control (e.g., enforce a permission in a new API that enables accessing car properties). *AutoAcRaptor* meets these challenges by conducting consistency analysis of AAOS-feature and access control checks across AAOS APIs. Our intuition is that Google and automakers should implement consistent auto feature checks to limit, block access to, or adjust functionality of Android APIs. They should similarly implement consistent traditional security checks (e.g., car permissions) when exposing underlying functionality to apps. Failure to do so may allow unprivileged third-party apps to access unintended resources and even manipulate them without having appropriate privileges. *AutoAcRaptor* leverages and adapts traditional consistency analysis techniques [9, 11, 12, 33] for our task.

Particularly, *AutoAcRaptor* uses convergence and similarity analysis to identify functionally-similar APIs.

Measurement. We run *AutoAcRaptor* on ten AAOS ROMs; two correspond to the latest AOSP codebases (versions 13 and 14), and eight are customized by General Motors (GM), Honda, Volvo and Polestar. Our analysis identifies 34 car system services (i.e., new AAOS-specific services) and five modified system services (i.e., Android system services containing new or modified AAOS entry point), on average per ROM.

Furthermore, our study shows that Google has *modified* 75, 73 traditional entry points on versions 13, 14, respectively to accommodate automotive functionality. The extent of vendor customization varies across automakers; Honda (v12L) introduces only one new entry point, while GM-SUV-22 (V10) introduces up to 249 entries.

Security Analysis. *AutoAcRaptor* identifies eight entry points with an inconsistent feature check and nine with an inconsistent access control enforcement on average per ROM. The number varies across automakers, reaching up to 14 and 9 in AOSP AAOS v13. AOSP codebases include approximately 4% potential inconsistencies, which are inherited by all automaker codebases. From these inconsistencies, we identified ten potential vulnerabilities[2]. We have reported the cases to the corresponding automakers. At the time of writing, five have been acknowledged and the rest are pending verification.

Contributions. We make the following contributions:

- We conduct the first systematic investigation of AAOS framework.
- We design and implement *AutoAcRaptor*, a pipeline that combines static analysis and light NLP to generate and evaluate access control specification of AAOS-specific entry points at the framework layer.
- We evaluate our tool on ten AAOS ROMs, quantifying the extent of AAOS customization and demonstrating the existence of auto feature and access control anomalies.

2 Background and Motivation

In this section, we cover background on AAOS and the different customization aspects adopted to accommodate Android for AAOS. We also present two cases that motivated our study.

2.1 Evolution of Android to AAOS

Android middleware has evolved towards implementing AAOS, by (1) introducing new car-specific system services (car-specific managers and car services), and

---

[2] Confirmed on automaker emulators.

by (2) customizing existing system services. Automotive apps invoke APIs in the (car/traditional) services using Binder IPC.

Introducing New Car-Specific Services: Car-specific system services and APIs are defined in a new library *car-lib* [15], which integrates three main components:

```
1    // Create Car instance
2    Car mCar = Car.createCar(context);
3    // Create CarPropertyManager instance
4    CarPropertyManager pm = mCar.getCarManager(Car.PROPERTY_SERVICE);
5    // Invoke public APIs in CarPropertyManager
6    pm.getProperty(propertyId, zoneId);
7    pm.getPropertyList(); }
```

Fig. 1. Invoking Car APIs.

1. **Car API**: This corresponds to the AAOS sdk. It defines *Manager* classes which encapsulate system service functionality and define corresponding public APIs. Apps can access Car API by creating an instance of the `Car` class (Fig. 1 shows an example).
2. **CarService**: AAOS defines new car-specific system services (e.g., `CarPowerManagerService`, `CarPropertyService`), which implement fundamental car functionality within AAOS, including navigation, infotainment, and audio control . The services are registered in and managed by a special system service `CarService` (registered in the `ServiceManager`), which can be queried by apps to access instances of specific car services. Specifically, it exposes a method `ICarImpl.getCarService(serviceName)` that accepts a name (string literal) of a car-specific service and returns corresponding binder instance.
3. **Car permissions**: AAOS defines new car permissions in a dedicated configuration file *automotive_android_manifest.xml*. The majority of the permissions (135 out of 145) are signature level. The permissions are enforced in AAOS entry points to mediate access to sensitive resources.

Modifying APIs in Android System Services: Adapting Android to AAOS also involves customizing existing Android system services by modifying some of their entry points. Modification is usually carried out by checking if the platform running is automotive (for example, by checking if `PackageManager.FEATURE_AUTOMOTIVE` is present) and accordingly providing different functionality. Such functionality can be categorized into two distinct behaviors:

Functionality Blocking: The functionality implemented by the API is partially or completely blocked in automotive builds. An example of a full blocking is depicted in AOSP's `DevicePolicyManager.setLocationEnabled(..)` (Fig. 2). As shown in Lines 2–5, when the platform is an automotive build, calls to disable the location will be ignored.

```
1  public void setLocationEnabled(boolean locationEnabled) {
2      if (mIsAutomotive && !locationEnabled) {
3          // Calls to disable location are ignored in  automotive builds
4          return;
5      }
6      mInjector.binderWithCleanCallingIdentity(() -> {
7          getLocationManager().setLocationEnabledForUser(locationEnabled);
8      }
9  }
```

Fig. 2. Simplified Implementation of `DevicePolicyManager. setLocationEnabled`.

An example of a partial blocking is depicted in AOSP's `DevicePolicy Manager.lockNow()` (Fig. 3), which allows device owners to lock the device immediately. As shown in Line 2 and Line 7, the API executes a few operations required to lock the device, e.g., evict credential encryption keys and set device Locked state. However, on automotive builds, it omits calls to turn off the screen as it would be a driving safety distraction (`powerManagerGoToSleep(..)` in Line 5 is not invoked).

```
1  public void lockNow(int flags,String callerPackageName,boolean parent){
2      mUserManager.evictCredentialEncryptionKey(callingUserId);
3      if (!mIsAutomotive) {
4          // Power off the display ONLY on non-automotive builds
5          mInjector.powerManagerGoToSleep(SystemClock.uptimeMillis(),
                   PowerManager.GO_TO_SLEEP_REASON_DEVICE_ADMIN, 0);
6      }
7      mInjector.getTrustManager().setDeviceLockedForUser(userToLock,true);
8      ...
```

Fig. 3. Simplified Implementation of `DevicePolicyManagerService.lockNow`.

Access Control Modification: Although less common, the access control enforced by the API may be altered to accommodate new AAOS security requirement. `AudioService.setMasterMute(..)` (Fig. 5) depicts an example. As shown in Line 7, the API enforces that the calling physical user should be `USER_SYSTEM` if the platform is automotive.

Adding APIs in Android System Services: Customizing Android system services further includes introducing new APIs, that service exclusive AAOS-specific functionality. `LockSettingService` defines a few of such APIs. For example, `addWeakEscrowToken(..)` (Fig. 4) only works on automotive platforms.

```
1   // API in LockSettingsService
2   public long addWeakEscrowToken(byte[] token, int userId) {
3       if (!mContext.getPackageManager().hasSystemFeature(FEATURE_AUTOMOTIVE)) {
4           throw new IllegalArgumentException("only for automotive devices.");
5       }
6       // actual functionality
7       ...
```

Fig. 4. Simplified Implementation of `LockSettingsService. addWeakEscrowToken`

2.2   Motivating Examples of Developer Confusion

We observe that AAOS developers may not have accurate knowledge of which Android APIs or code pieces should be blocked or adjusted to integrate AAOS. Figure 5 illustrates an *uncertain* automotive feature check. As indicated in the developer's comment, the feature check is added for safety but may not be necessary. The check is noted as *"to be removed"* in version 9 but is still lingering in the latest version 14.

```
1   public void setMasterMute(boolean mute) {
2       if (!isPlatformAutomotive()) {
3       // TODO: remove the isPlatformAutomotive check here.
4       // isPlatformAutomotive check is for safety but may not be necessary.
5           mute = false;
6       }
7       if ((isPlatformAutomotive() && userId==UserHandle.USER_SYSTEM)){
8           mAudioSystem.setMasterMute(mute);
```

Fig. 5. Uncertain Automotive Feature Check in `AudioService`

In contrast, Fig. 6 illustrates a *"to be added"* automotive feature check. As shown, the developers intend to customize the system service `DevicePolicyService` for auto-support in future versions. Such uncertainties obfuscate any attempt to understand the functional and security requirements of AAOS from source code.

```
1   public int getMinLockLength(boolean isPin, int complexity) {
2       complexity = PasswordMetrics.sanitizeComplexityLevel(complexity);
3       // TODO: b/131755827 add devicePolicyManager support for Auto
4       DevicePolicyManager dpM = mContext.getSystemService(Context.
            DEVICE_POLICY_SERVICE);
5       PasswordMetrics adminMetrics =
6       dpM.getPasswordMinimumMetrics(mContext.getUserId());
```

Fig. 6. Intended Future Automotive Support in `DevicePolicyService`

2.3   Motivating Examples of Inconsistent Checks

Inconsistent Auto Feature Checks. Through our analysis of AAOS code, we detected an API with a missing auto-feature check. As shown in Fig. 2, `DevicePolicyManager. setLocationEnabled` omits calling the method `LocationManager. setLocationEnabledForUser` when the supplied argument is `false` and the platform is an automotive build. However, the method `LocationManager. setLocationEnabledForUser` itself does not enforce the automative feature check. So, a calling app can bypass the feature check in `DevicePolicyManager. setLocatioEnabled` by directly invoking `setLocationEnabledForUser`. We reported the case to Google.

We refer the reader to Subsect. 4.7 for a description of cases with inconsistent access control checks.

3   System Design

In this section, we introduce a high-level overview of our tool, *AutoAcRaptor*, followed by an elaborate discussion of its key phases.

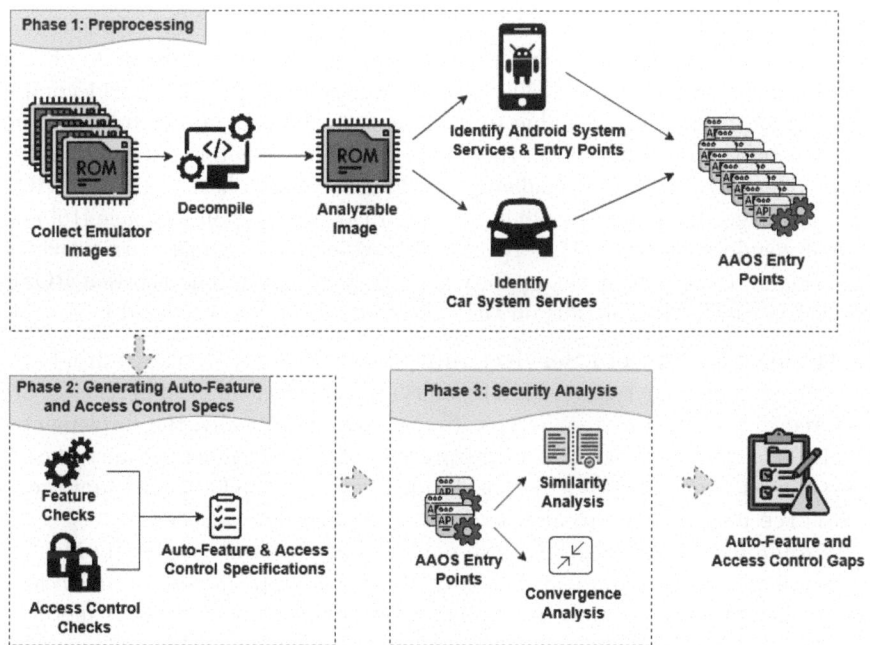

Fig. 7. Approach Overview of *AutoAcRaptor*.

3.1 Approach Overview

AutoAcRaptor consists of three phases, as depicted in Fig. 7. In the first step, *AutoAcRaptor* preprocesses AAOS ROMs to extract framework classes. Since the analysis spans both Google and automaker AAOS ROMs[3], *AutoAcRaptor* operates on compiled framework bytecode. It then analyzes the framework classes to locate car system services, modified Android system services, and corresponding entry points. In the second phase, it analyzes the extracted entry points to generate a specification that maps resources (i.e., potential sinks in an entry point) to auto feature and access control checks. In the third phase, *AutoAcRaptor* leverages the specification to perform a ROM-wide security evaluation. Specifically, *AutoAcRaptor* groups *functionally-similar* entry points and compares their auto feature/ access control specification. Finally, it flags inconsistent cases as potential anomalies.

3.2 Phase 1: Preprocessing

In the first step, *AutoAcRaptor* decompiles target AAOS ROMs and prepares them for upcoming analyses. It extracts framework and car-specific libraries, using a variety of tools (apktool [2], dex2jar [3], smali [8], etc..).

Collecting AAOS Images. Although the AAOS source code (Google version) is readily available in AOSP repositories [4], it is more difficult to obtain customized AAOS code. Specifically, (1) we cannot extract ROMs (for decompilation) from physical vehicles due to obvious cost and availability reasons, and (2) to the best of our knowledge, there are no public repositories for automaker source code. To address this challenge, we look for and collect available AAOS *emulator* images linked on the official Android Documentation webpage [16] and on some vendors' websites [1,5–7]. Since these emulated images are provided to developers to test their car applications, they closely mimic the real ROMs deployed in cars[4]. More details on the collected images are discussed in Sect. 4.

Identifying Car System Services and Entry Points. Unlike Android system services, which are registered in the Service Manager, car system services are registered in a central system service *CarService*. Specifically, the implementation of *CarService*, i.e., the class `ICarImpl`, creates a new binder instance of each car system service, and adds it to a list of local services (via `CarLocalServices.addService` method) and to another list maintaining active services.

AutoAcRaptor leverages the above registration pattern to statically identify car system services as follows: it first locates the class implementing *CarService*'s Stub. It then builds a precise call graph of its constructor method (`init` method) and interprocedurally traverses it to identify invocations to `CarLocalServices.addService`. Finally, it retrieves the concrete type of the car system service by resolving the arguments of `addService`.

---

[3] Source code is unavailable for automaker (emulator) ROMs.

[4] Although we cannot guarantee that an emulated ROM is an exact copy of the actual ROM deployed in cars.

To identify app-accessible entry points, *AutoAcRaptor* locates the interface of each resolved car system service and extracts its defined public methods.

Identifying AAOS Entry Points in Android System Services. *AutoAcRaptor* relies on the static patterns traditionally used by prior work to recover Android system services and their entry points [9,11,12,20,33](details are omitted for brevity). To pinpoint AAOS entry points in Android system services (i.e., entries modified/added to accommodate AAOS), *AutoAcRaptor* builds a control flow graph of each entry and inspects its conditional statements to locate those corresponding to automotive feature checks. Specifically, it leverages the following rules:

Rule 1: Direct automotive feature check. A conditional statement is classified as automotive if one operand in the predicate evaluates to an invocation of `PackageManager.hasSystemFeature` with the constant string *"android.hardware.type.automotive"* as an argument.

Rule 2: Implicit automotive feature check. *AutoAcRaptor* further tracks a special kind of control dependency that implies an automotive feature check. Consider the example in Fig. 8: Variable `mFeature` does not directly indicate an automotive feature check. However, `mFeature = true` must imply that `mPMS.hasSystemFeature(PackageManager.FEATURE_AUTOMOTIVE)` equals to `true`. To handle such cases, *AutoAcRaptor* tracks control dependencies caused by such equivalence checks, which are prevalent in AAOS.

```
1  boolean mFeature = false;
2  init(){
3    if(mPMS.hasSystemFeature(PackageManager.FEATURE_AUTOMOTIVE) mFeature = true;
4  }
5  entryPoint(..){
6    if(!mFeature) return;
7  }
```

Fig. 8. Implicit Automotive Feature Checks

3.3 Phase 2: Generating Auto-feature and Access Control Specs

At this stage, *AutoAcRaptor* statically analyzes the extracted entry points to build a specification that maps *resources* to observed auto feature and access control checks.

Resource Definition. A *resource* is defined as either (1) a reachable method invocation or field update instruction in an entry point or (2) the entry point itself.

Access Control and Auto Feature Checks. We perform a forward control-flow analysis on the API's interprocedural control flow graph (ICFG) and identify the conditional branches on which a resource is control dependent. A conditional branch is classified as an access control check if one of its operands corresponds to an invocation of known permission checks (e.g., `checkPermission`) or non-forgeable identifier checks (e.g., `UserHandle.getCallingUserId`). *AutoAcRaptor* similarly relies on Rule 1 (refer to Subsect. 3.2) to identify auto feature checks. Note that unlike access control checks where we only consider instructions in the `true` branch, we process both the `true` and `false` branches of auto feature checks. As stated earlier, resources in the true branch are exclusive to AAOS, while those in the false branch are blocked on AAOS.

Resource Reduction. As noted in prior work [17], not all resources control dependent on an access control are sensitive; consider a local variable update. We observe a similar phenomenon for auto feature checks; not all instructions guarded by an auto feature check are exclusive to AAOS. For example, a log statement is not relevant to AAOS even if it is control dependent on auto checks. We address this issue by relying on the following reduction strategies:

- **Frequency analysis:** *AutoAcRaptor* performs a frequency analysis to filter out commonly invoked resources across APIs, as they are less likely to be sensitive or AAOS-specific.
- **Removing data dependencies:** *AutoAcRaptor* conducts data flow analysis to identify and remove instructions exclusively used by the identified frequent instructions (e.g., `StringBuilder` methods used to construct log messages in `Log.d(..)`).
- **Resource name similarity analysis:** This strategy analyzes an API's resources to identify and exclude those that are less resembling the main functionality of the API through naming similarity analysis. Specifically, we use UniXcoder [21], a unified cross-modal pretrained NL-PL model for generating word embeddings. We use the model's tokenizer to tokenize the entry point name and a given resource name and generate corresponding embeddings. The latter are then compared using cosine similarity. If the calculated similarity is lower than an empirically selected threshold (see Sect. 4), the resource is not considered for further analysis.

The above strategies achieve a reduction of approximately 92% of instructions on average per entry point, narrowing down to the most relevant resources of an API.

Example of Generated Specifications. Figure 9 showcases the generated auto feature and access control specifications for the API resource `AudioService.setMasterMute(...)`. As shown, the analysis identifies an auto feature check and four access control checks.[5]

3.4   Phase 3: Security Analysis

To conduct the security analysis, *AutoAcRaptor* groups *functionally-similar* entry points and compares their auto feature and access control specification.

---

[5] A similar specification is generated for each reachable resource within the API's implementation. The specs are omitted for brevity.

$AutoFeature := 1$
$AccessControl_1 := [UserId = USER\_SYSTEM]$
$AccessControl_2 := [Perm = MODIFY\_AUDITO\_ROUTING]$
$AccessControl_3 := [Perm = INTERACT\_ACROSS\_USERS\_FULL]$
$AccessControl_4 := [Uid = SYSTEM]$

Fig. 9. Generated specifications for `AudioService.setMasterMute(...)`

Inconsistent specifications imply a potential anomaly. Recall that this solution addresses the lack of an oracle that can dictate whether a resource is *adequately* guarded by an auto-feature and/or access control check. *AutoAcRaptor* determines functional similarity through: Convergence and Similarity Analyses.

Convergence Analysis: *AutoAcRaptor* conducts convergence analysis across entry points to identify those that overlap in functionality [20,33]. This solution assumes that if the same resource is reachable using two entries, then they are similar. Note that, due to the large code size of AAOS frameworks, it is not trivial to identify converging entries. *AutoAcRaptor* achieves efficiency by using the specifications generated in Phase 2, which conveniently maps resources to observed auto feature and access control checks in different entries.

Similarity Analysis: Unlike convergence analysis where an exact common resource is required to identify functionally-overlapping APIs, this method adopts a more relaxed assumption. It relies on similarity to find such APIs – for example, it considers APIs exhibiting similar names (e.g., `CarPropertyService.getProperty` and `CarPropertyService.getPropertyList`) to implement similar functionality. This analysis can cover APIs missed by the convergence analysis. *AutoAcRaptor* uses UniXcoder to identify similar APIs, following the same method discussed in Phase 2 (in Resource Reduction).

4 Evaluation

We implement a prototype for *AutoAcRaptor* consisting of two components: (1) a static analysis component, built on top of WALA [31], and (2) a similarity analysis component, using UniXCoder. We evaluate *AutoAcRaptor* on a machine with Intel Core-i7 processor, featuring 12 cores with a maximum clock speed of 5.0 GHz, and 16 GB RAM.

4.1 Target ROMs

We collected two AAOS ROMs from AOSP and ten automaker-customized AAOS ROMs. The AOSP ROMs span versions 13 and 14, while the custom ones span versions 9 to 14. The latter are customized by General Motors (GM), Honda, Volvo, and Polestar [1,5–7].

When preprocessing the ROMs, we found that two ROMs, Volvo v10 and Polestar v9, could not be decompiled by apktool [2] or by smali [8]. Therefore, we excluded them from our analysis. Similarly, we found that another two ROMs, Honda v11 and GM-SUV-23 v11 were partially decompiled by the tools. However, we opted to include them in our analysis because we were able locate some AAOS system services and entry points. Table 1 lists the detailed statistics of the remaining ten ROMs (two AOSP and eight customized).

4.2   Research Questions

To evaluate our solution, we explore the following research questions:

- **RQ1:** What is the extent of Android customization carried out to accommodate AAOS?
- **RQ2:** What is the access control and auto-feature check landscape in AAOS entry points?
- **RQ3:** Does AAOS customization introduce access control and auto-feature check anomalies?
- **RQ4:** What is the accuracy of *AutoAcRaptor*?

4.3   RQ1: AAOS Customizations

Table 1. Detailed Statistics of Collected AAOS ROMs

OS Image	System Services		Entry Points						
	Android Services	Car Services	Total	Defining Service		AAOS Breakdown			
				Android Services	Car Services	Modified/Added by Google	Added by Automaker	Modified by Automaker	
GM-SUV-22 (V10)	165	31	3103	2958	145	200	249	217	
Polestar 2 (V10)	142	30	2859	2724	135	190	0	217	
GM-Cad-Lyriq-SUV-23 (V11)	159	30	3237	3070	167	220	5	231	
GM-SUV-23 (V11)	133	30	194	38	156	156	1	40	
Honda (V11)	124	29	159	24	135	135	2	31	
Volvo-XC40 (V11)	151	29	3213	3057	156	209	4	223	
GM-Cad-Lyriq-SUV-24 (V12L)	165	38	2167	1908	259	309	20	38	
Honda (V12L)	163	36	2133	1909	244	294	1	38	
AOSP (V13)	177	40	3515	3254	261	336	-	-	
AOSP (V14)	168	44	2210	1871	339	412	-	-	

System Services. Columns 2 and 3 in Table 1 list Android and Car system services, respectively. On average, *AutoAcRaptor* identifies 155 Android system services and 34 Car system services. The number of Car system services reaches up to 38 in the custom images, specifically, in GM-Cad-Lyriq-SUV- 24 (V12L) and up to 44 in the AAOS AOSP images (in version 14).

Entry Points. Columns 4 in Table 1 lists the total number of entry points in the identified system services, while columns 5 and 6 break down the entries by their defining class; Android system service (i.e., those defined also in non-AAOS builds), and Car system service. As shown, the number of entry points in car services varies from 135 in version 10 to 339 in version 14, exhibiting a steady increase throughout major versions.

The rest of the columns (columns 7–9) report a breakdown of AAOS entry points. Specifically, column 7 lists those entry points modified or added by Google for AAOS – this includes entries in both Android and Car system services; whereas the last two columns show the number of entries added and modified by automakers. We identify entry points modified by automakers by comparing against their reference AOSP ROM. For example, we compare Polestar 2 (V10) to AOSP V10.[6]

Note that entries in columns 7 and 9 might overlap, as an entry point can be modified by both Google and the automaker.

The number of modified/added APIs by Google for automotive support are largely consistent within a major version in the collected ROMs, with the exception of the two aforementioned partially corrupt ROMs (Honda v11 and GM-SUV-23 v11). We speculate that the slight differences are due to decompilation errors or due to minor version differences.

Automaker Customization Landscape. As shown in the last two columns, the number of added entries varies across vendors, ranging from none in Polestar 2, to 249 in GM-SUV-22. In addition, the automakers modify a high number of AOSP entries (defined in both Android and Car system services), ranging from 31 in Honda v11 and reaching up to 231 in GM-Cad-lyriq-SUV-23.

4.4   RQ2: Security Landscape

Table 2 reports the number of AAOS entry points that enforce an access control check (columns 2–5) and/or an auto-feature check (column 6). On average, around 60% of AAOS entry points enforce access control. As further shown, most added entries by automakers (column 4) do not enforce access control, while a larger portion of Google and automaker-modified entries (columns 3 and 5, respectively) enforces access control. Note that a lack of access control enforcement does not necessarily imply a security flaw when the underlying resources are not sensitive.

The last column lists entry points implementing auto feature checks. We observe that the checks are only present in AOSP AAOS and they are largely consistent across all ROMs.

4.5   RQ3: Access Control and Auto-feature Check Anomalies

AutoAcRaptor uses two strategies to identify access control and auto-feature check anomalies:

1) Convergence Analysis. Column 2 in Table 3 reports the number of API pairs detected by *AutoAcRaptor* as converging on at least one common resource. Observe that the analysis only considers pairs where at least one API is automotive (i.e., an API that (1) enforces an automotive feature check, or (2) is defined in Car services, or (3) is modified/added by an automaker).

---

[6] Note that we are not considering older AOSP ROMs (version < 13) in further analysis.

Table 2. Number of Entry Points with Access Control and Auto-Feature Checks

OS Image	Access Control Checks				Auto-Feature Checks
	Car Services	Modified/Added by Google	Added by Automaker	Modified by Automaker	
GM-SUV-22 (V10)	73	54	33	105	47
Polestar 2 (V10)	69	54	0	105	47
GM-Cad-Lyriq-SUV-23 (V11)	94	52	1	89	44
GM-SUV-23 (V11)	89	0	0	17	47
Honda (V11)	85	0	0	18	47
Volvo-XC40 (V11)	93	52	1	98	44
GM-Cad-Lyriq-SUV-24 (V12L)	145	50	4	28	43
Honda (V12L)	141	50	0	29	43
AOSP (V13)	155	72	-	-	68
AOSP (V14)	211	72	-	-	68

Column 3 in Table 3 lists the number of pairs with an access control or feature check anomaly. As mentioned earlier, this corresponds to the cases where two converging APIs enforce different access control (e.g., one enforces a permission while the other does not), and/or only one implements an auto feature check. On average, *AutoAcRaptor* reports 144 unique pairs with overlapping resources per image, among which about 9% are flagged as anomalous. Such inconsistencies signal potential security flaws and warrant further investigation.

2) API-Name Similarity Analysis. We count the number of API pairs that share semantically-similar names. Again, *AutoAcRaptor* ensures that at least one API in each pair is automotive. The second column of Table 4 reports the count of such APIs whereas the third reports those with different access control or auto feature checks. On average, *AutoAcRaptor* reports that 55 pairs are similar, and 12% are flagged as potentially anomalous.

Table 3. Convergence Analysis Results

OS Image	Number of API Pairs	
	Total	Access Control & Auto-Feature check Anomalies
GM-SUV-22 (V10)	163	9
Polestar 2 (V10)	160	9
GM-Cad-Lyriq-SUV-23 (V11)	135	14
GM-SUV-23 (V11)	39	0
Honda (V11)	38	16
Volvo-XC40 (V11)	134	15
GM-Cad-Lyriq-SUV-24 (V12L)	183	12
Honda (V12L)	177	0
AOSP (V13)	197	9
AOSP (V14)	221	8

4.6   RQ4: *AutoAcRaptor* Accuracy

We evaluate *AutoAcRaptor*'s accuracy based on its true positive (TP) and false positive (FP) rates. Due to the lack of ground truth, we resort to manual analysis to determine the positives of *AutoAcRaptor*. We consider a reported anomaly TP, if the pair of APIs perform the same sensitive functionality (identified through manual analysis) but enforces different access control and/or auto-feature checks. Conversely, we consider a reported anomaly as FP if the API pair with inconsistent checks actually perform *dissimilar functionality*, hence the inconsistency is reasonable. We perform this assessment on a representative ROM: GM-Cad-Lyriq-SUV-24 (V12L), and manually validate all reported cases in the convergence and similarity analyses.

Table 4. Similarity Analysis Results

OS Image	Number of API Pairs	
	Total	Access Control & Auto-Feature check Anomalies
GM-SUV-22 (V10)	46	1
Polestar 2 (V10)	45	1
GM-Cad-Lyriq-SUV-23 (V11)	45	5
GM-SUV-23 (V11)	39	4
Honda (V11)	30	4
Volvo-XC40 (V11)	44	5
GM-Cad-Lyriq-SUV-24 (V12L)	62	8
Honda (V12L)	62	8
AOSP (V13)	70	10
AOSP (V14)	112	30

As discussed in Sect. 3, *AutoAcRaptor*'s performance is highly dependent on the similarity score by UniXCoder. Thus, *AutoAcRaptor*'s TPs and FPs are dependent on the threshold chosen for the similarity score. Table 5 shows *AutoAcRaptor*'s TP and FP rates under multiple threshold values, varying from 0.6 to 0.9. As shown in the table, the threshold 0.8 achieves the best trade-off, leading to lower FPs and higher TPs, thus we select this value in our analysis. *AutoAcRaptor* achieves a TP rate of about 71% and a FP rate of about 29% in the convergence analysis and achieves a TP rate of about 75% and a FP rate of about 25% in the similarity analysis. We manually analyzed about 10% of the randomly selected false positives and found the primary root causes: 1) for convergence analysis, *AutoAcRaptor* correctly identified a common convergence point but it is not the primary functionality of the APIs, and 2) for similarity analysis, the APIs are identified as similar based on their names but have no similarity in their behavior, that is, their functionality is significantly different.

To evaluate false negatives, we analyzed about 5% of system services (10 in total), and corresponding entry points that were not reported as converging or similar by *AutoAcRaptor*. The analysis, which spanned unique 7140 API pairs, took 7.5 working days by one author. We did not find any pair that warranted further inspection (i.e., contained at least one convergence point, or was similar), therefore we concluded that *AutoAcRaptor*'s false negative rate is negligible.

Table 5. Impact of Various Similarity Thresholds

Threshold	Convergence		Similarity
	TP	FP	TP
0.6	0.67	0.33	0.39
0.7	0.71	0.29	0.41
0.8	0.71	0.29	0.75
0.9	0.71	0.29	0.71

4.7   Case Studies of Observed Anomalies

To demonstrate the security impact of our tool, we randomly picked a few reported access control and auto feature check inconsistencies and attempted to build end-to-end PoCs. We note though that not all gaps are exploitable. The reasons are twofold. First, triggering an inconsistency may require certain conditions that are hard to meet (e.g., the API requires a hardware feature not available on the emulator). Second, the inconsistency lies in proprietary functionality that is difficult to analyze and to understand its execution output.

Nonetheless, we were able to confirm ten cases on corresponding AAOS emulators. All have been reported to the respective vendors. Table 6 summarizes the anomalies. Next, we discuss three representative cases:

Table 6. Summary of Verified Anomalies

OS Image	Entry Point	Observed Anomaly
AOSP (V13)	CarPropertyService.getProperty	Misplaced permission check
GM-Cad-Lyriq-SUV-24 (V12L)	CarPropertyService.registerListener	Missing permission check
AOSP (V13)	DevicePolicyManagerService. setPasswordMinimumLowerCase	Missing auto feature check
GM-SUV-22 (V10)	PhoneProjectionBinderService. isProjectionForeground	Missing permission check
GM-Cad-Lyriq-SUV-24 (V12L)	CarPowerManagementService. unregisterListener	Missing access control check
Honda (V12L)	AccountManagerService.getPreviousName	Missing access control check
AOSP (V14)	ConnectivityService. startNattKeepaliveWithFd	Missing permission check
AOSP (V14)	PowerManagerService.goToSleep	Missing auto feature check
AOSP (V14)	UserManagerService.setuserRestriction	Missing auto feature check
AOSP (V14)	LocationManagerService. setLocationEnabledForUser	Missing auto feature check

Case 1: The `CarPowerManagementService` in GM-Cad-Lyriq-SUV-24 (V12L) defines two APIs: `finished` and `unregisterListener`. The former ensures that the caller has (`Car.PERMISSION_CAR_POWER`), and verifies that the method is being called from either a system or the same process. If checks are successful, an internal method `finishedImpl` is invoked, which our tool flags as a main *resource*. The latter API, `unregisterListener`, performs a single check to ensure that the caller has (`Car.PERMISSION_CAR_POWER`), and transitively invokes the same resource `finishedImpl`. It thus misses the process checks. We reported the case to the respective automaker (General Motors), who has acknowledged it and promised to address it in their next release.

Case 2: `startNattKeepalive` and `startNattKeepaliveWithFd` are two APIs from `ConnectivityService` in AOSP (V14). Both invoke `startNattKeepalive` internal method. However, the former requires a signature-level permission (`PACKET_KEEPALIVE_OFFLOAD`), whereas the latter lacks this permission check, resulting in an inconsistency. We reported the case to the Google Android team.

Case 3: Another case that we observed was in a car-specific API. Figure 10 shows a partial bypass of the access control check guarding car property IDs in Line 3. Since the access control is placed after the log operation, it discloses the car configuration data - as the log message is dumped within the calling app's process scope. In other APIs of this service, this information is protected. We reported it to Google, who promised to remediate it in future versions.

```
1  public CarPropertyValue getProperty(int prop, int zone) {
2      if (mConfigs.get(prop) == null) {
3          Slogf.e(TAG, "propId not in config list:0x" + toHexString(prop));
4          return null;
5      }
6      CarServiceUtils.assertPermission(mContext, mHal.getReadPermission(prop));
7      return mHal.getProperty(prop, zone);
```

Fig. 10. Partial Bypass of Access Control Check Guarding Car Property IDs.

5   Related Works

Android Automotive Framework. Although AAOS is a relatively new framework, several recent studies [22,26,27,29] have focused on its privacy and security aspects. For example, one study [29] analyzes 14 third-party AAOS apps and finds that 78% of them have privacy policies that are inconsistent with the permissions they request. Another study [27] highlights privacy-related attacks, such as fingerprinting and location inference in AAOS. Additionally, PriDrive [22] examines data collection practices through network traffic analysis in AAOS. Pricar [26] proposes a privacy-preserving data-sharing framework, while another study [25] investigates the security of in-vehicle infotainment systems. Further research has also analyzed AAOS's architecture [28] and security measures [13], along with a survey of discovered vulnerabilities. These studies either identify specific vulnerabilities or evaluate the architectural security of AAOS. In contrast, our work presents the *first systematic study* of auto-feature and access control gaps within AAOS.

Other Automotive Attack Surfaces. Several studies investigate the security of other automotive platforms. [14] analyzes the external attack surface of modern cars, revealing vulnerabilities in diagnostic tools, media players, Bluetooth, and cellular networks that can be exploited for remote control and data theft. It offers recommendations for improved security. Similarly, [24] focuses on the security risks of smartphone integration in In-Vehicle Infotainment (IVI) systems like MirrorLink, uncovering vulnerabilities that allow attackers to exploit compromised smartphones to send malicious messages to the vehicle's internal network. [18] studies attacks on the widely used Control Area Network (CAN)-bus, discussing ongoing challenges. VeCure [35] provides a security framework against CAN-bus attacks, while CAN-CID [30] offers a context-aware system for real-time detection of CAN-based attacks.

Android Access Control Analysis. Our work is closely related to approaches analyzing the traditional Android software stack [9,11,12,17,19,33] for access control issues. Stowaway [19] uses dynamic analysis to create permission specifications, while PScout [11] enhances them through static analysis. Axplorer [12] categorizes protected resources, and Arcade [10] develops protection maps using path-sensitive analysis. Other works [9,17,23,33,34] focus on inconsistencies across paths to the same resource, with Kratos [33] comparing Java layer paths and AceDroid [9] normalizing access control checks. More recent works (Poirot [17] and Bluebird [34]) use probabilistic inference to model uncertainty while generating access control specification.

Our tool, *AutoAcRaptor* shares similarities with these works but is specifically designed for AAOS. Traditional convergence analyses (e.g., Kratos [33] and Acedroid [9]) rely on (1) analyzing `ServiceManager` to extract traditional system services, and (2) converging APIs to detect inconsistencies. Poirot [17] and Bluebird [34] improve upon these, but still miss car-specific services and non-converging yet functionally similar APIs. *AutoAcRaptor*, in contrast, is specifically tailored to identify car-specific entry points and auto-feature checks, and utilizes similarity analysis to cover non-converging yet functionally similar APIs which can lead to inconsistencies.

6   Conclusion

We propose an automated analysis pipeline *AutoAcRaptor* that conducts program analysis and lightweight NLP to systematically evaluate auto-feature and access control specifications for AAOS framework entry points. We run *AutoAcRaptor* on ten ROMs and identify 23 anomalies per ROM. The findings highlight that AAOS developers lack appropriate security specifications and emphasize the need for further scrutiny in AAOS frameworks.

Acknowledgements. This research was supported, in part by NSERC under grant RGPIN-07017 and by the Canada Foundation for Innovation under project 40236. This work benefited from the use of the CrySP RIPPLE Facility at the University of Waterloo. Any opinions, findings and conclusions in this paper are those of the authors only and do not necessarily reflect the views of our sponsors.

References

1. Android emulator (2024). https://developer.volvocars.com/in-car-apps/android-emulator-xc40/
2. Apktool (2024). https://apktool.org/
3. dex2jar (2024). https://github.com/pxb1988/dex2jar
4. Git repositories on android (2024). https://android.googlesource.com/
5. Gm developers (2024). https://developer.gm.com/in-vehicle-apps
6. Honda android automotive OS emulator (2024). https://global.honda/en/cars-apps/index.html
7. Polestar developer portal (2024). https://www.polestar.com/global/developer#emulator
8. smali (2024). https://github.com/google/smali
9. Aafer, Y., Huang, J., Sun, Y., Zhang, X., Li, N., Tian, C.: Acedroid: normalizing diverse android access control checks for inconsistency detection. In: Proceedings of the 2018 Network and Distributed System Security Symposium (2018)
10. Aafer, Y., Tao, G., Huang, J., Zhang, X., Li, N.: Precise android API protection mapping derivation and reasoning. In: Proceedings of the 2018 ACM SIGSAC Conference on Computer and Communications Security, pp. 1151–1164 (2018)
11. Au, K.W.Y., Zhou, Y.F., Huang, Z., Lie, D.: Pscout: analyzing the android permission specification. In: Proceedings of the 2012 ACM Conference on Computer and Communications Security (CCS 2012), pp. 217–228. Association for Computing Machinery (2012). https://doi.org/10.1145/2382196.2382222
12. Backes, M., Bugiel, S., Derr, E., Mcdaniel, P., Octeau, D., Weisgerber, S.: On demystifying the android application framework: re-visiting android permission specification analysis. In: USENIX Security Symposium (2016)
13. Chatzoglou, E., Kambourakis, G., Kouliaridis, V.: A multi-tier security analysis of official car management apps for android. Future Internet **13**(3), 58 (2021)
14. Checkoway, S., et al.: Comprehensive experimental analyses of automotive attack surfaces (2011). https://www.usenix.org/conference/usenix-security-11/comprehensive-experimental-analyses-automotive-attack-surfaces. Accessed 20 Dec 2023
15. cs.android.com: Car-lib. https://cs.android.com/android/platform/superproject/main/+/main:packages/services/Car/car-lib/. Accessed 20 Dec 2023
16. Android Developers: Test using the android automotive OS emulator. https://developer.android.com/training/cars/testing/emulator. Accessed 20 Dec 2023
17. El-Rewini, Z., Zhang, Z., Aafer, Y.: Poirot: probabilistically recommending protections for the android framework. In: Proceedings of the 2022 ACM SIGSAC Conference on Computer and Communications Security (CCS 2022), pp. 937–950. Association for Computing Machinery (2022). https://doi.org/10.1145/3548606.3560710
18. Fakhfakh, F., Tounsi, M., Mosbah, M.: Cybersecurity attacks on can bus based vehicles: a review and open challenges. Library Hi Tech **40**(5), 1179–1203 (2022)

19. Felt, A.P., Chin, E., Hanna, S., Song, D., Wagner, D.: Android permissions demystified. In: Proceedings of the 18th ACM Conference on Computer and Communications Security, pp. 627–638 (2011)
20. Gorski, S.A., et al.: Acminer: extraction and analysis of authorization checks in android's middleware. In: Proceedings of the Ninth ACM Conference on Data and Application Security and Privacy, CODASPY 2019, pp. 25–36. Association for Computing Machinery, New York (2019). https://doi.org/10.1145/3292006.3300023
21. Guo, D., Lu, S., Duan, N., Wang, Y., Zhou, M., Yin, J.: Unixcoder: unified cross-modal pre-training for code representation. arXiv preprint arXiv:2203.03850 (2022)
22. Gözübüyük, B., Tang, B., Shin, K., Pesé, M.: Analyzing privacy implications of data collection in android automotive OS. arXiv (2024). https://arxiv.org/abs/2409.15561
23. III, S.A.G., Thorn, S., Enck, W., Chen, H.: FReD: identifying file re-delegation in android system services. In: 31st USENIX Security Symposium (USENIX Security 2022), Boston, MA, pp. 1525–1542. USENIX Association (2022). https://www.usenix.org/conference/usenixsecurity22/presentation/gorski
24. Mazloom, S., Rezaeirad, M., Hunter, A., McCoy, D.: A security analysis of an in-vehicle infotainment and app platform (2016). https://www.usenix.org/conference/woot16/workshop-program/presentation/mazloom. Accessed 20 Dec 2023
25. Mazloom, S., Rezaeirad, M., Hunter, A., McCoy, D.: A security analysis of an {in-vehicle} infotainment and app platform. In: 10th USENIX Workshop on Offensive Technologies (WOOT 2016) (2016)
26. Pesé, M.D., Schauer, J.W., Mohan, M., Joseph, C., Shin, K.G., Moore, J.: Pricar: privacy framework for vehicular data sharing with third parties. In: 2023 IEEE Secure Development Conference (SecDev), pp. 184–195. IEEE (2023)
27. Pesé, M.D., Shin, K.G.: Survey of automotive privacy regulations and privacy-related attacks (2019)
28. Pesé, M., Shin, K., Bruner, J., Chu, A.: Security analysis of android automotive. SAE Technical Paper 2020-01-1295 (2020). https://doi.org/10.4271/2020-01-1295
29. Pesé, M.: A first look at android automotive privacy. SAE Technical Paper 2023-01-0037 (2023). https://doi.org/10.4271/2023-01-0037
30. Rajapaksha, S., Kalutarage, H., Al-Kadri, M.O., Madzudzo, G., Petrovski, A.V.: Keep the moving vehicle secure: context-aware intrusion detection system for in-vehicle can bus security. In: 2022 14th International Conference on Cyber Conflict: Keep Moving!(CyCon), vol. 700, pp. 309–330. IEEE (2022)
31. Wala github repository. https://github.com/wala/WALA. Accessed 20 Dec 2023
32. SamMobile: Porsche cars to feature android automotive, google play store integration soon (2023). https://www.sammobile.com/news/porsche-cars-android-automotive-google-play-store-soon/. Accessed 20 Dec 2023
33. Shao, Y., Chen, Q.A., Mao, Z.M., Ott, J., Qian, Z.: Kratos: discovering inconsistent security policy enforcement in the android framework. In: Network and Distributed System Security Symposium (2016)
34. Vyas, P., Waheed, A., Aafer, Y., Asokan, N.: Auditing framework APIs via inferred app-side security specifications. In: 32nd USENIX Security Symposium (USENIX Security 2023), Anaheim, CA, pp. 6061–6077. USENIX Association (2023). https://www.usenix.org/conference/usenixsecurity23/presentation/vyas
35. Wang, Q., Sawhney, S.: Vecure: a practical security framework to protect the can bus of vehicles. In: 2014 International Conference on the Internet of Things (IOT), pp. 13–18. IEEE (2014)

Poster: SPECK: From Google Textual Guidelines to Automatic Detection of Android Apps Vulnerabilities

Roberto Rossini, Simeone Pizzi, Samuele Doria, Mauro Conti, and Eleonora Losiouk[(✉)]

Department of Mathematics, University of Padua, 35121 Padua, PD, Italy
{simeone.pizzi,mauro.conti,eleonora.losiouk}@unipd.it,
sdoria@math.unipd.it

Abstract. This paper addresses vulnerabilities in Android apps introduced by developers failing to follow Google's security guidelines. Existing tools do not cover all guidelines, leaving many violations undetected. We propose SPECK, a static rule-based taint analysis system that detects violations of all 31 Google guidelines, outperforming current tools. Analyzing 500 popular apps, we find every app has at least one violation, highlighting the need for comprehensive detection.

1 Introduction

The origin of the Android success comes from being an open-source platform that allows developers to freely build and publish apps. At the same time, this openness also attracts attackers looking for vulnerabilities to be exploited. Beyond OS-level vulnerabilities and malicious apps, a critical but often overlooked attack surface stems from developers' misuse of Android APIs or lack of security expertise [12]. To mitigate such threats and support developers, Google provides: (i) security guidelines [3], (ii) Lint [5], a code analysis tool, (iii) and a security-focused course on Google Play Academy [4]. Unfortunately, the textual format of these guidelines might prevent their adoption during development.

We first assessed whether existing tools detect the vulnerabilities described in Google's official guidelines. To this end, we designed and implemented (IVA), an Android app that intentionally includes violations of all the guidelines, and use it to benchmark the detection capabilities of existing tools. We found that they detect at most half of the violations.

To address this gap, we designed **SPECK**[1] (Security and Privacy chECK of Android apps vulnerabilities), a rule-based static analyzer that automatically detects Google guidelines' violations. We formalized 31 detection rules from Google's textual guidelines and implemented them in SPECK, also providing developers with actionable feedback by pointing to the exact lines of code responsible for each violation.

In this work, we address the following research questions:

---

[1] https://github.com/samudoria/SPECK.

M. Egele et al. (Eds.): DIMVA 2025, LNCS 15748, pp. 167–172, 2025.
https://doi.org/10.1007/978-3-031-97623-0_10

- **RQ1:** How many violations of the Google textual guidelines can be detected by the current state-of-art static analysis tools?
- **RQ2:** How many violations can SPECK detect?
- **RQ3:** How fast can SPECK detect violations in Android apps?
- **RQ4:** How well does SPECK detect violations with respect to state-of-art tools?

We evaluated SPECK on the 500 most downloaded apps from the Google Play Store. Remarkably, 19 of the 31 formalized rules are violated in over 50% of the apps, often multiple times.

2   Google Textual Guidelines and State-of-Art Tools

To assess how well current static analysis tools detect violations of Google's security and privacy guidelines (RQ1), we developed a custom app, IVA, which systematically violates all the 31 guidelines (Sect. 2.1) and serves as ground truth for evaluating tool effectiveness (Sect. 2.2).

2.1   Intentionally Vulnerable App

Instead of relying on real-world or open-source apps, where obfuscation or complexity could hide guideline violations, we built IVA from scratch. This ensures: (i) full control over the source code, (ii) guaranteed coverage of all guideline violations, (iii) no external libraries, removing false positives from third-party code. We thoroughly analyzed Google's textual guidelines and embedded one violation per rule into IVA. Manual verification confirmed each violation was implemented correctly and in isolation.

2.2   RQ1 - How Many Violations of the Google Textual Guidelines Can Be Detected by the Current State-of-Art Static Analysis Tools?

We ran ten static analysis tools on IVA, namely: Amandroid [14], Androwarn [2], Lint [5], Quark-Engine [8], RiskInDroid [13], SUPER [10], Mariana Trench [6], Mobile Security Framework (MobSF) [7], Androbugs [1], and Quick Android Review Kit [9], and recorded which guideline violations were found. No tool detected all violations. The top-performing tools–Lint [5], MobSF [7], and Androbugs [1]–each detected 11 out of 31 violations, achieving a detection rate of 34.4%. On the other hand, Androwarn [2] was the worst-performing tool, detecting no violation. On average, tools detected approximately 6.7 violations, corresponding to a detection rate of just 20.9%. This confirms the lack of comprehensive, automatic support for developers seeking guideline compliance.

Given the limitations of the existing tools, we developed SPECK, a system that automatically inspects Android apps for violations of 31 rules derived from Google's security guidelines (full implementation details can be found in our public repository[2]). SPECK detects violations by statically inspecting the app's source code, looking for patterns and API calls, and by applying taint analysis rules that focus on the data flows within the app. Concerning the latter, we integrated FlowDroid [11], a taint analysis framework for Android. SPECK extends FlowDroid to support custom sources, sinks, and data flow patterns necessary for detecting issues.

3   Evaluation

Here, we present the results we obtained after running SPECK against the 500 most downloaded real-world apps from the Google Play Store. First, we measure SPECK precision in detecting violations through a manual verification (Sect. 3.1), then we answer to RQ2 (Sect. 3.2) and RQ3 (Sect. 3.3). Finally, we compare SPECK against the state-of-the-art (Sect. 3.4).

3.1   Rules Precision

To assess SPECK precision in detecting guideline violations, we identified two independent reviewers—external to the project—to manually verify a sample of 2,778 reported violations across ten randomly selected apps. Each reviewer inspected the relevant source code and labeled violations as true positives or false positives. Disagreements were resolved through a secondary review by two of the authors.

Among the 31 formalized rules, 26 generated results suitable for manual evaluation. The remaining five either reported no violations or involved checks (e.g., manifest-only rules) for which precision is not meaningful. SPECK achieved a high overall precision, averaging 92% across the 26 evaluated rules. Notably, 13 rules reached perfect precision (100%), typically those based on manifest analysis or direct API invocations. The lowest precision (60%) was observed for Rule 24, which relies on taint analysis to detect improper hostname verification and is prone to false positives due to the complexity of tracking indirect data flows.

3.2   RQ2 - How Many Violations Can SPECK Detect?

To answer this research question, we measured the number of apps violating a rule and the number of times the apps violate it. As shown in Fig. 1 and Fig. 2, 19 rules are violated by more than 50% of the apps, while other rules can have zero violations. Moreover, the most violated rules have more than 10.000 violations each. Our analysis shows that specific rules are frequently violated. This may be because developers do not consider these violations relevant, or because they stem from legitimate use cases. Conversely, some rules were never violated in our dataset. These may reflect guidelines that are easier to adhere to or less prone to accidental misuse. In general, rules with high violation counts often require

---

[2] https://github.com/samudoria/SPECK.

an understanding of developer intent. In such cases, SPECK issues warning messages for all detected violations to ensure that potentially unintentional issues are not overlooked.

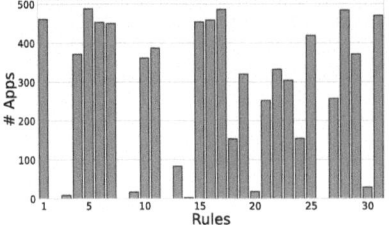

Fig. 1. Number of apps violating the rules.

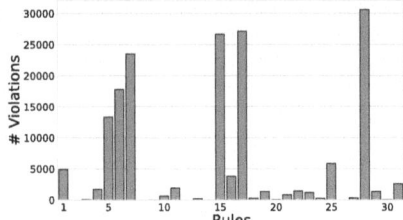

Fig. 2. Number of violations by rule.

3.3 RQ3 - How Fast Can SPECK Detect Violations in Android Apps?

To answer the third research question, we measured the average execution time each rule requires to analyze an app. Figure 3 shows the distribution of these times across the 500 apps. The results reveal three performance: 12 rules execute almost instantly; 12 rules complete within 1–20 s; the remaining 7 rules require over 300 s, primarily due to FlowDroid's taint analysis and call graph modeling. Most rules with near-zero processing time rely on Manifest analysis, which involves inspecting a single file. Some API-based rules also show minimal overhead, as they only require parsing a limited number of files that import relevant API classes. Finally, rules involving code analysis have the highest processing times—often exceeding 300 s—primarily due to the generation of call graphs.

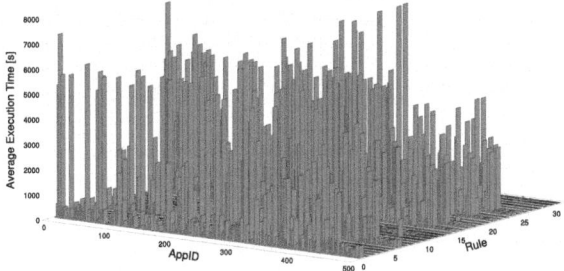

Fig. 3. Average execution time required by each rule to analyze an app.

3.4   RQ4 - How Well Does SPECK Detect Violations with Respect to State-of-Art Tools?

To further assess the precision of SPECK, we compared its performance against a subset of the tools, selecting the four that implement detection logic for the highest number of Google's guidelines. We used the set of true positives from the 2,778 manually verified violations (described in Sect. 3.1) to directly compare SPECK with the following tools: MOBSF [7], ANDROBUGS [1], QARK [8], and SUPER [10]. Specifically, we executed each tool on the 10 apps containing the validated violations and checked whether the other tools also flagged any of the true positives identified by SPECK.

For example, SPECK reported 185 true violations of Rule 5 across the 10 apps, but MOBSF, ANDROBUGS, and QARK failed to detect any of them, while SUPER did not cover Rule 5 at all. Overall, SPECK outperformed all four tools in the number of correctly identified violations. During the evaluation, we also encountered tool-specific limitations: MOBSF failed to complete the analysis of two apps within a five-hour timeout; ANDROBUGS crashed due to a `dex2jar` issue on one app; and QARK failed to decompile two apps.

4   Conclusions

In this paper, we present SPECK, the first static analysis tool designed to verify adherence to the complete set of Google's textual guidelines. Existing research tools detect specific vulnerabilities but do not offer comprehensive checks for full compliance with Google's security and privacy guidelines We formalized the guidelines into 31 actionable rules and implemented SPECK as a rule-based static analyzer that automatically detects violations. For each issue, SPECK highlights the exact line of code involved, helping developers address vulnerabilities early in the development process.

Although SPECK is currently limited by FlowDroid's overhead and by the missing analysis of native libraries, its relevance is highlighted by our analysis of 500 popular apps from the Google Play Store: over 50% violated at least 19 of the rules, often multiple times. Furthermore, SPECK currently outperforms the current state-of-the-art static rule-based analyzers. These findings underscore the pressing need for automated tools like SPECK to support secure Android development.

Acknowledgement. We would like to thank Julien Branlant, Michele Agnello, Alberto Molon, Nemanja Lalic and Ahmad Zubair Zahid for their support to the project.

This work was supported by the European Commission under the Horizon 2020 Programme (H2020), as part of the LOCARD project (Grant Agreement no. 832735), and by Google through a Google Research Program Scholar award.

References

1. Androbugs (2025). https://github.com/androbugs2/androbugs2. Accessed 19 June 2025
2. Androwarn (2025). https://github.com/maaaaz/androwarn. Accessed 19 June 2025
3. Google Guidelines (2025). https://developer.android.com/training/articles/security-tips. Accessed 19 June 2025
4. Google play academy (2025). https://playacademy.exceedlms.com/student/path/63550-security-by-design. Accessed 19 June 2025
5. Improve your code with lint checks (2025). https://developer.android.com/studio/write/lint. Accessed 19 June 2025
6. Mariana trench (2025). https://github.com/facebook/mariana-trench. Accessed 19 June 2025
7. Mobile security framework - MobSF (2025). https://github.com/MobSF/Mobile-Security-Framework-MobSF. Accessed 19 June 2025
8. Quark-engine (2025). https://github.com/quark-engine/quark-engine. Accessed 19 June 2025
9. Quick android review kit (2025). https://github.com/linkedin/qark. Accessed 19 June 2025
10. Super android analyzer (2025). https://github.com/SUPERAndroidAnalyzer/super. Accessed 19 June 2025
11. Arzt, S., et al.: FlowDroid: precise context, flow, field, object-sensitive and lifecycle-aware taint analysis for android apps. ACM SIGPLAN Not. **49**(6), 259–269 (2014)
12. Fischer, F., et al.: Stack overflow considered harmful? The impact of copy & paste on android application security. In: 2017 IEEE Symposium on Security and Privacy (SP), pp. 121–136. IEEE (2017)
13. Merlo, A., Georgiu, G.C.: RiskInDroid: machine learning-based risk analysis on android. In: De Capitani di Vimercati, S., Martinelli, F. (eds.) SEC 2017. IAICT, vol. 502, pp. 538–552. Springer, Cham (2017). https://doi.org/10.1007/978-3-319-58469-0_36
14. Wei, F., Roy, S., Ou, X.: Amandroid: a precise and general inter-component data flow analysis framework for security vetting of android apps **21**, 1–32 (2018)

OS and Network

Taming the Linux Memory Allocator
for Rapid Prototyping

Ruiyi Zhang[(⊠)], Tristan Hornetz, Lukas Gerlach, and Michael Schwarz

CISPA Helmholtz Center for Information Security, Saarbrücken, Germany
ruiyi.zhang@cispa.de

Abstract. Microarchitectural attacks pose an increasing threat to system security. They enable attackers to extract sensitive information such as cryptographic keys, website usage patterns, or keystrokes. Software-level defenses, such as constant-time implementations, mitigate some attack vectors but impose significant challenges on developers. Operating-system-level mitigations, such as page coloring and memory isolation, address these threats but require intricate kernel modifications and time-consuming workflows, making prototyping new defenses complex.

In this paper, we present MAPAlloc (Microarchitectural Prototyping Allocator), a flexible, cross-architecture framework for rapidly prototyping memory allocation-based defenses and attacks on Linux systems. Using a simple domain-specific language, MAPAlloc allows for precise control over physical memory allocation on x86, ARMv8, and RISC-V. MAPAlloc enables quick implementation and evaluation of mitigations such as page coloring and novel techniques like layered page coloring, increasing the number of cache colors from 32 to 256 on modern CPUs. We demonstrate MAPAlloc's versatility through case studies that prevent Prime+Probe and DRAMA attacks and reverse-engineer the AMD Zen 4 complex cache-indexing function for use in layered page coloring. Additionally, we prototype a Prime+Probe attack with an incomplete non-linear slice function from previous work by limiting the physical memory using MAPAlloc. Without MAPAlloc, such defense and attack prototypes require complicated modifications of the Linux kernel, making them hard to develop and test. Thus, MAPAlloc is an essential framework for simplifying research in microarchitectural security.

1 Introduction

Microarchitectural attacks are increasingly becoming a significant and practical threat. These attacks not only threaten security by inferring cryptographic keys [6,31] but also threaten privacy, such as in website fingerprinting [55,68], keystroke logging [52,62], and compromising the privacy guarantees of differential privacy algorithms [25]. Academia and industry have dedicated substantial effort to addressing this issue, targeting both hardware and software levels. While hardware mitigations look promising, retrofitting them to existing

M. Egele et al. (Eds.): DIMVA 2025, LNCS 15748, pp. 175–194, 2025.
https://doi.org/10.1007/978-3-031-97623-0_11

hardware CPUs is often not possible [49,51,63]. Moreover, it typically takes years until mitigations appear in hardware, if they appear at all [12,56]. A more flexible approach involves software-level defenses against microarchitectural attacks. Constant-time implementation [23], commonly found in cryptographic libraries [41,46,59,64], can avoid a broad class of microarchitectural attacks, including ones that are based on timing or memory access patterns. However, constant-time code is challenging to implement, making it impractical for all but the most high-risk targets. Moreover, it shifts the burden to the developer instead of fixing it at a more central point, such as the operating system.

Consequently, previous works proposed multiple mitigations on the operating-system level [4,29,32,54]. These mitigations influence memory allocation to better isolate attacker and victim applications. Page coloring [5] assigns non-overlapping cache sets to mutually untrusted applications, preventing eviction-based cache attacks such as Prime+Probe and Evict+Reload. CATT [4] mitigates the effect of Rowhammer on the kernel by separating kernel and user-space memory such that these memory regions never end up in adjacent DRAM rows. Similarly, ZebRAM [32] modifies the allocator to have guard rows in DRAM, making Rowhammer flips ineffective. Such mitigations all require complex changes to the kernel, requiring considerable expertise in kernel data structures and functionality. Additionally, as microarchitectural attacks typically require native code execution, the workflow of compiling and rebooting into a new kernel for tests is time-consuming and tedious.

In this paper, we present MAPAlloc (Microarchitectural Prototyping Allocator), a rapid-prototyping framework that provides extensive control over the *physical* memory allocation of the Linux kernel. MAPAlloc supports an expressive Domain-Specific Language (DSL) for memory constraints to enable precise per-process control over the physical pages a process can access. MAPAlloc can easily provide proof-of-concept implementations for mitigations such as page coloring by simply specifying the cache-set function using the DSL and the process to which the restriction should be applied. Moreover, MAPAlloc is implemented as a cross-architecture kernel module, supporting x86, ARMv8, and RISC-V CPUs. Thus, MAPAlloc can be loaded at runtime on a wide range of systems. MAPAlloc is designed for rapid prototyping, allowing the evaluation of defense strategies such as emulating hardware coloring effects. However, it is not a production-ready mitigation. Any real defense could similarly bypass or modify the memory allocator, just as MAPAlloc does. Therefore all mitigations prototyped using MAPAlloc will work in real systems.

To demonstrate the flexibility of MAPAlloc, we demonstrate how it can use DRAM addressing functions [45] to separate DRAM rows and cache-set functions to isolate cache sets [54], quickly reproducing known mitigations. In case studies, we show that our prototyped mitigation using MAPAlloc effectively prevents Prime+Probe and DRAMA attacks. Additionally, we introduce *layered page coloring*, an improved variant of page coloring that combines cache sets with cache slices to provide more flexibility in assigning non-overlapping cache sets. We introduce a graph-based algorithm to calculate the number of available colors when combining arbitrary linear and non-linear microarchitectural hash functions. We show that on modern CPUs, layered page coloring can increase the number of colors from 32 to 256. Further, we reverse-engineer the cache set-indexing function on AMD Zen 4 CPUs, showing that this complex function can also be used for layered page coloring on the L2 and L3 caches. Thus, layer page coloring is a viable defense on modern CPUs that could realistically be implemented in the Linux kernel.

As a side effect, MAPAlloc can also simplify microarchitectural attack prototyping. For example, existing non-linear cache-slice functions for modern CPUs have only been reverse-engineered for systems with at most 4 GB of DRAM [15]. However, by limiting memory allocations to the first 4 GB of physical memory using MAPAlloc, attacks can still be tested on systems with more memory. Similarly, MAPAlloc can reduce the time it takes Rowhammer attacks to find exploitable rows by limiting rows to a specific DIMM or even bank. Such artificial constraints on the system can ease attack prototyping, as shown in previous work [11]. Note that MAPAlloc does not enhance real attacks, as it operates as a privileged Linux kernel module. It only assists in prototyping by applying controlled memory constraints.

Contributions. The main contributions of this work are:

1. We present MAPAlloc, a generic framework that allows for customizable refinement of physical page allocations across different architectures, enabling researchers to quickly prototype mitigations (such as page coloring) and attacks.
2. We introduce layered page coloring with linear and non-linear microarchitectural hash functions, showing that such a combination can increase the number of security domains for page coloring.
3. We reverse-engineer the complex set addressing and cache-slice functions on AMD Zen 4 and demonstrate their use in page coloring.
4. In three case studies, we demonstrate the use cases of MAPAlloc by building page coloring for the last-level cache and the DRAM and artificially limiting the physical memory to prototype Prime+Probe with an incomplete non-linear cache-slice function.

Structure. The paper is organized as follows. Section 2 provides relevant background. Section 3 presents the design and implementation of our frameworks. In Sect. 4, we evaluate MAPAlloc by building prototypes for (layered) page coloring against eviction-based attacks and DRAMA attacks, and we prototype a Prime+Probe attack with an incomplete non-linear cache-slice function. Section 5 evaluates the functionality and performance of MAPAlloc. We discuss related works and limitations in Sect. 6. Section 7 concludes.

Availability. Our framework is available at https://github.com/cispa/MAPAlloc.

2 Background

2.1 Memory Allocators

Memory allocators are critical components of modern operating systems, responsible for managing the allocation and deallocation of memory in units called pages, each typically 4 KB in size. The main allocator in Linux, known as the buddy allocator [8], groups free memory pages into buddies of various sizes, which are powers of two. For example, an order-0 buddy is one 4 KB page, while an order-3 buddy is 8 contiguous 4 KB pages (32 KB). The allocator uses free lists to keep track of available memory blocks for each size. When a process requests memory, the allocator can split larger blocks into smaller ones or merge adjacent blocks to form larger ones, reducing fragmentation. Memory allocation requests include flags, known as Get Free Pages (GFP) flags [8], which specify additional information like whether the memory is for user or kernel space. These flags help the allocator decide how to fulfill the request efficiently. Additionally, the per-CPU page (PCP) allocator [9] manages free pages for each CPU, speeding up memory allocation by reducing contention. Together, these mechanisms ensure efficient memory management in the Linux kernel.

2.2 Cache Eviction and Attacks

Modern processors use cache memory to store frequently accessed data and instructions, significantly improving performance by reducing the time needed to access data from the main memory. Unlike the larger 4 KB pages in main memory, the cache is organized into cache sets, each containing multiple 64-byte chunks, known as cache lines. Despite their speed, caches have limited size, making cache eviction necessary to create space for new entries when the cache is full. Well-known cache side-channel attacks, such as Prime+Probe [44] and Evict+Reload [18], exploit the cache eviction process to infer sensitive information. In Prime+Probe, the attacker fills cache sets with their own data and later measures access times to determine which cache lines were evicted by the victim's memory accesses, revealing the victim's access patterns. Conversely, Evict+Reload relies on shared memory. The attacker evicts a specific cache line by filling the cache set, waits for the victim to access it, and then measures the reload time to infer whether the victim accessed the memory.

2.3 DRAMA and Rowhammer

Unlike side channel attacks that target CPU caches, DRAMA attacks [45] target internal caching structures of DRAM modules called row buffers. DRAM is structured into multiple components, including channels, DIMMs, ranks, and banks, with the row being the smallest independently accessed unit. Each bank consists of numerous rows, with a corresponding row buffer for read and write operations. When data is accessed, the corresponding row is loaded into the row buffer, which can then be read. Loads into the row buffer have variable latency depending on the row buffer state. If the row currently loaded is already cached in the row buffer, access times are typically 50–70 cycles faster. By measuring these access times, an attacker can monitor the state of the row buffer and implicitly the victim's memory access patterns. Similar to cache sets, a row can contain memory from different applications.

While DRAMA attacks can leak victim memory access patterns, Rowhammer attacks [30,48,50] can corrupt data. Similar to DRAMA attacks, Rowhammer operates on the DRAM. However, in contrast to DRAMA, Rowhammer exploits the physical properties of the DRAM cells. As DRAM cells are dynamic memory, they leak charge over time and must be refreshed regularly to prevent memory corruption. By repeatedly accessing (hammering) a row of memory, the power leakage of DRAM cells is accelerated, and bit flips, typically in the adjacent rows, are induced. These bit flips can corrupt data, allowing attackers to manipulate the memory content of other processes. While multiple mitigations on hardware [3,12,38], software [2,4] and combined [27] levels have been proposed, Rowhammer remains challenging to mitigate. This is due to the closed nature of DRAM modules and memory controllers and the fact that Rowhammer is a physical phenomenon that even differs between identical setups [16]. Multiple previous mitigation techniques, such as increasing refresh rates and ECC DRAM, have been shown to be only partially effective [12].

2.4   Page Coloring

Page coloring is a memory allocation technique employed for performance [67] or security [54] purposes. It allows for assigning "colors", which are labels, to memory pages. The color of a page can be determined by arbitrary attributes, such as physical address or allocating process. These colors can then serve as an additional attribute during allocation to distribute memory pages more precisely. For example, page coloring can help to optimize cache utilization by considering the cache sets as colors. Thus, pages with different colors can be used for frequently accessed memory, as the non-overlapping cache sets prevent mutual eviction. This increases the cache hit rate and, thus, an application's performance. In the context of security, page coloring can be used to prevent side-channel attacks by isolating microarchitectural elements between processes. For example, if pages from processes in different security domains are mapped to disjoint cache sets, this prevents eviction-based cache side channels between the processes.

3   Framework

In this section, we present a high-level overview of MAPAlloc and its practical implementation for memory allocation manipulation. MAPAlloc allows researchers to influence the *physical* memory allocation for a specific application using a simple DSL. Once configured, MAPAlloc allocates pages only with the physical addresses that fulfill the constraints given by the DSL. We evaluate the proof-of-concept program on x86, ARMv8, and RISC-V CPUs.

3.1 Design

MAPAlloc consists of two core components: a DSL specifying constraints on physical addresses and a custom allocator that allocates pages that fulfill these constraints. While MAPAlloc has to run in the kernel, it exposes a user-space interface for providing the DSL-defined constraints. Constraints are on a per-process basis, specified via the process ID.

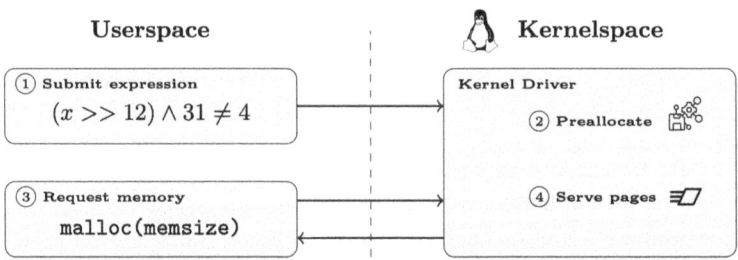

Fig. 1. The design of MAPAlloc. A client requests a memory policy, which the kernel driver preallocates. Afterward, the client can request memory allocations from this pool.

Figure 1 shows the overview of MAPAlloc. After providing DSL-defined constraints to MAPAlloc, MAPAlloc preallocates physical pages fulfilling the constraints in kernel mode. These preallocated pages are kept in a pool to quickly provide them to the application if needed. Hence, allocating pages with MAPAlloc does not require costly searches for fitting pages, as the pool decouples this search from the allocation. As a result, allocations are always fast, independent of the constraint's complexity. We allow for the configuration of this pool size to ensure an ideal tradeoff between memory overhead and allocation speed.

To ensure that all memory allocations are served via MAPAlloc and its pool, MAPAlloc hooks the Linux kernel's memory allocation functions. This captures all memory allocations, including explicit allocations via `mmap` and `brk/sbrk` and implicit allocations triggered by page faults. Constraints can also apply to already allocated pages. For example, since user-space applications never directly interact with physical pages, the operating system (or, in this case, MAPAlloc) can transparently migrate existing pages to ones that meet specific constraints. MAPAlloc provide APIs to enforce migration by iterating through the relevant pages, updating the respective page-table entries, and flushing the TLB to ensure coherence.

3.2 DSL

We provide a lightweight DSL for specifying constraints. The current proof-of-concept implementation supports an arbitrary number of constraints involving arithmetic operations, bit operations, and comparisons. These constraints are combined using logic operations. The DSL's grammar is shown in the appendix in Fig. 5. The advantage of a DSL is that constraints can be complex and can be extended with functions in a backward-compatible manner. MAPAlloc parses the grammar to apply the constraints to the memory allocations. While the current implementation does not yet support stateful constraints, this is not a limitation of the design. Future contributions are encouraged to extend the DSL with this capability.

3.3   Implementation

To make the usage of MAPAlloc as simple as possible, we implement it as a Linux kernel module instead of a kernel patch. While this reduces the flexibility of MAPAlloc and impacts the performance, it greatly enhances the maintainability. Furthermore, using kernel modules is often possible when a custom kernel is difficult to install, such as on ARM and RISC-V development boards or smartphones. The kernel module is written in C and hooks the Linux kernel's memory management functions to manipulate memory allocations. We do not rely on architecture-specific functions, making MAPAlloc compatible with a wide range of architectures. We successfully verify its functionality on x86, ARMv8, and RISC-V.

Our implementation hooks the `__alloc_pages` function of the Linux kernel. This function is the lowest-level function responsible for returning a physical page. It handles all single-page allocations, both implicit and explicit allocations. Thus, we only have to hook here to cover the majority of cases for getting a user-accessible physical page. We use a `kretprobe` to hook into this function before it returns. Due to our implementation as a kernel module, we can neither change nor replace the function generically. Thus, we have to resort to such a probe.

We opt for the return probe, as this allows using the results of the kernel function while being able to change the return value. In the handler of the return probe, we have access to the page allocated by the kernel. We can either directly use this page if it fulfills the constraints, replace it with a page from our pool, or reject it, forcing the kernel to allocate a new page.

Additionally, we hook `vm_mmap_pgoff`, which is used for `mmap` functionality. Hooking this function allows for intercepting explicit multi-page allocations, replacing them with pages that fulfill the constraints.

3.4   Usage

Listing 1 shows a sample usage of MAPAlloc. The main workflow is to create a new process via `fork`, provide the DSL-based constraints to MAPAlloc for the new process, and then replace the constrained process with the target process via `exec*`. This workflow ensures that all pages of the process fulfill the constraints. While MAPAlloc can also be applied to a running process, it only affects the pages allocated after applying MAPAlloc. However, we can also support this use case by migrating existing physical pages of the target process to ones that fulfill the constraints.

```
1 const char* constraint = "((x >> 12) & 31 >= 1) && "
2                          "((x >> 12) & 31 <= 4)";
3 if(mapalloc_init()) {
4   printf("Error initializing MAPAlloc, did you load the module?\n");
5   return 1;
6 }
7 pid_t pid;
8 if((pid = fork()) != 0) {
9   mapalloc_constrain(pid, constraint);
10  execve([...])
11 }
```

Listing 1: Example usage of MAPAlloc for starting a new process that only receives physical pages mapping to a subset of cache sets.

4   Case Studies

In this section, we demonstrate different use cases for MAPAlloc. We demonstrate that MAPAlloc makes it easy to prototype page coloring, both for cache sets and DRAM rows. Additionally, we introduce layered page coloring for non-linear microarchitectural hash functions, showing that combining microarchitectural hash functions enables finer-grained coloring. We also show that MAPAlloc helps prototyping attacks by artificially limiting the physical address space, allowing the use of incomplete cache-slice functions.

4.1   Eviction Set Prevention

In this case study, we focus on the color attributes of the cache slice and the cache sets. With MAPAlloc, we evaluate the page coloring prototype against eviction-based cache side-channel attacks.

$$i(a) = a_6, \ a_7, \ a_8, \ a_9 \oplus a_{28} \oplus a_{29}, \ a_{10} \oplus a_{27} \oplus a_{30}, \ a_{11} \oplus a_{26} \oplus a_{31}$$
$$a_{12} \oplus a_{25} \oplus a_{32}, \ a_{13} \oplus a_{24} \oplus a_{33}, \ a_{14} \oplus a_{23} \oplus a_{34}, \ a_{15} \oplus a_{22} \oplus a_{35}, \ a_{16} \oplus a_{21} \oplus a_{36}$$
$$h(a) = a_{17}, \ a_{18}, \ a_{19}$$

Fig. 2. Set index function i and slice hash function h for EPYC 9124 (Zen 4). As both functions do not share bits, they can be trivially combined. The total number of combined colors is the product of the colors provided by set and slices.

The cache set index function is typically an identity function over parts of the physical address. For a cache with S sets, bits $a_6 \ldots_{6+\log_2(S)}$ of the physical address a determine the set index. While an attacker can control bits within a page, the bits $a_{12} \ldots_{6+\log_2(S)}$ can still contribute to page coloring. Coloring based on the cache set is trivial to implement using MAPAlloc, as the constraint is a simple bitmask on the physical address.

Layered Page Coloring. As regular page coloring is limited to 32 or 64 colors on current x86 CPUs, we introduce layered page coloring, combining multiple indexing functions to create more colors. One such additional indexing function is the cache-slice function that partitions the cache into multiple slices. However, the cache set and slice functions have overlapping bits, making this combination non-trivial. The increased number of colors depends on the number of overlapping bits used in both functions and how the attacker-controllable bits $a_{6\ldots11}$ affect the slice hash function. In such cases, a linear algebra-based approach based on the kernel of the hash function can be used [19]. We reimplement this approach, which allows efficient computations of memory partitions based on multiple *linear* hash functions. On AMD EPYC 9124, the cache-index function i and slice hash h do not interfere with each other, as shown in Fig. 2. Therefore, it is trivial to compute a good partition that splits both cache sets and slices into different colors.

As the case of non-linear function has yet to be solved generically, we model the problem using graphs to calculate the number of available colors. While our approach scales exponentially, the problem we solve is small enough to be practical. We refer to the combination of cache slice and cache set as *extended set*. We evaluate which extended set the attacker can reach for all physical addresses when controlling the page offset. The address and the reachable extended set are connected nodes in a graph. On the full graph, we calculate the number of components, i.e., parts of the graph that are not connected. This number is equivalent to the number of colors available for page coloring. Additionally, each component contains the extended set for a color that can be used in MAPAlloc. We calculate the number of available colors for an Intel Coffee Lake with 6 slices based on the non-linear slice function provided by Gerlach et al. [15]. This non-linear slice function doubles the number of usable colors to 64.

Table 1. The result of our coloring isolation. *Color Num* is the number of colors provided by using set partition or slice partition, which is not controllable by an attacker who can control bits 6–12. *EVC_Time* indicates how long an attacker needs to build eviction sets via the tool evsets [60]. ✗ indicates that eviction set building failed.

CPU	Coloring Element	Color Num	EVC_Time	After coloring
Intel Core i3-5010U	Cache Set	64	16s	✗
AMD EPYC 7252	Cache Set	64	30s	✗
AMD EPYC 9124	Cache Set	32	✗	✗
AMD EPYC 9124	Cache Slice	8	✗	✗

To evaluate the page coloring as a prototype mitigation, we use a global variable as the target. We allocate a 128MB memory buffer and select candidate addresses from this buffer to construct an eviction set, using the eviction-set generation from Vila et al. [60]. Afterward, we calculate the color of the target variable based on the cache sets or cache slices function. We assume an L3 Prime+Probe attacker can access all the remaining page colors except for the one the target variable has. Hence, we deploy this memory constraint to avoid allocating addresses with the same color and repeat the eviction set construction. We evaluate this policy on various machines, as shown in Table 1. The tool fails to find eviction sets on AMD EPYC 9124, where effective eviction construction is naturally complex, as the non-inclusive L3 cache is divided into 256 partitions (colors). All test machines run Ubuntu 22.04. The Intel machine uses the Linux kernel 5.15.0, and the AMD EPYC machines use the Linux kernel 6.8.0.

Under the mitigation policy of MAPAlloc, our results show that a Prime+Probe attacker cannot build eviction sets for the target set on any test machine. We validate the effectiveness of this mitigation by repeating the building process 100 times. None of these repetitions found an eviction set. On AMD EPYC 9124, each eviction address must have the same color as the target, chosen from 256 possible colors. As a proof-of-concept to demonstrate the isolation of coloring, we successfully build eviction sets by refining allocations to reside in the same slice and set as the victim variable.

4.2   Mitigating DRAMA and Rowhammer

In this section, we demonstrate the use of page coloring to prevent DRAMA and Rowhammer attacks. Our approach to preventing DRAMA uses reverse-engineered DRAM mapping functions to implement a page coloring scheme that avoids row conflicts between victim and attacker.

Layered page coloring can be employed, extending our approach to mitigate DRAMA or Rowhammer and cache-based side-channel attacks.

DRAMA. DRAMA attacks [45] exploit the conflict of the DRAM row buffer. The row buffer is shared within a DRAM bank; the bank to which a memory address belongs is determined by the DRAM mapping function. This function uses implementation-dependent physical address bits to map to a channel, DIMM, rank, and bank. We use open-source tooling [21] to reverse-engineer the mapping functions on Intel CPUs with different memory configurations. Similar to the cache-slice function, if bits within the range of a 4 KB page are used in the mapping function, MAPAlloc cannot rely on them to create coloring. Therefore, we only impose constraints on bits higher than 12.

Our goal in protecting against DRAMA attacks is to create two disjoint pools of addresses D and N. It must hold that for DRAM bank b

$$\forall a \in D, \text{bank}(a) = b \quad \text{and} \quad \forall a \in N, \text{bank}(a) \neq b$$

If these constraints are fulfilled, addresses in D and N map to different banks, thereby preventing row buffer conflicts. We can achieve the desired coloring by first allocating D, which is possible as we know the DRAM mapping function, and then blocking all addresses mapping to b for further allocations. In practice we choose $|N| = 4$ KB and $|D| = 1$ MB. To evaluate our coloring scheme, we test it on multiple different machines as listed in Table 2. Our results show that independent of the DRAM mapping function, we can prevent row buffer conflict for the address set N via coloring. Therefore, an attacker cannot mount DRAMA attacks on addresses in N, as they cannot access addresses aliasing to the same row buffer.

Table 2. The result of our DRAM banks coloring isolation. ✗ denotes that the attacker fails to create row buffer conflicts.

CPU	Banks Mapping	Color Num.	After Coloring
Intel Core i3-5010U	$a_{14} \oplus b_{17}, a_{15} \oplus b_{18}, a_{16} \oplus b_{19}$	8	✗
Intel Core i9-9980HK	$a_{14} \oplus b_{18}, a_{15} \oplus b_{19}, a_{16} \oplus b_{20}, a_{17} \oplus b_{21}$	16	✗

Rowhammer. Page coloring mitigations against Rowhammer have been proposed in previous work [32,37]. One mitigation idea implemented in [32] is to add *guard rows* next to rows one wants to protect against Rowhammer. More precisely, one computes the row index r_v of victim data via its physical address and then adds n guard rows at $r_v \pm n$. The number of guard rows depends on a DRAM-dependent parameter called the blast radius [33]. The blast radius is the maximal distance from a hammered row at which bit flips are reliably observed. MAPAlloc can easily add the guard rows by adding constraints that only allow sensitive rows to be n rows away from attacker-controlled rows. If implemented naively, this approach drastically reduces the available memory depending on the choice of n. However, the memory overhead can be reduced if the guard rows are filled with data that is not sensitive to bit flips. Furthermore, as MAPAlloc is implemented in the kernel, we can transparently apply error correction to the guard rows. Errors in the guard rows can be detected and corrected by the error correction code, while sensitive rows are protected by the fact that they are out of the blast radius. Similar layered mitigations using error correction [10,26] have already been explored in related work.

A related approach [37] proposes subarray groups to partition the DRAM more efficiently. However, we do not implement this approach because MAPAlloc currently does not support hypervisor-based applications, and subarray groups only make sense in a hypervisor context.

4.3 Prime+Probe with Incomplete Non-linear Slice Function

While linear cache-slice functions have been thoroughly explored and reverse-engineered [20,24,36,39,53], non-linear cache-slice functions remain limited in availability. Although they have been reverse-engineered for some CPUs [15, 22,40,66], the results are often incomplete because the functions only work for systems with limited memory. For example, the cache-slice functions for Intel

Coffee Lake and Alder Lake as reported by Gerlach et al. [15] only work for systems where the highest physical address does not have any bits above bit 31 set. Thus, this limits the applicability to systems with less than 8 GB of memory.

For prototyping attacks, MAPAlloc can be used to artificially limit the available physical memory of the system. Thus, MAPAlloc ensures that the incomplete slice function is still valid for all allocated pages. We evaluate this by mounting a Prime+Probe attack on an Intel Xeon E-2176M (Coffee Lake) with 6 cores with the slice function of Gerlach et al. [15]. We rely on MAPAlloc to restrict the physical memory for attacker and victim to 2 GB to ensure we can always apply the slice function.

We target the AES T-Table implementation in OpenSSL. To align with prior research, we use OpenSSL 1.0.1e [14,17,34,47]. We base our implementation on the one from Gruss et al. [17] but replace the cache-slice function and apply MAPAlloc to restrict the memory. With these adaptions, we recover, on average, more than 97% of the key correctly. This result shows that MAPAlloc helps to test microarchitectural attacks that would otherwise only work if the system is physically changed, i.e., if DRAM is physically removed.

5 Evaluation

In this section, we evaluate the functionality and performance of MAPAlloc across architectures.

5.1 Functionality

We test the functionality of MAPAlloc on various microarchitectures among x86, ARM, and RISC-V architectures. Table 3 lists the machines used for testing the proof-of-concept constraint outlined in Listing 1. The slice functions and DRAM addressing functions used in case studies are determined via open-source reverse-engineering tools [15,45].

In our tests, we first deploy the memory policy to constrain the page color for upcoming allocations. Then, we allocate memory via mmap, global variables, and heap variables to verify if their physical addresses fulfill the specified constraints. Our results demonstrate that MAPAlloc can manage memory allocation and maintain page coloring on all tested machines.

5.2 Performance

In the following section, we evaluate the performance of MAPAlloc.

Pre-allocation. When initializing, MAPAlloc pre-allocates pages in kernel mode to avoid expensive search operations at runtime. The cost of this operation primarily depends on the number of pages we consider and check against our constraints. In our implementation, this number is set to 87.5% of the system's

Table 3. CPUs tested for MAPAlloc.

CPU	μarch	Release	Last-level Cache	Known Functions	
				Slice	DRAM
Intel Core i5-2520M	Sandy Bridge	2011	Inclusive	●	●
Intel Core i3-5010U	Broadwell	2015	Inclusive	●	●
Intel Xeon E3-1505M	Skylake	2015	Inclusive	●	●
Intel Xeon E-2176M	Coffee Lake	2018	Inclusive	●	●
Intel i3-8130U	Kaby Lake R	2018	Inclusive	●	●
Intel Celeron N4500	Jasper Lake	2021	Non-inclusive	●	●
AMD EPYC 7252	Rome	2019	Non-inclusive	●	○
AMD EPYC 9124	Genoa	2023	Non-inclusive	◑	○
Broadcom BCM2711	Cortex-A72	2019	Inclusive	N/A	○
Allwinner D1	RISC-V C906	2021	N/A	N/A	○

The icons represent the knowledge of a function. ● : known or found by open-sourced tools [15, 45], ○ : unknown, ◑ : discovered in this work

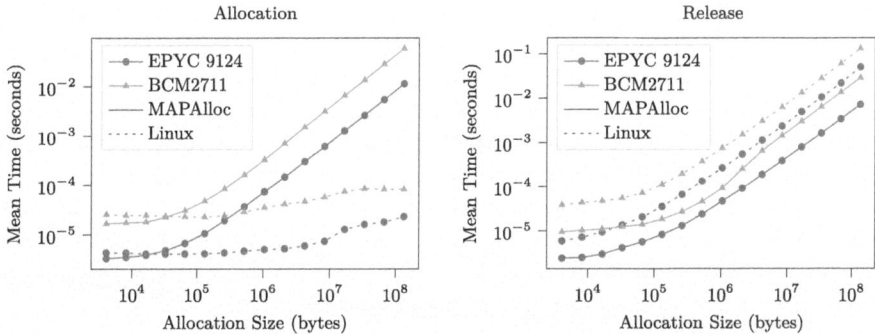

Fig. 3. Mean time to allocate and release memory blocks of different sizes with MAPAlloc and Linux ($n = 4096$ per sample, Linux v6.8.0). Less is better.

unused memory. For a freshly booted system, the pre-allocation time depends mainly on the amount of available DRAM. On an AMD EPYC 9124 with 16 GB of DRAM, pre-allocation takes roughly 4.5 s. On the Intel Core i3-5010U with only 8 GB of DRAM, it takes about 1.8 s. Since pre-allocating pages with MAPAlloc is a one-time initialization effort, it does not directly influence the runtime of user applications. We confirm this by running the popular 7-zip LZMA data compression benchmark [42] alongside another coloring process that pre-allocates pages. After 10 iterations, the average impact remains at 0.4%.

Memory Allocation. To assess the performance of our memory allocator, we measure the mean time required for allocating and releasing differently-sized mappings. The results of this evaluation for the AMD EPYC 9124 (x86_64) and the Broadcom BCM2711 (ARMv8) are shown in Fig. 3. For small allocations of 8 or fewer pages, MAPAlloc's page allocator performs slightly better than Linux. This is likely due to the low complexity of our implementation as compared to Linux's allocation routines. However, our prototype implementation does not implement bulk allocations but instead remaps every page individually. Furthermore, MAPAlloc zeroes the pages during initialization instead of during cleanup. Hence, we observe significantly longer allocation times than with Linux for larger memory blocks, with the time growing roughly proportionally to the allocation size. On the other hand, releasing pages with MAPAlloc is consistently faster than with Linux, up to 7 times faster. Note that when freeing pages, MAPAlloc only clears the user page table entries and adds the pages back to the pool. In contrast, Linux might run more expensive management tasks, such as attempting to merge smaller page buddies into larger ones.

Microarchitectural Effects. One of the applications we propose for MAPAlloc is studying the runtime effects of page coloring schemes on software. To demonstrate MAPAlloc's effectiveness for this purpose, we investigate the 7-zip compression benchmark while preventing access to parts of the cache. See Fig. 4 for the results of this evaluation. As expected, we observe a decline in throughput on the AMD EPYC 9124 and Intel Core i3-5010U, as we reduce the number of available cache sets. This effect is most pronounced in the compression benchmark. For example, restricting the available cache sets to only 512 out of 16384 on the AMD EPYC 9124 degrades the compression throughput to 62.2% of the throughput with the full cache. While we also observe performance degradation in the decompression benchmark, this is less significant, with the throughput being reduced to only 97.5%percent. Hence, we conclude that LZMA compression on x86_64 significantly benefits from a large cache, whereas this is not necessarily the case with LZMA decompression.

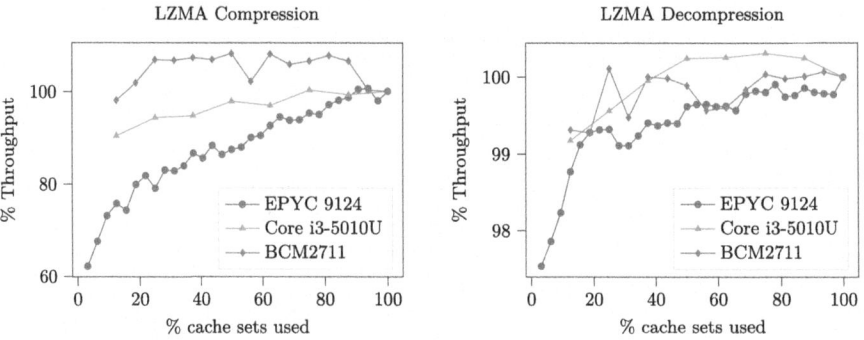

Fig. 4. Mean LZMA de/compression throughput with limited cache access, normalized by baseline performance (7-zip v23.01, 1 thread).

⟨*expression*⟩ ::= ⟨*term*⟩
  | ⟨*expression*⟩ || ⟨*term*⟩

⟨*term*⟩ ::= ⟨*factor*⟩
  | ⟨*term*⟩ && ⟨*factor*⟩

⟨*factor*⟩ ::= ⟨*bitwise-term*⟩
  | ⟨*factor*⟩ | ⟨*bitwise-term*⟩

⟨*bitwise-term*⟩ ::= ⟨*bitwise-factor*⟩
  | ⟨*bitwise-term*⟩ ^ ⟨*bitwise-factor*⟩

⟨*bitwise-factor*⟩ ::= ⟨*bitwise-shift*⟩
  | ⟨*bitwise-factor*⟩ & ⟨*bitwise-shift*⟩

⟨*bitwise-shift*⟩ ::= ⟨*arithmetic*⟩
  | ⟨*bitwise-shift*⟩ « ⟨*arithmetic*⟩
  | ⟨*bitwise-shift*⟩ » ⟨*arithmetic*⟩

⟨*arithmetic*⟩ ::= ⟨*term1*⟩
  | ⟨*arithmetic*⟩ + ⟨*term1*⟩
  | ⟨*arithmetic*⟩ - ⟨*term1*⟩

⟨*term1*⟩ ::= ⟨*factor1*⟩
  | ⟨*term1*⟩ * ⟨*factor1*⟩
  | ⟨*term1*⟩ / ⟨*factor1*⟩
  | ⟨*term1*⟩ % ⟨*factor1*⟩

⟨*factor1*⟩ ::= ⟨*primary*⟩
  | ~ ⟨*factor1*⟩

⟨*primary*⟩ ::= ⟨*literal*⟩
  | (⟨*expression*⟩)

⟨*relational-expression*⟩ ::= ⟨*expression*⟩ == ⟨*expression*⟩
  | ⟨*expression*⟩ != ⟨*expression*⟩
  | ⟨*expression*⟩ < ⟨*expression*⟩
  | ⟨*expression*⟩ > ⟨*expression*⟩
  | ⟨*expression*⟩ <= ⟨*expression*⟩
  | ⟨*expression*⟩ >= ⟨*expression*⟩

⟨*literal*⟩ ::= ⟨*number*⟩
  | ⟨*boolean*⟩

⟨*number*⟩ ::= ⟨*digit*⟩+

⟨*boolean*⟩ ::= true
  | false

⟨*digit*⟩ ::= 0-9

Fig. 5. Grammar for the memory constraint DSL.

6   Discussion

In this section, we discuss the applicability to virtual machines, hardware mechanisms with similar capabilities, implementation limitations, and related work.

6.1   Virtual Machines

Our implementation of MAPAlloc is limited to non-virtualized environments. MAPAlloc neither supports running in the hypervisor to constrain virtual machines nor inside virtual machines. Conceptually, nothing hinders implementing MAPAlloc in a hypervisor to apply constraints to entire virtual machines. However, similar approaches for virtual machines have already been implemented by prior work [61]. Therefore, we leave integrating support for arbitrary memory constraints into such hypervisors for future work.

Running MAPAlloc inside a virtual machine is more challenging. While MAPAlloc does run inside a virtual machine, it is functionally inoperable. The reason is that physical addresses inside a virtual machine, so-called guest physical addresses, are not actual physical addresses. Guest physical addresses are virtual addresses on the host translated to real physical addresses using extended page tables. Thus, the virtual machine does not know physical addresses. Constraints enforced by MAPAlloc in virtual machines degenerate to constraints on virtual addresses and can not be used to prototype defenses or attacks. This limitation can be overcome by adding a hypervisor part to MAPAlloc that could provide the actual physical address to MAPAlloc running inside the virtual machine.

6.2   Hardware Mitigations

Numerous hardware-based isolation techniques offer an orthogonal alternative to software approaches like MAPAlloc. For instance, Intel's Cache Allocation Technology (CAT) partitions the cache structure among different threads, providing isolation. Intel CAT was considered as a potential mitigation against cache side-channel attacks [35]. However, recent work [43,65] has demonstrated its inefficiency. On AMD EPYC server CPUs, the new cache range reservation feature [1] enables the hypervisor to lock specific L3 cache ways for a designated system memory range. Although this feature was not originally intended for security purposes, it effectively isolates the specified system memory range from the rest of the memory in the L3 cache. Configuring this range involves two model-specific registers shared across an entire core complex, which limits its application to multiple processes. Ultimately, neither Intel nor AMD provides a hardware feature that mitigates DRAM attacks like DRAMA.

6.3   Related Work

Related work on partitioning memory, also called cache coloring, has been explored for a wide range of applications. Initially, cache coloring aimed to improve performance by optimizing cache utilization [7,28,58]. These approaches ensure a good cache hit rate for frequently used memory pages. Security applications of memory coloring have been explored by Hofmann et al. [19]. These applications mainly include mitigating cache-based side channel attacks [54,67]. Similar to cache coloring, separating DRAM banks into multiple domains has been proposed as a Rowhammer mitigation in previous work [32,37].

6.4 Implementation Limitations

In the following, we discuss limitations specific to our implementation of MAPAlloc. We note that overcoming these limitations is merely an engineering effort.

Corner Cases. Our implementation of MAPAlloc considers the common cases of memory allocation in the Linux kernel. However, due to the complexity of the Linux kernel, we are likely missing corner cases that are not handled. As MAPAlloc is not designed as a security mechanism but rather as a rapid prototyping framework, we consider such corner cases as unproblematic.

Exempted Pages. While MAPAlloc can enforce constraints on most pages, there are some pages where constraints cannot be enforced. We identify two categories of such pages. First, kernel pages mapped to the user space, such as vDSO [13]. As we do not enforce constraints on kernel pages, these pages are entirely out of scope for MAPAlloc. Second, implicit or explicit shared memory, if not all applications sharing the memory have compatible constraints. Such shared memory can also happen without a developer knowing, e.g., if the operating system uses page deduplication [57].

7   Conclusion

We introduced MAPAlloc, a versatile, cross-architecture framework for efficiently prototyping defenses and attacks on Linux that require control over physical memory allocations. Using a simple domain-specific language, MAPAlloc enables precise control over physical memory allocation across x86, ARMv8, and RISC-V architectures. MAPAlloc enables rapid implementation and evaluation of mitigations, such as page coloring. Additionally, MAPAlloc allows for quickly extending such techniques, as we show with layered page coloring, which significantly expands the number of cache colors from 32 to 256 on modern CPUs. We achieve this increase in isolation domains by reverse-engineering AMD Zen 4's cache indexing function and combining it with our reverse-engineered cache-slice function. By simplifying and accelerating the prototyping process, MAPAlloc bridges the gap between theoretical concepts and practical implementations, helping to improve the robustness of system security.

Acknowledgments. We thank our shepherd and the anonymous reviewers for their valuable feedback and suggestions that helped improve the paper.

References

1. AMD64 Architecture Programmer's Manual (2024)
2. Aweke, Z.B., Yitbarek, S.F., Qiao, R., Das, R., Hicks, M., Oren, Y., Austin, T.: ANVIL: software-based protection against next-generation Rowhammer attacks. ACM SIGPLAN Notices (2016)
3. Bennett, T., Saroiu, S., Wolman, A., Cojocar, L.: Panopticon: a complete in-dram rowhammer mitigation. In: Workshop on DRAM Security (DRAMSec) (2021)
4. Brasser, F., Davi, L., Gens, D., Liebchen, C., Sadeghi, A.R.: CAn't touch this: software-only mitigation against Rowhammer attacks targeting kernel memory. In: USENIX Security Symposium (2017)
5. Bray, B.K., Lunch, W.L., Flynn, M.J.: Page allocation to reduce access time of physical caches (1990). http://i.stanford.edu/pub/cstr/reports/csl/tr/90/454/CSL-TR-90-454.pdf
6. Brumley, D., Boneh, D.: Remote timing attacks are practical. In: USENIX Security Symposium (2003)
7. Bugnion, E., Anderson, J.M., Mowry, T.C., Rosenblum, M., Lam, M.S.: Compiler-directed page coloring for multiprocessors. ACM SIGPLAN Notices (1996)
8. Corbet, J.: Some kernel memory-allocation improvements (2015). https://lwn.net/Articles/658081/
9. Corbet, J.: Remote per-CPU page list draining (2022). https://lwn.net/Articles/884448/
10. Dio, A.D., Koning, K., Bos, H., Giuffrida, C.: Copy-on-flip: hardening ECC memory against rowhammer attacks. In: NDSS (2023)
11. Easdon, C., Schwarz, M., Schwarzl, M., Gruss, D.: Rapid prototyping for microarchitectural attacks. In: USENIX Security (2022)
12. Frigo, P., et al.: TRRespass: exploiting the many sides of target row refresh. In: S&P (2020)
13. Frysinger, M.: vdso(7) — linux manual page (2024)
14. Ge, Q., Yarom, Y., Cock, D., Heiser, G.: A survey of microarchitectural timing attacks and countermeasures on contemporary hardware. J. Cryptogr. Eng. (2016)
15. Gerlach, L., Schwarz, S., Faroß, N., Schwarz, M.: Efficient and generic microarchitectural hash-function recovery. In: S&P (2024)
16. Gerlach, L., Thomas, F., Pietsch, R., Schwarz, M.: A large-scale rowhammer reproduction study using the blacksmith fuzzer. In: ESORICS (2023)
17. Gruss, D., Maurice, C., Wagner, K., Mangard, S.: Flush+Flush: a fast and stealthy cache attack. In: DIMVA (2016)
18. Gruss, D., Spreitzer, R., Mangard, S.: Cache template attacks: automating attacks on inclusive last-level caches. In: USENIX Security Symposium (2015)
19. Hofmann, J., Fournet, C., Köpf, B., Volos, S.: Gaussian elimination of side-channels: linear algebra for memory coloring. In: ACM CCS (2024)
20. Hund, R., Willems, C., Holz, T.: Practical timing side channel attacks against kernel space ASLR. In: S&P (2013)
21. IAIK: DRAMA Reverse-Engineering Tool and Side-Channel Tools (2016). https://github.com/IAIK/drama
22. Inci, M.S., Gulmezoglu, B., Irazoqui, G., Eisenbarth, T., Sunar, B.: Seriously, get off my cloud! Cross-VM RSA Key Recovery in a Public Cloud. Cryptology ePrint Archive, Report 2015/898 (2015)
23. Intel Corporation: Guidelines for Mitigating Timing Side Channels Against Cryptographic Implementations (2020). https://www.intel.com/content/www/

us/en/developer/articles/technical/software-security-guidance/secure-coding/
mitigate-timing-side-channel-crypto-implementation.html

24. Irazoqui, G., Eisenbarth, T., Sunar, B.: Systematic reverse engineering of cache slice selection in intel processors. In: Euromicro Conference on Digital System Design (2015)

25. Jin, J., McMurtry, E., Rubinstein, B.I.P., Ohrimenko, O.: Are we there yet? Timing and floating-point attacks on differential privacy systems. In: S&P (2022)

26. Juffinger, J., Lamster, L., Kogler, A., Eichlseder, M., Lipp, M., Gruss, D.: CSI: rowhammer-cryptographic security and integrity against rowhammer. In: IEEE S&P (2022)

27. Juffinger, J., Lamster, L., Kogler, A., Eichlseder, M., Lipp, M., Gruss, D.: CSI: rowhammer-cryptographic security and integrity against rowhammer. In: IEEE S&P (2023)

28. Kessler, R.E., Hill, M.D.: Page placement algorithms for large real-indexed caches. TOCS (1992)

29. Kim, T., Peinado, M., Mainar-Ruiz, G.: StealthMem: system-level protection against cache-based side channel attacks in the cloud. In: USENIX Security Symposium (2012)

30. Kim, Y., et al.: Flipping bits in memory without accessing them: an experimental study of DRAM disturbance errors. In: ISCA (2014)

31. Kocher, P.C.: Timing attacks on implementations of Diffe-Hellman, RSA, DSS, and other systems. In: CRYPTO (1996)

32. Konoth, R.K., et al.: ZebRAM: comprehensive and compatible software protection against rowhammer attacks. In: OSDI (2018)

33. Lang, Z., Jattke, P., Marazzi, M., Razavi, K.: Blaster: characterizing the blast radius of rowhammer. In: Workshop on DRAM Security (DRAMSec) (2023)

34. Lipp, M., Gruss, D., Spreitzer, R., Maurice, C., Mangard, S.: ARMageddon: cache attacks on mobile devices. In: USENIX Security Symposium (2016)

35. Liu, F., et al.: Catalyst: defeating last-level cache side channel attacks in cloud computing. In: HPCA (2016)

36. Liu, F., Yarom, Y., Ge, Q., Heiser, G., Lee, R.B.: Last-level cache side-channel attacks are practical. In: S&P (2015)

37. Loughlin, K., Rosenblum, J., Saroiu, S., Wolman, A., Skarlatos, D., Kasikci, B.: Siloz: leveraging dram isolation domains to prevent inter-vm rowhammer. In: SOSP (2023)

38. Marazzi, M., Solt, F., Jattke, P., Takashi, K., Razavi, K.: Rega: scalable rowhammer mitigation with refresh-generating activations. In: S&P (2023)

39. Maurice, C., Le Scouarnec, N., Neumann, C., Heen, O., Francillon, A.: Reverse engineering intel complex addressing using performance counters. In: RAID (2015)

40. McCalpin, J.D.: Mapping addresses to l3/cha slices in intel processors. Technical report (2021)

41. OpenSSL: OpenSSL: The Open Source toolkit for SSL/TLS (2019). http://www.openssl.org

42. Pavlov, I.: 7-zip (2023). https://7-zip.org/. v23.01

43. Wieczorkiewicz, P., Branco, R., Lee, B.: On the Effectiveness of Intel's CAT as a Side-Channel Mitigation Technology (2024). https://langsechq.gitlab.io/spw24/papers/LangSec2024-Branco-CAT-paper.pdf

44. Percival, C.: Cache missing for fun and profit. In: BSDCan (2005)

45. Pessl, P., Gruss, D., Maurice, C., Schwarz, M., Mangard, S.: DRAMA: exploiting DRAM addressing for cross-CPU attacks. In: USENIX Security Symposium (2016)

46. Pornin, T.: BearSSL: A smaller SSL/TLS library (2022). https://www.bearssl.org
47. Purnal, A., Turan, F., Verbauwhede, I.: Prime+Scope: overcoming the observer effect for high-precision cache contention attacks. In: CCS (2021)
48. Qiao, R., Seaborn, M.: A new approach for rowhammer attacks. In: International Symposium on Hardware Oriented Security and Trust (2016)
49. Qureshi, M.K.: CEASER: mitigating conflict-based cache attacks via encrypted-address and remapping. In: IEEE MICRO (2018)
50. Razavi, K., Gras, B., Bosman, E., Preneel, B., Giuffrida, C., Bos, H.: Flip feng shui: hammering a needle in the software stack. In: USENIX Security Symposium (2016)
51. Saileshwar, G., Qureshi, M.: MIRAGE: mitigating conflict-based cache attacks with a practical fully-associative design. In: USENIX Security Symposium (2021)
52. Schwarz, M., et al.: KeyDrown: eliminating software-based keystroke timing side-channel attacks. In: NDSS (2018)
53. Seaborn, M.: L3 cache mapping on Sandy Bridge CPUs (2015). http://lackingrhoticity.blogspot.com/2015/04/l3-cache-mapping-on-sandy-bridge-cpus.html. Accessed 26 June 2015
54. Shi, J., Song, X., Chen, H., Zang, B.: Limiting cache-based side-channel in multi-tenant cloud using dynamic page coloring. In: DSN-W (2011)
55. Shusterman, A., et al.: Robust website fingerprinting through the cache occupancy channel. In: USENIX Security Symposium (2019)
56. Sun, K., Branco, R., Hu, K.: A New Memory Type Against Speculative Side Channel Attacks (2019)
57. Suzaki, K., Iijima, K., Yagi, T., Artho, C.: Memory deduplication as a threat to the guest OS. In: EuroSys (2011)
58. Taylor, G., Davies, P., Farmwald, M.: The TLB slice—a low-cost high-speed address translation mechanism. In: ISCA (1990)
59. The Mbed TLS Contributors: Security (2024). https://mbed-tls.readthedocs.io/en/latest/project/long-term-plans/#security
60. Vila, P., Köpf, B., Morales, J.: Theory and practice of finding eviction sets. In: S&P (2019)
61. Volos, S., Fournet, C., Hofmann, J., Köpf, B., Oleksenko, O.: Principled microarchitectural isolation on cloud CPUs. In: ACM CCS (2024)
62. Weber, D., Thomas, F., Gerlach, L., Zhang, R., Schwarz, M.: Indirect meltdown: building novel side-channel attacks from transient execution attacks. In: ESORICS (2023)
63. Werner, M., Unterluggauer, T., Giner, L., Schwarz, M., Gruss, D., Mangard, S.: ScatterCache: thwarting cache attacks via cache set randomization. In: USENIX Security Symposium (2019)
64. wolfSSL: wolfSSL: Embedded TLS Library (2023). https://www.wolfssl.com/
65. Yan, M., Sprabery, R., Gopireddy, B., Fletcher, C., Campbell, R., Torrellas, J.: Attack directories, not caches: side channel attacks in a non-inclusive world. In: S&P (2019)
66. Yarom, Y., Ge, Q., Liu, F., Lee, R.B., Heiser, G.: Mapping the Intel Last-Level Cache. Cryptology ePrint Archive, Report 2015/905 (2015)
67. Ye, Y., West, R., Cheng, Z., Li, Y.: Coloris: a dynamic cache partitioning system using page coloring. In: PACT (2014)
68. Zhang, R., Kim, T., Weber, D., Schwarz, M.: (M)WAIT for it: bridging the gap between microarchitectural and architectural side channels. In: USENIX Security (2023)

Linux Hurt Itself in Its Confusion! Exploiting Out-of-Memory Killer for Confusion Attacks via Heuristic Manipulation

Lorenzo Bossi, Daniele Mammone, Michele Carminati, Stefano Zanero, and Stefano Longari[✉]

Dipartimento di Elettronica, Informazione e Bioingegneria, Politecnico di Milano, Milan, Italy
lorenzo1.bossi@mail.polimi.it, {daniele.mammone,michele.carminati, stefano.zanero,stefano.longari}@polimi.it

Abstract. The Linux kernel's Out-of-Memory (OOM) killer ensures system stability by terminating processes when memory is exhausted, but its heuristic-based design was not built for adversarial contexts. This paper introduces OOM Confusion Attacks, a novel class of Denial of Service (DoS) attacks that exploit the OOM killer to execute privileged process termination, targeting critical services rather than attacker processes. By orchestrating memory exhaustion through numerous unprivileged processes, these attacks may kill target applications, block service recovery, and destabilize systems. We demonstrate the feasibility of OOM Confusion Attacks on default Linux configurations commonly used by cloud providers, formulate and quantify the resource constraints for success, and evaluate application exposure to OOM Confusion Attacks. Additionally, we identify race conditions that can be exploited to block the recovery of privileged services. To mitigate these threats, we propose strategies to increase the resilience of critical applications.

Keywords: Denial of Service attacks · Out-of-Memory killer · Memory Exhaustion Attacks

1 Introduction

The Linux kernel's Out-of-Memory (OOM) killer is a critical mechanism designed to maintain system stability when memory resources are exhausted. When the system runs out of memory, and no further allocations can be fulfilled, the OOM killer selects and terminates a process to free memory and allow the system to continue operating [25]. The selection happens through a heuristic scoring system, where each process is assigned a score based on various factors such as memory consumption and process priority. The necessity to design such a measure arose due to the possibility - and common practice - to employ memory

---

L. Bossi and D. Mammone—Contributed equally to this work.

© The Author(s), under exclusive license to Springer Nature Switzerland AG 2025
M. Egele et al. (Eds.): DIMVA 2025, LNCS 15748, pp. 195–215, 2025.
https://doi.org/10.1007/978-3-031-97623-0_12

overcommitment, where the kernel allows applications to allocate more memory than is physically available, based on the expectation that not all allocated memory will be used simultaneously. This, of course, optimizes memory utilization and improves performance under normal operating conditions but introduces the risk that the system may run out of memory if applications simultaneously attempt to use more memory than the system can provide, leading to instability, unresponsiveness, or kernel panic. The OOM killer was conceived as a last-resort measure for mitigating such issues and was not designed with adversarial contexts in mind, nor was it intended as a security feature. Even so, it would still be invoked in the event of a memory exhaustion attack. Memory exhaustion attacks are a subset of resource exhaustion attacks [15,19], which aim at overwhelming a system's resources and conducting it in a non-operational state. Classical memory exhaustion attacks indiscriminately deplete a system's memory. The OOM killer, in case of a standard memory exhaustion attack, would mostly act against the attacker in an attempt to restore the system's stability. However, being designed without an adversarial context in mind, a knowledgeable attacker may instead exploit the OOM killer authorization level for their own objectives [9].

This paper introduces a novel class of Denial of Service (DoS) attacks, defined as OOM Confusion Attacks. An OOM Confusion Attack overloads the system's RAM by executing many unprivileged processes, each allocating a small amount of memory. This method exploits the OOM killer's heuristic rules, putting the system in a condition where the OOM killer acts as a confused deputy and kills the target victim processes instead of those of the attacker, potentially disrupting critical services and causing a system-wide DoS. We study the feasibility of OOM Confusion Attacks, quantifying the requirements and exploring the thresholds in terms of memory consumption, process creation, and resource exhaustion necessary for the attacker to accomplish its goals. Furthermore, we examine the potential system-level and operational impacts of these attacks. We test the effectiveness of these attacks on the default Linux configuration provided by cloud providers, demonstrating that our attack can induce the system in a state where it is impossible to restart killed services for several hours. Furthermore, we demonstrate and evaluate the susceptibility of common applications as targets for OOM Confusion Attacks. Additionally, under certain conditions, the attack can exploit a race condition to block the restart of a privileged previously terminated service. Beyond understanding the attack, this work also proposes strategies to harden systems against such exploitation. By studying the OOM killer's behavior under adversarial conditions, we aim to identify ways to hinder successful attacks, making them less feasible.

Our contributions can be summarized as follows:

- We define a novel class of Denial-of-Service attacks, termed OOM Confusion Attacks, which aim to exploit the OOM killer present in most Linux distributions as a confused deputy to terminate the targets of the attacker.
- We formalize the requirements to successfully implement an OOM Confusion Attack, demonstrating its feasibility in the latest Linux distributions.

- We evaluate the susceptibility of standard applications and Linux configurations provided by many hosting services against OOM Confusion Attacks.

2 Linux Memory Management Primer

Linux employs a sophisticated memory management framework to handle the allocation and deallocation of memory for processes [24]. This framework includes physical memory (RAM), virtual memory, and swap space to maximize system performance. The key mechanisms used by the Linux kernel to manage memory include caching, where frequently accessed data is cached in memory for faster access, paging, where memory - divided into fixed-size pages is moved between RAM and disk (swap space) as needed, and reclamation, where the kernel attempts to reclaim memory by identifying and freeing inactive or unnecessary pages. Moreover, the Linux kernel implements the **Memory Overcommitment** mechanism. It allows processes to successfully request more memory than the physically available [25]. The requested memory is then mapped to physical memory only when it is actually used, avoiding under-utilization of physical RAM, as most processes do not use all the memory they request at once. However, it might happen that all processes actually use the memory they requested simultaneously, exhausting the physical space on memory. In this case, the kernel cannot free memory through standard methods and has to choose between being deadlocked on memory or killing a process to free some memory. When a system is in this state, we say it has arisen a **OOM Condition**. Since user-space processes are allowed to allocate more space than the physical memory available, critical processes like kernel threads might end up without physical memory available, possibly leading to deadlocks. By killing a user-space process when memory usage reaches a threshold, the kernel can ensure itself and the most critical processes have enough space to continue running.

Note that in critical systems which require strong memory guarantees, memory overcommitment is usually disabled. By doing so, every process is guaranteed to be able to use the memory it has allocated without waiting for the kernel to free some space for it. In this case, if the available memory is exhausted, the following requests for memory will fail. However, in most Linux distributions for desktop and server environments, overcommitment is enabled by default, as it is considered a more flexible solution [25]. In both cases, a careful evaluation of memory requirements should be made to ensure the system is suitable for the expected workload. A trade-off must be made between memory under-utilization and the risk of degraded performance and process termination.

2.1 The Linux Out-of-Memory Killer

As introduced before, when an OOM condition arises, the kernel fails to reclaim memory with Page Reclaiming and swapping mechanisms. In this state, the kernel pauses further memory allocation requests and invokes the OOM Killer.

This mechanism is designed to exit from OOM conditions and restore the system to an operational state. When invoked, the OOM selects a candidate process for termination. This action is necessary to prevent the entire system from freezing due to unfulfilled memory requests and to prioritize the survival of critical system processes and the overall system's health.

Victim Selection and Badness Score. A key challenge for the OOM Killer is determining which process to terminate. To do so, the kernel assigns a *badness* score to each process based on various factors. The process with the highest score is then selected for termination[1].

The main idea behind the *badness* score is to make a balanced measure of process memory usage reduced by estimating the process relevance. Note that the selection routine involves evaluating each process individually. The termination occurs only after the evaluation is complete, and the procedure is not atomic, meaning that by the time the selected process is killed, it might no longer have the highest score. However, hanging all processes for the whole routine would be too inefficient. The formula for calculating the *badness* score has evolved over various Linux versions, becoming progressively simpler and more predictable. Initially, the *badness* score considered numerous factors, including the system's memory capacity, memory utilization, whether the process was running as root, the number of child processes, the process runtime, whether it was making direct hardware access, the process nice value (a priority modifier for CPU time allocation), and the so-called OOM Score Adjustment (*oom_score_adj*), a variable enabling system administrators to influence the killing process. However, such a complicated formula made tuning the OOM score adjustment value very difficult and often led to unpredictable kills. Moreover, these calculations could be computationally expensive, especially on systems with many running processes. To cope with these problems, the *badness* formula has been gradually simplified until in the kernel update v4.17-rc1 (April 2018), it got into the current form [36]:

$$B_{proc} = pages_{proc} + \frac{oom\_score\_adj}{1000} \cdot pages_{total} \qquad (1)$$

where $pages_{proc}$ are the memory pages currently used by the process, and $pages_{total}$ are the total memory RAM and swap pages in the system.

This puts, by default, every process on the same level of relevance and expects the system administrator to hint at the relevance and the expected memory usage of each process by assigning the OOM score adjustment. The system administrator now has better control of the OOM killer behavior, as the formula is much easier to understand. Furthermore, a simplified *badness* reduces the time needed to resolve OOM situations, improving system responsiveness under heavy memory pressure. The drawback is that the OOM score adjustment assignment is now fundamental to ensure correct OOM management. Without a correctly

---

[1] Certain processes, such as kernel threads and the init system - responsible for initializing the system, starting essential services, and managing processes - are excluded by default to preserve system stability, as terminating these could result in a system-wide failure state, needing a manual system reboot.

set adjustment, the kernel now, by default, treats every process the same way, creating an opportunity for attackers to manipulate OOM Killer behavior.

Score Adjustment. The system administrator can express its preference on which process to kill by assigning an OOM score adjustment value to each process. The OOM score adjustment is a value in $[-1000, +1000]$, which should represent a proportion of the system memory discounted or added to the process memory usage. That means a process with a big adjustment is likely to be killed, while one with a low adjustment is unlikely to be killed. If the adjustment is set to -1000, the process is considered unkillable and excluded from the victim selection. Only the root user can assign negative values or decrease the current value for a process. When a new process is created, the score adjustment is inherited by default from the parent process.

3   Related Works

While there has been significant research on the performance of the Linux Out-of-Memory killer [6,14,29], focusing on its decision-making algorithms and efficiency in reclaiming resources, and other works have studied methodologies and mitigation of memory exhaustion attacks [12,20,26,33], there is limited analysis of how the OOM killer itself can be abused as part of a memory exhaustion attack strategy. For example, since the invocation of the OOM killer can be slow due to excessive swapping [23] and reclaiming [28], there are many implementations of user-space daemons trying to predict the OOM condition and take preventive actions [1,14]. One of the most popular examples is Facebook's implementation of a user-space OOM daemon [14], showing that the kernel OOM Killer can be inefficient and memory management can be offloaded to userland, offering finer control over process termination and mitigation strategies. The abuse of the OOM killer to disrupt system availability is not a completely novel concept. CVE-2004-0807 [11] and CVE-2020-10781 [10] allowed malicious users to exhaust system memory by opening multiple connections to a forking server, leading to the invocation of the OOM killer when memory is depleted. CVE-2019-14891 [8] enabled attackers to exploit concurrent OOM killer invocations to terminate Kubernetes' container management process and conduct network escape attacks. CVE-2020-10781 [10] also allowed an attacker to trigger the OOM killer by exhausting memory through reading ZRAM-related files. CVE-2014-7300 [7] permitted a system user to kill GNOME Shell by abusing PrtSc requests. Although these CVEs involve the abuse of the OOM killer, they all focus on exploiting specific vulnerabilities in particular servers providing distinct services. In contrast, our work conducts a generalized study of the internal mechanisms of the OOM killer and demonstrates how it can be abused to terminate a specific process on the target system. Our approach is independent of any specific application used to trigger its functionality. Furthermore, we propose a series of mitigations for this class of attacks that apply system-wide—providing protection for any remote server configuration rather than only addressing specific vulnerabilities in individual applications.

4   Motivation and Attacker Model

The Linux kernel is one of the most impactful and widely used software components in modern computing. As the core of Linux-based operating systems, it powers a diverse range of devices and systems, from smartphones and desktops [4] to supercomputers and cloud infrastructures [34].

It is also one of the most studied pieces of software in security, both due to the impact of potential vulnerabilities and to its open-source nature. In fact, the concept of Denial of Service (DoS) via memory exhaustion attacks is a well-established security topic. However, to the best of the authors' knowledge, no study has been conducted on the adversarial exploitation of the OOM killer as a confused deputy, either to target a specific process to kill or execute a hard-to-recover DoS attack on a target machine.

Even if exhausting memory can represent a significant harm to the system's availability, killing processes can be a further threat, especially for processes executing critical tasks and for processes that are not restarted on failure. Furthermore, in adversarial scenarios, attackers may deliberately terminate specific processes to disrupt critical functionalities of dependent systems without requiring to maintain a system-wide failure [8].

While an attacker might not be able or willing to perform a prolonged memory exhaustion attack, exploiting OOM Confusion Attacks allow them to target a specific process for termination, possibly in a timed manner, without keeping the memory fully utilized for a prolonged period. Alternatively, the attacker could utilize the OOM Confusion Attack to ensure that a particular process is always killed while its child processes or other instances successfully complete their tasks, avoiding a full system freeze.

4.1   Attacker Model

The main goal of the attack is to bring the system into a condition where the OOM killer is forced to terminate a specific target process. In this scenario, the attacker must recreate a state in which the OOM killer's heuristics select a specific process to kill among the available ones. The method an attacker may use for OOM Confusion Attack depends on the interaction scenario with the system. However, the key point in each case is that the attacker needs to start some processes exhausting system memory. Here, we report some of the most relevant strategies.

Multi-user Environments. In this scenario, multiple users share the same Linux system, and each has access to its unprivileged user [31]. In this case, the attacker may conduct a OOM Confusion Attack to target processes spawned by other users. These setups are prevalent in academic systems, enterprise servers, and shared computing clusters, where the operating system and hardware are shared among users. This is the most privileged position since the attacker has a high degree of control over the creation of malicious processes.

Vulnerable unprivileged applications are a potentially more common but less controllable context. In this scenario, an attacker does not have direct access to the system, e.g., it does not have SSH access to a system user, but can interact with an exposed vulnerable application, which allows Remote Code Execution (RCE) when exploited [5,21,40]. Modern security best practices dictate that applications should not run with root privileges to minimize the potential impact of exploitation. While the attacker might not be able to escalate privileges, they can spawn processes with the same privileges as the parent vulnerable application. The attacker may achieve the same results as multi-user environments but has less degree of control on the system.

Induced Process Spawning is another, although less common, scenario where an attacker may not have direct access to execute commands on a target system, but can exploit vulnerabilities or misconfigurations to induce a process on the machine to spawn numerous processes [11]. A practical example is the exploitation of a forking server, e.g., Exim [13], Apache HTTP Server [16], Samba [32]. This is a common pattern among web servers, which forks a child to handle incoming requests. When the maximum number of requests that can be processed in parallel (and thus the maximum number of simultaneous child processes), is not correctly limited, it is possible to create an excessively large number of processes. An attacker may abuse this by sending a large number of simultaneous requests, forcing the server to spawn excessive child processes, eventually leading to memory exhaustion and invoking the OOM killer.

5   OOM Confusion Attack Design

The objective of this attack is to force the OOM Killer to terminate a victim process owned by a different user on the system. The attack assumes the attacker has no administrative privileges; otherwise, the attack would be unnecessary.

First, the attacker identifies a victim process they want to kill and gets its *badness* score. Recalling that the OOM Killer always terminates the process with the highest badness, the attacker fills the memory with a series of processes whose badness is lower than the victim's. However, when the victim is not the process with the highest badness on the system, it is necessary to iteratively kill every process with badness higher than the victim's and fill with children the previously occupied memory. After some iterations of this process, the victim will be the process with the highest badness on the system and will be killed by the OOM. To achieve this goal, the attacker needs two components: a process, which we define as a "launcher" or "parent" that continuously launches "children" processes filling the memory, and with a badness lower than the victim's one. Moreover, the badness of the children must be higher than the launcher's one. Otherwise, it may be killed before the kill of the victim process.

The attacker needs, therefore, to keep track of two variables to implement an effective attack: the amount of fillable memory to ensure that the OOM killer is invoked, the maximum amount of spawnable processes to understand the amount of memory they need to assign to each one, and the *oom_score_adj* of each child, to keep its badness lower than the victim's one.

Spawnable Processes. The limit on the number of processes depends on the scenario the attacker is targeting; in particular, it depends on the cgroup of the process which is starting the attacker's processes [39]. cgroup [18] is a mechanism inside the Linux kernel to limit the resources available to a process, e.g., the maximum number of spwnable child processes. The common practice is to limit the maximum number of processes to a percentage of the kernel parameter *threads_max*, which represents the maximum number of threads that can be created on the whole system at any given time and whose formula is [37]:

$$threads\_max = \frac{ram_{total} - reserved}{8 \cdot thread\_size}$$

where ram_{total} is the total RAM of the system in pages, *reserved* is the number of memory pages reserved for the kernel, and *thread_size* is the default number of pages occupied by a new thread. For instance, systemd [30] groups together all processes belonging to a specific user session, allowing system administrators to define resource limits for each user. The default configuration limits the number of processes per user at $max_{procs} = 33\% \cdot threads\_max$. This means that an attacker with user-level access in a multi-user environment would be constrained to 33% of *threads_max*. Nevertheless, this limit differs for processes not associated with a user session, e.g., process demons defined in a service unit. Here, $max_{procs} = 15\% \cdot threads\_max$.

Target Memory Utilization. After computing the maximum number of processes spwnable, the attacker computes the necessary memory to fill to induce the OOM Killer to kill the victim process. The total memory to fill should be a value in the range $pages_{tofill} \in [pages_{free}, pages_{total} - reserved]$, where $pages_{free} = ram_{free} + swap_{free}$ is the number of free pages on the system, considering both RAM and swap, $pages_{total} = ram_{total} + swap_{total}$ is the total number of pages on the system, considering both RAM and swap, and *reserved* is the memory reserved to the kernel.

The number of pages to assign to each process - if the attacker controls it - is then easily computed by dividing the total pages to fill by the number of spawnable processes. Finally, it is possible to compute the badness of children's processes:

$$B_{child} = pages_{child} + \frac{adj_{child}}{1000} \cdot pages_{total} \qquad pages_{child} = \frac{pages_{tofill}}{max_{procs}}$$

where adj_{child}, the *oom_score_adj* of each child, must be chosen carefully by the attacker to make the attack feasible, as we explain later.

Attack Feasibility Evaluation. The attack is feasible when the following constrains hold:

$$B_{child} < B_{victim} \tag{2}$$

$$N_{procs} \cdot pages_{child} \geq pages_{tofill} \tag{3}$$

$$N_{procs} < max_{procs} - running_{procs} \tag{4}$$

$$\text{if } (victim \neq parent): B_{parent} < B_{victim} \tag{5}$$

where B_{child} is the badness of each children, B_{victim} is the badness of the victim process, B_{parent} is the badness of the parent process, N_{procs} is the number of children to create, $running_{procs}$ is the number of processes currently running in the attacker's cgroup. Equation 2 enforces the badness of each child process must be lower than the victim's badness, otherwise children will be killed before the victim. Equation 3 enforces the total memory cumulatively filled by children to be greater or equal to the target memory to fill $pages_{tofill}$. This constraint imposes that the children triggers the OOM Killer, until the victim process is terminated. Equation 4 imposes that it is possible to create enough children without exceeding the maximum number of processes allowed by cgroups. Equation 5 ensures that the badness of the parent is lower than that of the victim when they are different processes. This prevents the parent from being terminated before the victim. However, this equation does not apply when the parent itself is the victim, as in such a case, their badness values would be equal. To ensure the attack is feasible, it is necessary to find values that satisfy all the constraints for N_{procs}, $pages_{child}$, and B_{child}. To find a feasible B_{child}, it is necessary to find a feasible adj_{child} value. If such values can be found, then the attack is feasible. For simplicity, we assume that B_{parent} is fixed. The other symbols in the set of constraints are constants and can be retrieved from the system settings or computed. The latter applies to $pages_{tofill}$, which represents the total number of pages that the attack must allocate to induce the OOM Killer to terminate the victim process.

Algorithm 1. The iterative algorithm computing the constraints for feasibility

Input: B_{parent} - the badness of the parent
  $[procs]$ - the list of running processes, descending sorted by badness
  $pages_{free}$ - the memory pages available in memory
Output: N_{procs} - the number of childs to create
  adj_{child} - the child's adjustment
  $pages_{child}$ - the pages each child has to fill
$pages_{tofill} \leftarrow pages_{free}$
for $proc$ in $procs$ **do**
  $result \leftarrow find\_values(pages_{tofill})$
  if $result \equiv not\_found$ **then**
    return $result$
  end if
  $N_{procs}, adj_{child}, pages_{child} \leftarrow result$
  $pages_{tofill} \leftarrow pages_{tofill} + proc.pages$
  if $proc \equiv victim$ **then**
    $pages_{child} \leftarrow \frac{pages_{tofill}}{N_{procs}}$
    return $N_{procs}, adj_{child}, pages_{child}$
  end if
end for

Algorithm 1 is the pseudo-code of the iterative process computing $pages_{tofill}$. Initially, this value is initialized to $pages_{free}$. At each iteration, the algorithm

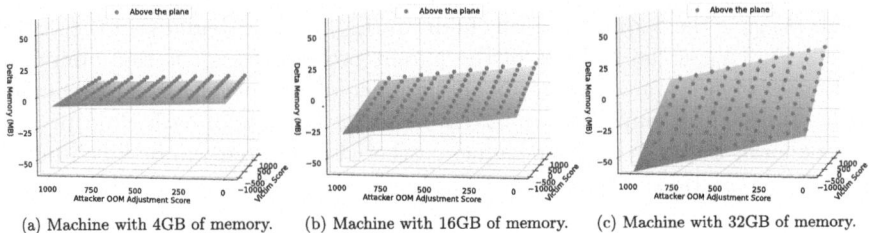

(a) Machine with 4GB of memory. (b) Machine with 16GB of memory. (c) Machine with 32GB of memory.

Fig. 1. Attack feasibility plot. The space above the plane represents the combinations where the attack is feasible.

picks as a target the process with the highest badness value and computes whether it's possible to kill it, finding the values that satisfy all the constraints given the current $pages_{tofill}$ value. If these values cannot be found, the procedure terminates, and the attack is considered unfeasible. If these values exist, and the current target is the victim, the procedure terminates, returning the parameters for the attack. Otherwise, the algorithm sums to $pages_{tofill}$ the pages occupied by the current target and starts a new iteration targeting the next process with the highest badness value. Note that if the estimation is done in advance, the memory usage of processes in the list may fluctuate during the attack, and new processes can be spawned.

In Fig. 1, we plot Eqs. 2 and 3, showing the feasibility of an attack depending on the relationship between the adjustment scores of the attacker and the victim adjustment, and the $\Delta_{memory} = pages_{child} - pages_{victim}$. The values are plotted considering systems where pages have the default dimension of 4 KB.

6   Experimental Evaluation

Our experiments aim to demonstrate the feasibility of OOM Confusion Attacks and their potential impact on a target system. We summarize as follows the questions we aim to answer, which can be mapped to the experiments:
a) What is the victim system response to an OOM Confusion Attack?
b) What is the impact of OOM Confusion Attacks on commonly used processes?
c) Does the current implementation of the OOM killer further exacerbate the impact of OOM Confusion Attacks?
d) How much do the default settings of commercial services and common processes expose them to OOM Confusion Attacks?
e) What are the requirements for executing OOM Confusion Attacks in scenarios with limited system knowledge?

6.1   Evaluation Setup

The setup remains consistent across experiments unless stated otherwise. The test environment uses virtual machines running Linux Ubuntu 24 LTS servers, accessed via SSH. Since the attacker is expected to operate without elevated privileges, in each system, we create a new unprivileged user for the attacker. We collect the logs generated by the invocation of the OOM killer to study the attack. However, the attack places significant stress on the machine, which eventually leads to the kernel ring buffer being overrun, resulting in the loss of log entries. To address this, we use a privileged user on the machine to forward logs through the serial port, ensuring they are saved on the host machine. In cases where this setup is not feasible, such as when testing on remote machines, a background process with a priority adjustment of −1000 runs dmesg in watch mode, forwarding the logs to a file in real-time. Once the environment is ready, the unprivileged user executes the steps outlined in Sect. 5 to determine the optimal parameters for the attack. The attacker then executes a *launcher* process, meant to spawn small processes (*children*) that are the ones with the goal of exhausting the memory. Note that since the attack may terminate the SSH connection, it is crucial to ensure that the launcher process continues running even after the session ends by using the nohup command.

Launcher. The attack is conducted through a custom launcher process. The parent process repeatedly forks new child processes, each configured to consume a controlled amount of memory. Each child process remains idle after memory allocation, keeping CPU usage low and allowing the parent process to spawn other children easily. The children's process is put in an idle state, waiting for I/O operations on a named pipe created by the parent. After all the child processes are successfully spawned, the launcher enters an infinite idle state. This ensures that it does not free its allocated memory, which would undermine the attack. Importantly, the launcher is designed to consume less memory than the child processes, reducing the likelihood that it will be targeted and terminated by the OOM killer before completing its task.

Children. The primary goal of the child processes is to allocate a defined amount of memory and then idle indefinitely without releasing it, creating sustained memory pressure on the system. Each child process reserves memory for a fixed-size buffer. After the allocation is complete, the child process reads from its standard input, which is connected to the named pipe created by the launcher. Since no data is ever written to the pipe, the child process hangs indefinitely in a blocked state, consuming memory without performing any further actions.

6.2 System Response

To evaluate the system's response to an OOM Confusion Attack, we run the attack in multiple configurations. By analyzing the logs, we can generalize the system behavior by identifying five main phases, as shown in Fig. 2a, confirming theoretical hypotheses and highlighting some unforeseen events.

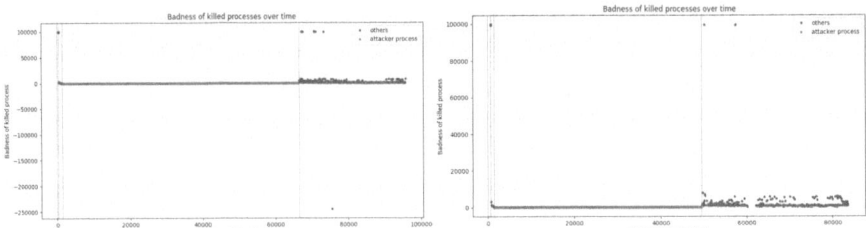

(a) Badness of the killed processes in the first 26 hours of a prolonged attack on a Ubuntu 24.10 server with 6GB of RAM and no swap space. (b) Badness of the killed processes in the first 24 hours of a prolonged attack on a Ubuntu 24.10 server with 6GB of RAM and 3GB of swap space.

Fig. 2. Comparison of process badness over time with and without swap space.

Attack Initiation: the attacker is spawning the processes, and the kernel has not invoked the OOM Killer yet.

Killing Processes with Larger Badness: Once the OOM killer is invoked, it kills every process in the badness ranking above those of the attacker.

Killing Only Attacker Processes: Once the OOM killer reaches the attacker processes, it attempts to remove them. For a significant (yet variable) period of time, the system is unable to perform other relevant tasks due to the excessive memory pressure. This is the phase of maximum memory pressure and performance degradation for the whole system.

Attempt to Restart Services: Eventually, the system stabilizes once enough attacker processes have been killed. Theoretically, this phase may never be reached since, as long as the attacker keeps spawning children's processes, the memory pressure may be too high for the system to react. However, it was eventually always reached in our experiments, suggesting that the degradation of performances, also affecting the attacker's actions, tends to lead toward convergence. In this phase, the system attempts to respawn killed processes with a restart policy. However, the memory in the system remains insufficient, causing the OOM killer to kill them again. Note that at the end of the attack, the memory is left almost completely full, making every further allocation trigger again the OOM condition. Interestingly, we notice in Fig. 2a that there is an outlier benign process being killed with a badness of -250000. This theoretically impossible event led us to discover the additional vulnerability presented in Sect. 6.4, where due to the non-atomic assignment of an adjustment score, some critical processes with low scores may be killed on startup.

The Importance of Swap Space. Despite the common belief that swap slows down the system under memory pressure, we found that it generally facilitates system recovery in OOM Confusion Attacks. In fact, the attacker needs to fill the swap space to invoke the OOM killer. However, while increasing RAM would lead to the increase of the *threads_max* parameter (see Sect. 5), allowing the attacker to distribute the total memory over more processes, this does not happen when only swap is increased. Hence, in the context of OOM Confusion Attacks, the benefit of forcing the attacker to use bigger processes due to the necessity to fill the memory space before the OOM killer is invoked is larger than the cost of the I/O operations caused by swapping, as visible in Fig. 2b.

6.3   Process Exposure to OOM Confusion Attacks

The objective of the experiment is to evaluate whether common processes present in Linux default configurations may be targets of OOM Confusion Attacks. To obtain such information, we map their memory and adjustment score and execute the attack as described in Sect. 6.1. The experiment results are presented in Table 1. The memory allocation is an estimate obtained through multiple readings while the system is idle and no specific tasks are running (the value in parenthesis represents the logged value at the time of death of the process). We execute the attack across setups, comparing RAM alone vs. RAM with swap (e.g., 6 GB vs. 3 GB + 3 GB). Most processes have an adjustment score of 0 or above. As a general pattern found also in the analysis of various other services and applications, processes with positive score adjustments are generally associated with user-level applications. An example is the systemd user instance, which manages user-specific services and typically has a +100 OOM score adjustment. Many system-related services and background processes are assigned a default OOM score adjustment value of 0. This includes a broad range of services, such as networking, hardware management, and security. These processes are not as critical as core system daemons, however targeting them may limit the system functionality. For example, some web servers - for instance, Nginx - have a zero badness score by default, as killing them would compromise only the server availability. In contrast, critical daemons, which are responsible for the whole system's correct functioning and supporting other services, are assigned negative OOM score adjustment values to ensure they are protected from being terminated by the OOM Killer. For instance, dbus-daemon has a OOM score adjustment of −900, and the sshd daemon is assigned a OOM score adjustment of −1000, rendering them unfeasible to target through OOM Confusion Attacks. Confirming the theoretical analysis, an attack initiated from user space with an adjustment score of 0 enables the attacker to target all processes with an adjustment score of 0 or higher, regardless of the memory they occupy. Notably, the time required for the attack to terminate processes varies significantly depending on the scenario. As anticipated in the previous experiment, the availability of swap space proves more effective in delaying the attacker's objectives compared to increased RAM. However, the most significant delay occurs during the time taken to fill the memory and trigger the OOM killer (indicated in the table by the "first kill" row). In contrast, the order in which processes are terminated has a less substantial impact.

Table 1. Memory usage, adjustment, and Time-To-Kill (TTK) of various common Linux applications.

category	sample services	memory	score adj.	restart mode	3 GB	3-3 GB	6 GB	8-8 GB	16 GB	16-16 GB	32 GB
user session	systemd (user)	2–30 MB (4.7 MB)	+100	no°	9 s	32 s	15 s	4 m41 s	1 m37 s	7 m4 s	3 m55 s
	sd-pam (user)	2–30 MB (2.9 MB)	+100	no°	9 s	32 s	15 s	4 m45 s	1 m41 s	7 m6 s	4 m
logging	rsyslogd	5–20 MB (5 MB)	0	on-failure	14 s	34 s	17 s	4 m57 s	1 m50 s	7 m7 s	4 m26 s
	journald	20–100 MB (20.1 MB)	−250	no	×	×	×	×	×	×	×
scheduling	cron	1–5 MB (2.3 MB)	0	on-failure	17 s	48 s	35 s	5 m37 s	1 m59 s	8 m16 s	4 m48 s
backup services	rsync	10–100 MB (2.5 MB)	0	on-failure	17 s	48 s	35 s	5 m34 s	1 m58 s	8 m12 s	4 m48 s
disk management	udisksd	5–50 MB (5.4 MB)	0	no	13 s	34 s	16 s	4 m50 s	1 m45 s	7 m19 s	4 m12 s
power management	upowerd	5–30 MB (4.1 MB)	0	no	11 s	41 s	31 s	5 m14 s	1 m52 s	7 m49 s	4 m36 s
updates	fwupd	10–200 MB (8.9 MB)	0	no	10 s	34 s	15 s	4 m49 s	1 m44 s	7 m9 s	4 m8 s
	unattended-upgrades	10–100 MB (13 MB)	0	no	10 s	33 s	15 s	4 m47 s	1 m43 s	7 m7 s	4 m6 s
containerization	dockerd	50–200 MB (179.2 MB)	−500	always	×	×	×	×	×	×	×
	systemd-machined	10–200 MB (3.7 MB)	0	no	16 s	46 s	35 s	5 m16 s	1 m54 s	8 m2 s	4 m43 s
policy enforcement	polkitd	5–50 MB (6 MB)	0	no	12 s	34 s	16 s	5 m13 s	1 m52 s	7 m48 s	4 m33 s
network	ModemManager	5–50 MB (5 MB)	0	on-abort	13 s	35 s	17 s	4 m53 s	1 m45 s	7 m21 s	4 m16 s
	systemd-resolve	5–50 MB (4.9 MB)	0	always	14 s	35 s	18 s	4 m57 s	1 m48 s	7 m32 s	4 m18 s
	dnsmasq	1–50 MB (1 MB)	0	no°	19 s	52 s	38 s	5 m42 s	2 m3 s	8 m18 s	4 m53 s
ssh connection	sshd daemon	2–10 MB (7.5 MB)	−1000	on-failure	×	×	×	×	×	×	×
	sshd user instance	2–10 MB (4.2 MB)	0	no	16 s	43 s	34 s	5 m25 s	1 m56 s	7 m59 s	4 m46 s
web servers	nginx	2–30 MB (1.6 MB)[a]	0	no	17 s[β]	51 s[β]	37 s[β]	5 m40 s[β]	2 m3 s[β]	8 m26 s[β]	4 m52 s[β]
	apache	2–30 MB (3.7 MB)	0	on abort	15 s[β]	47 s[β]	32 s[β]	5 m26 s[β]	2 m[β]	8 m13 s[β]	4 m38 s[β]
MQTT broker	mosquitto	2–100 MB (4.3 MB)	0	no	15 s	46 s	22 s	5 m6 s	1 m53 s	7 m51 s	4 m31 s
database	postgres	0.1–1 GB+(3.8 MB)	−900	no	15 s[α]	43 s[α]	22 s[α]	5 m18 s[α]	1 m51 s[α]	7 m47 s[α]	4 m30 s[α]
shell	bash	2–10 MB (3.9 MB)	0*	no°	16 s	36 s	23 s	5 m29 s	1 m55 s	8 m5 s	4 m45 s
first kill					9 s	32 s	15 s	4 m41 s	1 m37 s	7 m4 s	3 m55 s

[a] memory consumed by each worker.
[α] indicates that the time refers to the kill of all the workers spawned by the named service.
[β] indicates that the time refers to the kill of all the workers spawned by the named service and the parent process.
° indicates that the service is not handled by systemd (hence would not restart).
* indicates that while the default is 0, the process inherits the parent's adjustment.

Table 2. Metrics for normal and race conditions of various services.

	After Startup			During Race		
Service	oom_score_adj	oom_score	badness	oom_score_adj	oom_score	badness
SSHD	−1000	0	MIN_LONG	0	667	2046
Docker	−500	338	−718716	0	667	2044
Systemd-Journald	−250	508	−82840	0	670	2059

6.4 Race Condition Analysis

Linux lacks a mechanism to specify the OOM score adjustment of a process before its startup. When a process is created, it inherits the OOM score adjustment of its parent. The adjustment can later be modified either by the child during its startup procedure or by the parent process. However, this introduces a race condition [27] between the process creation and the application of the correct OOM score adjustment. An attacker could exploit this brief window to terminate a process that would otherwise be infeasible to kill with our approach.

In standard contexts, the race condition does not significantly affect the attacker's capabilities: if the parent has a higher adjustment value than its child, the attacker can target the parent process directly. Otherwise, it is easier to wait for the parent to adjust the child's OOM score adjustment. However, an interesting scenario arises when the child process is created by a privileged process,

Table 3. Attacker's minimum process memory requirements in various hosting platforms configurations.

	Provider	Distribution	RAM	Swap	threads-max	cgroup%		Minimum Memory	
						user	system	user	sys. daemon
A	local	Ubuntu24.10LTS	3 GB	3 GB	22634	33%	15%	484 KB	1759 KB
B	local	Ubuntu24.10LTS	6 GB	None	46260	33%	15%	364 KB	829 KB
C	aws	Ubuntu24.04LTS	0.5 GB	None	3215	33%	15%	346 KB	757 KB
D	aws	LinuxAmazon2023	0.5 GB	0.5 GB	3084	33%	15%	771 KB	1682 KB
E	azure	Ubuntu24.04.1LTS	1 GB	None	6698	33%	15%	342 KB	753 KB
F	oracle	OracleLinux8.10	1 GB	2 GB	6842	50%	50%	843 KB	843 KB
G	hetzner	Ubuntu24.04.1LTS	4 GB	None	30288	33%	15%	363 KB	800 KB
H	vultr	Ubuntu24.10LTS	1 GB	2.5 GB	7080	33%	15%	1416 KB	3107 KB
I	d.ocean	Ubuntu24.04LTS	1 GB	None	7402	33%	15%	358 KB	787 KB
J	linode	Ubuntu24.04.1LTS	1 GB	0.5 GB	7126	33%	15%	593 KB	1301 KB

such as the kernel or the init system. The OOM killer excludes init and kernel processes as potential victims, making their adjustment value irrelevant. Nevertheless, this value is often set to 0 by default. Init systems like *systemd* [30], among handling the system startup, manage the lifecycle of user-defined services. If the service does not apply the correct adjustment during its startup, *systemd* allows users to specify the adjustment score for service processes in the service's configuration file, known as the unit file. Until the correct adjustment value is set, the process has systemd's adjustment, which is 0. An attacker could exploit this timeframe, during which the child's adjustment is 0, to terminate the process. To ensure it is not a vulnerability of the specific service start-up implementation, we considered both adjustment assignment methods: either an adjustment is assigned by the service start-up process (sshd), or an adjustment is assigned using systemd unit file (dbus-daemon). Since the kernel expose only the *oom_ score*, which is a rescaled and approximate version of the badness score [38], we verify our hypothesis by patching the kernel to print the badness along with the *oom_ score*, and monitoring the badness score when restarting its systemd service at constant time intervals. Then, we compare its badness with adjustment 0 and the badness with the correct adjustment. We tested several services, and we found the 3 reported in Table 2 that changed their adjustment after being started by systemd. In all of the three cases, the badness during the race-condition time frame is significantly higher.

6.5   Default Configurations

To estimate the exposure of service providers to risks associated with OOM Confusion Attacks, we analyze the default configuration of systems offered by commercial hosting platforms and the default configuration of common services and processes present in fresh Linux installations.

Hosting Platforms. We analyzed virtual machines across various hosting platforms to assess whether different configurations were set by default and to evaluate the presence of custom processes that could impact the feasibility of the attack. Upon executing - successfully - a non-targeted version of the attack, we did not observe significant differences in configurations regarding enforced process limits or custom processes that substantially affected the attack's feasibility. However, this exploration highlights the critical role of RAM and swap size on attack viability. In Table 3, we present the minimum size achievable by the attacker processes for each setup, along with the relevant variables required to obtain it. It is important to note that the basic configuration is disadvantageous for the attacker. However, if a legitimate user installs additional applications on the servers, the attacker's capabilities could remain unchanged—if the new application is killed while reaching the target—or could improve. The latter occurs when the newly installed application has a lower *badness* score than the target, as it would assist the attacker in consuming available memory more effectively. We present the cases where the attacker initiates the attack from an active user session or from an exploited system daemon, which, as discussed in Sect. 5 have different maximum process limits. Contrary to expectations, the results show that machines with very limited RAM but with swap enabled are actually more resilient than much larger machines without swap. For example, when comparing the test layouts D and G, attacker processes in layout D are twice as big as those in layout G, despite layout G having eight times more RAM and four times more combined RAM and swap space. Furthermore, given the considerations made in Sect. 5, we can observe that in most cases, the attacker has a smaller memory consumption than the processes shown in Table 1, indicating that in similar adjustment score conditions, all processes would be targetable.

Restart Mode Configuration. While, as shown in the previous experiments, the vast majority of processes may be targets of OOM Confusion Attacks, even if executed from user space, such processes may restart automatically, lowering the attack impact. *Systemd*, in particular, allows to set a restart mode which specifies whether and under what conditions the service should be automatically restarted [35]. The available restart modes include *no, always, on-failure, on-abort, on-success, on-abnormal,* and *on-watchdog.* A termination due to the OOM killer would not trigger only on *no, on-success,* and *on-watchdog.* There are also other settings, such as the delay before restarting or how many attempts to restart the process should be done. Unless explicitly defined, the default behavior is to not restart the process, preventing runaway restart loops in cases where the service repeatedly fails. In Table 1, the Default Restart Mode column shows the default settings found on the installation of the various applications. Some critical services, since they may need to be resilient to failure, have a restart mode explicitly defined. However, only a few have a restart mode set, indicating that the majority, if targets of the attack, would not automatically restart. We also analyzed the restart mode configuration of all services running under *systemd* in a default installation of Ubuntu24.10LTS server. Of the 524 services handled,

34 were set as *always*, 28 as *on-failure*, 7 as *on-abort*, and 456 as *no*, indicating that only a small subset of services restarts unless explicitly stated by the user.

6.6   Induced Process Spawning Attacks Use Case: Exim

In the induced process spawning scenario, the attacker interacts only through the system's interface, and has therefore limited knowledge and capabilities. To understand the requirements necessary for OOM Confusion Attack feasibility, we examine Exim [13], a widely used mail transfer agent (MTA) on Unix-like systems. Exim handles email delivery using a forking model, spawning a new child process for each connection. These child processes persist while the session is active and consume memory based on processed data, making Exim an ideal attack target in this scenario (Sect. 4.1). The attack exploits Exim's forking model by remotely opening numerous email connections, ensuring each child process consumes controlled memory. The goal is to trigger the OOM killer to terminate the Exim parent process, potentially disrupting mail delivery. To evaluate attack feasibility, the attacker needs the badness and adjustment scores of the children and target process, the number of spawnable children, and the required memory to fill. Some of this information is accessible—assuming no custom modifications of adjustment scores were made on the target system, which is uncommon (Sect. 6.5). The attacker can set up a local test environment with Exim to monitor forked processes' memory allocation and badness and determine default limits on children from both the application and cgroup controls. Note, however, that Exim's default child process limit is 20, far below what is typical in real environments. Public project analysis shows this limit is often increased (Table 4a). Additionally, in default settings, the parent and target processes do not restart automatically. The required memory to fill remains unknown to the attacker. They may first test feasibility by spawning as many large children as possible directly on the victim server or proceed directly, ignoring precise memory calculations, as memory does not modify the attack requirements, and the attacker would have no control over it. To evaluate attack feasibility, we conducted an experiment by launching the Exim server on a virtual machine and connecting to it from the host, with the objective of killing the parent Exim process. The results are shown in Table 4b. Each child process had a badness score of 4792 and a PSS (Proportional Set Size) value of 4063 KB. Each experiment was repeated 10 times to ensure result consistency and estimate the mean values of the attack. Every attack attempt was successful. Note that the difference between the theoretical number of children processes to spawn (computed through Eq. 5), is consistent with the experimental ones. As shown in Table 4b, all configurations except the default settings would be vulnerable to the attack. Notably, the suggested system requirements, where mentioned, never exceeded 8 GB.

Table 4. Experiments on OOM Confusion Attacks against Exim attacks.

(a) Examples of connection limits
in public Exim projects.

Project	Max	Host
Exim default	20	Inf
MXroute [22]	5000	Inf
Baruwa [2]	Inf	60
Wikimedia [17]	4000	5-50
MailCleaner [3]	Inf	Inf

(b) Results of attacks against Exim in various scenarios.

RAM	Free Memory	Theory	Exp.	TTK
1GB	772656 KB	194	188	3.74s
2GB	1786880 KB	443	439	8.87s
4GB	3733504 KB	923	922	18.3s
8GB	7758204 KB	1924	1939	43.7s

7   Discussion and Mitigations

While our experiments demonstrate the feasibility of OOM Confusion Attacks, presenting their effect on the system, the exposure of common applications, and the lack of mitigations implemented in default configurations of Linux distributions found in common hosting services, the final objective of our work is that of identifying mitigations against the attack. System administrators can mitigate the risk of OOM Confusion Attacks by conducting a threat evaluation to identify potential targets and attack vectors, and then tuning the system accordingly by adding RAM and swap space to the system, assigning custom score adjustments, or limiting the number of processes and memory usage.

Adding RAM to the system is an effective strategy only when the limit on the number of spawned processes is not dependent on *threads_max*. Otherwise, the main beneficial effect of adding RAM is to increase the time required to complete the attack, as shown in Table 1.

Adding Swap Space is generally a good strategy, as it increases the amount of memory the attacker needs to fill to conduct an OOM Confusion Attack without affecting the constraints in the number of processes. Furthermore, it extends the time required to perform the attack.

Assigning score adjustments is an effective method of preventing attacks on critical processes. To configure adjustments effectively, system administrators should consider the relevance of each process, its expected memory utilization, and its exposure to untrusted actors. Assigning lower adjustment values to critical processes and potential attack targets ensures they are terminated only when absolutely necessary. Assigning higher values to processes exposed to untrusted actors, such as network-exposed services or processes spawned from user interaction, mitigates the risk of them being abused to perform an OOM Confusion Attack. Even when a score adjustment is assigned to a single process, by default, it is automatically propagated to its child processes. For instance, Exim does not provide a mechanism to assign a custom oom_score_adj to the child processes, causing them to inherit whatever adjustment is assigned to the main process. Consequently, in the scenario presented in Sect. 6.6, assigning a low score adjustment to the main Exim process would not protect it from the attack, as the adjustment would also be propagated to the children spawned by the attacker.

Limiting the number of processes that a user or a service can spawn helps system administrators reduce the risk of OOM Confusion Attacks , by forcing the attacker to spawn larger processes to consume enough memory to trigger the OOM Killer, resulting in a higher badness score for those processes. However, this solution requires a trade-off between system security and usability. Furthermore, limiting the number of processes a service can spawn, might expose it to thread exhaustion attacks.

Memory constraints through process grouping can sometimes be counterproductive in preventing OOM Confusion Attacks. When such limits are enforced using cgroup v2, the default behavior is to invoke the OOM killer on the processes inside the cgroup when the limit is reached, making all process within the cgroup potential targets. For example, in the scenario presented in Sect. 6.6, setting a Memory consumption limit on the Exim service would protect other processes on the system but would also make the Exim main process easier to target, as it would reduce the amount of memory required to invoke the OOM killer. Therefore it is fundamental to isolate the processes potentially controllable by an attacker to the potential targets. In such case, cgroups can prevent processes inside the group from exhausting system memory and targeting processes outside that group.

The Race Condition presented in Sect. 6.4 deserves an additional note. The straightforward solution to remove the race condition would be to assign a −1000 adjustment to systemd and the kernel processes. Since the process cannot be killed anyway, this would not directly impact the functionality of the OOM killer. The indirect effect, however, is that such services would have to explicitly assign 0 as the default adjustment score to the spawned processes after they are created, which would obtain - initially - a privileged adjustment score. This solution may create the risk of an attacker being capable of spawning rapidly many processes, which would, however, have an adjustment score of −1000, being therefore unkillable.

8 Conclusions

In this paper, we introduced OOM Confusion Attacks, a novel class of Denial of Service attacks that exploit the Linux OOM killer as a confused deputy. These attacks leverage the heuristic-based decision-making of the OOM killer to manipulate it into targeting victim processes instead of the attacker's processes. We conducted an in-depth evaluation of the requirements for OOM Confusion Attacks both through a formal iterative analysis to assess the feasibility of the attack and a comprehensive set of experiments demonstrating its impact on target systems. These experiments also evaluated the exposure of common applications and commercial hosting services to OOM Confusion Attacks. Additionally, we identify race conditions that can be exploited to block the recovery of privileged services. A final discussion has been presented to propose strategies and mitigations to increase the resilience of systems, and in particular critical services, against OOM Confusion Attacks. Future work should further investigate the feasibility of attacks in different scenarios, and evaluate the impact of proposed mitigations on the usability of the system.

Acknowledgments. This work was partially supported by the MICS (Made in Italy – Circular and Sustainable) Extended Partnership and received funding from Next-Generation EU (Italian PNRR – M4 C2, Invest 1.3 – D.D. 1551.11-10-2022, PE00000004, CUP D43C22003120001) and project SERICS (PE00000014) under the MUR National Recovery and Resilience Plan funded by the European Union - NextGenerationEU.

References

1. Android low memory killer daemon. https://source.android.com/docs/core/perf/lmkd?hl=it
2. Baruwa. https://github.com/baruwaproject/baruwa2
3. Alinto: Mailcleaner community edition. https://github.com/MailCleaner/MailCleaner8
4. Backlinko. https://backlinko.com/iphone-vs-android-statistics
5. Biswas, S., Sohel, M., Sajal, M., Afrin, T., Bhuiyan, T., Hassan, M.: A study on remote code execution vulnerability in web applications. In: International Conference on Cyber Security and Computer Science (ICONCS 2018), pp. 50–57 (2018)
6. Chen, W., Pi, A., Wang, S., Zhou, X.: OS-augmented oversubscription of opportunistic memory with a user-assisted OOM killer. In: Proceedings of the 20th International Middleware Conference, pp. 28–40 (2019)
7. CVE-2014-7300. https://nvd.nist.gov/vuln/detail/CVE-2014-7300
8. CVE-2019-14891. https://nvd.nist.gov/vuln/detail/CVE-2019-14891
9. CVE-2019-6637. Available from MITRE, CVE-ID CVE-2019-6637. https://cve.mitre.org/cgi-bin/cvename.cgi?name=CVE-2019-6637
10. CVE-2020-10781. https://cve.mitre.org/cgi-bin/cvename.cgi?name=2020-10781
11. CVE-2004-0807. https://access.redhat.com/security/cve/CVE-2004-0807
12. Du, Z., et al.: Medusa: unveil memory exhaustion DoS vulnerabilities in protocol implementations. In: Proceedings of the ACM on Web Conference 2024, pp. 1668–1679 (2024)
13. Exim internet mailer. https://www.exim.org/
14. Facebook: OOMD-meta incubator. https://github.com/facebookincubator/oomd
15. Fink, G.A., Griswold, R.L., Beech, Z.W.: Quantifying cyber-resilience against resource-exhaustion attacks. In: 2014 7th International Symposium on Resilient Control Systems (ISRCS), pp. 1–8. IEEE (2014)
16. The Apache Software Foundation: Apache prefork model documentation. https://httpd.apache.org/docs/2.4/mod/prefork.html
17. Wikimedia Foundation. https://github.com/wikimedia/operations-puppet
18. cgroups - Linux man pages. https://man7.org/linux/man-pages/man7/cgroups.7.html
19. Groza, B., Minea, M.: Formal modelling and automatic detection of resource exhaustion attacks. In: Proceedings of the 6th ACM Symposium on Information, Computer and Communications Security, pp. 326–333 (2011)

20. Hareesh, M., Yaswanth, K., Sreeja, M., Kalady, S.: Accurate fork bomb detection by process name. In: 2017 International Conference on Intelligent Computing, Instrumentation and Control Technologies (ICICICT), pp. 1504–1508. IEEE (2017)
21. Hassan, M.M., Mustain, U., Khatun, S., Karim, M., Nishat, N., Rahman, M.: Quantitative assessment of remote code execution vulnerability in web apps. In: Kasruddin Nasir, A.N., et al. (eds.) InECCE2019. LNEE, vol. 632, pp. 633–642. Springer, Singapore (2020). https://doi.org/10.1007/978-981-15-2317-5_53
22. Jarland Donnell, R.A.: Scripts for updating mxroute directadmin servers. https://github.com/mxroute/da_server_updates
23. Kay, T.: Linux swap space. Linux J. **2011**(201), 5 (2011)
24. Kudrjavets, G., Kumar, A., Thomas, J., Rastogi, A.: What do you mean by memory? When engineers are lost in the maze of complexity. In: Proceedings of the 46th International Conference on Software Engineering: Software Engineering in Practice, pp. 405–407 (2024)
25. Kudrjavets, G., Thomas, J., Kumar, A., Nagappan, N., Rastogi, A.: When malloc() never returns NULL—reliability as an illusion. In: 2022 IEEE International Symposium on Software Reliability Engineering Workshops (ISSREW), pp. 31–36. IEEE (2022)
26. Nakagawa, G., Oikawa, S.: Fork bomb attack mitigation by process resource quarantine. In: 2016 Fourth International Symposium on Computing and Networking (CANDAR), pp. 691–695. IEEE (2016)
27. Netzer, R.H., Miller, B.P.: What are race conditions? Some issues and formalizations. ACM Lett. Programm. Lang. Syst. (LOPLAS) **1**(1), 74–88 (1992)
28. Owda, H., Shah, M.A., Musa, A.I., Tamimy, M.I.: A comparison of page replacement algorithms in Linux memory management. Memory **1**, 2 (2014)
29. Patare, P., Govindan, V.: Efficient handling of low memory situations in Linux. Int. J. Eng. Res. Technol. **4** (2015)
30. Poettering, L.: The systemd system and service manager. https://github.com/systemd/systemd
31. Saggu, J.S.: Multi-user operating system. https://www.naukri.com/code360/library/multi-user-operating-system
32. Samba: Windows interoperability suite of programs for Linux and Unix. https://www.samba.org/
33. Shah, K.D., Patel, K.V.: Security against Fork Bomb attack in Linux based systems. Int. J. Res. Advent Technol. **7**, 125–128 (2021)
34. StackScale: The 500 most powerful supercomputers use Linux (2024). https://www.stackscale.com/blog/most-powerful-supercomputers-linux/
35. Systemd unit files documentation. https://www.freedesktop.org/software/systemd/man/latest/systemd.service.html
36. Torvalds, L.: Badness score definition. https://github.com/torvalds/linux/blob/master/mm/oom_kill.c#L202
37. Torvalds, L.: threads-max default value setting. https://github.com/torvalds/linux/blob/master/kernel/fork.c#L997
38. Torwalds, L.: Linux OOM_score definition. https://github.com/torvalds/linux/blob/master/fs/proc/base.c#L585
39. Vissol, C.: Systemd and cgroup. https://medium.com/@charles.vissol/systemd-and-cgroup-7eb80a08234d
40. Xiao, F., Yang, Z., Allen, J., Yang, G., Williams, G., Lee, W.: Understanding and mitigating remote code execution vulnerabilities in cross-platform ecosystem. In: Proceedings of the 2022 ACM SIGSAC Conference on Computer and Communications Security, pp. 2975–2988 (2022)

Overlapping Data in Network Protocols: Bridging OS and NIDS Reassembly Gap

Lucas Aubard[1]([⊠]), Johan Mazel[2], Gilles Guette[3], and Pierre Chifflier[2]

[1] Inria, Rennes, France
lucas.aubard@inria.fr
[2] ANSSI, Paris, France
{johan.mazel,pierre.chifflier}@ssi.gouv.fr
[3] IMT Atlantique, Cesson Sévigné, France
gilles.guette@imt-atlantique.fr

Abstract. IPv4, IPv6, and TCP have a common mechanism allowing one to split an original data packet into several chunks. Such chunked packets may have overlapping data portions and, OS network stack implementations may reassemble these overlaps differently. A Network Intrusion Detection System (NIDS) that tries to reassemble a given flow data has to use the same reassembly policy as the monitored host OS; otherwise, the NIDS or the host may be subject to attack. In this paper, we provide several contributions that enable us to analyze NIDS resistance to overlapping data chunks-based attacks. First, we extend state-of-the-art *insertion* and *evasion* attack characterizations to address their limitations in an overlap-based context. Second, we propose a new way to model overlap types using Allen's interval algebra, a spatio-temporal reasoning. This new modeling allows us to formalize overlap test cases, which ensures exhaustiveness in overlap coverage and eases the reasoning about and use of reassembly policies. Third, we analyze the reassembly behavior of several OSes and NIDSes when processing the modeled overlap test cases. We show that 1) OS reassembly policies evolve over time and 2) all the tested NIDSes are (still) vulnerable to overlap-based evasion and insertion attacks.

Keywords: Intrusion detection system · Evasion · Insertion · IP · TCP

1 Introduction

Some Internet protocols use chunking[1] mechanism. It was introduced to answer a potential discrepancy between medium or underlying protocol capacities. When chunking occurs, the receiver must reassemble all the chunks to retrieve the

---

[1] We use the term chunking as a generic way to refer to "splitting an original data chunk into several". Thus, it both refers to the "fragmentation" mechanism for IPv4 and IPv6 and the "segmentation" mechanism for TCP.

ⓒ The Author(s), under exclusive license to Springer Nature Switzerland AG 2025
M. Egele et al. (Eds.): DIMVA 2025, LNCS 15748, pp. 216–236, 2025.
https://doi.org/10.1007/978-3-031-97623-0_13

initial data packet. However, the chunking mechanism can lead to overlaps. The most common case is a chunk retransmission with the same data that starts and finishes at the same byte offsets. Nevertheless, other types of overlaps exist, i.e., partial overlaps, and the data can be different on the overlapping portion. IPv4 and TCP RFC specifications [7,14] neither forbid data overlaps nor specify the behavior an implementation must adopt (e.g., prefer data from the older chunk). IPv6 RFC specification initially did not forbid overlaps in the first drafts, but has banned them since 2017 [12].

Network packet analysis is one of the possible techniques commonly used to detect intrusions. Some widely deployed Network Intrusion Detection Systems (NIDS) are signature-based, meaning that they match suspicious patterns or signatures of known attacks on the reassembled flow data. Therefore, NIDSes must reassemble consistently the network traffic with the monitored hosts to detect attacks. Ptacek and Newsham [27] introduced a set of IP and TCP ambiguities that may lead to NIDS misassemblies with supervised hosts and thus, NIDS circumvention. The ambiguities exist because 1) NIDSes receive a copy of the network traffic from and to the hosts, and 2) NIDSes and monitored hosts are distinct machines. Thus, NIDSes cannot easily determine how a host processes a specific packet when data overlap occurs.

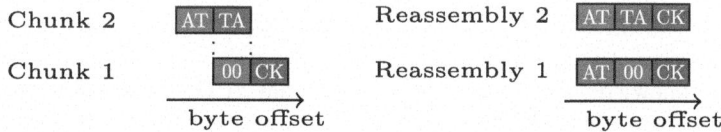

Fig. 1. Data overlap ambiguity illustration.

Figure 1 is an illustration of the data overlap issue. The reassembled data here differs depending on which chunk the reassembly policy favors. Someone with bad intentions may exploit the multiple reassembly possibilities to hide a malicious payload. If the NIDS reassembles with reassembly strategy 1 while the host reassembles with strategy 2, the former cannot see the malicious payload and raise any security alert.

Several works showed that reassemblies depend on the IPv4 [19,24,27,30], IPv6 [9,13,21] and TCP [19,25,27,30] implementations. Since attackers may use the overlapping ambiguity to bypass their security functionalities, NIDSes like Suricata [5] and Snort [28] introduced a feature allowing users to associate each supervised host (through IP address) to a specific reassembly policy [3,4,6]. Other NIDSes like Zeek (formerly Bro [26]) have chosen a different approach: implementing only one reassembly policy but allowing users to enable overlap-related alerts. The set of implemented IPv4 and TCP reassembly policies in Suricata and Snort are based on the works of Novak and Sturges [24,25] published in 2005 and 2007. These works are the latest that tested OS policies for IPv4 and TCP. Thus, we identify two main problems. The first one is that the

OSes Novak and Sturges tested have since been updated, and their protocol implementation may have changed. The second problem arises from the manual approach the related works [9,13,24,25,27,30] used to design their test cases. There is thus no certainty on test case coverage exhaustiveness. Reassembly policies implemented within Suricata and Snort may be based on out-of-date and/or partial policy descriptions, giving a wrong sentiment of security regarding overlap-based attacks. So, the knowledge of modern OS reassembly policies must be updated (see Sect. 4.1), and the exhaustive coverage of overlap test cases must be ensured (see Sects. 3.1 and 5).

After Ptacek and Newsham's seminal work [27], an entire research area has focused on finding any sequence of packets (i.e., not only based on the overlap ambiguity) that protocol implementations process differently. Research has especially intensified from the growing deployment of censorship systems (CS) since the technique has been found successful in their circumvention [10,11,19,20,32,33]. The methods used to find such sequences have moved from manual [19,20,32] (in which authors have to discover the ambiguities all by themselves) to semi-automatic ones using fuzzing [10,36,37] or symbolic execution [33,34]. Until 2020, some works [19,20,32,33] reported the (more and more) relative success of overlapping chunk-based attacks. We argue that the recent works using semi-automatic approaches did not perform extensive (i.e., complete) testing using overlap-based strategies, mainly because fuzzing and symbolic execution methods are good at finding novel chunk sequence examples but not at exhaustive testing.

To our knowledge, no work has verified that NIDSes' overlap reassembly policies are consistent with OSes' since Ptacek and Newsham unveiled the issue in 1998. We fill that gap in this paper. The question that will guide us throughout is: *Do NIDSes reassemble overlap test cases differently as the OSes, and therefore, is it possible to use data overlaps to attack a NIDS or the hosts it supervises?* The contributions are the following:

- In Sect. 2, we extend *insertion* and *evasion* definitions to overlap-based attack context. This enables us to cover all the related attack scenarii and to characterize the requirements for the overlapping data portion.
- In Sect. 3.1, we model chunk sequences with Allen's spatio-temporal reasoning [8] and, thus, ensure overlap coverage exhaustiveness (in contrast to 10 over 13 related-works, see Sect. 5).
- In Sect. 4, we describe IPv4, IPv6, and TCP reassembly policies of a large range of OSes, including recent ones (e.g., Windows 11, the Linux-based Debian 12, FreeBSD 14.1, OpenBSD 7.6, Solaris 11.4) and of three widely deployed NIDSes (i.e., Snort, Suricata and Zeek). We find that:
  i) OS reassembly policies evolve over time. In particular, Windows and Linux-based OSes have changed their IPv4 reassembly policies (regarding state-of-the-art), while TCP policies have barely been modified.
  ii) Snort, Suricata, and Zeek reassemble IPv4, IPv6, and TCP chunks in a partially consistent way with OSes. This opens the way to insertion and evasion attacks. A CVE [22] was assigned to some of the disclosed problems.

2   Problem Definition and Threat Model

This section first defines insertions and evasions in the specific context of overlap-based attacks. It then details the attacker capabilities needed to exploit OS and NIDS reassembly discrepancies.

2.1   Problem Definition

An attacker may use overlap data ambiguity to exploit NIDS and host reassembly divergence. We extend Ptacek and Newsham [27] and Wang et al. [33] *insertion* and *evasion* packet-based attack definitions to fit the context of IP and TCP chunk overlaps. The main shortcomings that we address are:

- Ptacek and Newsham consider a data chunk as being either *totally* accepted or dropped but not *partially* accepted. With the overlap from Fig. 1, the "ATTACK"/"AT00CK" reassembly divergence is impossible.
- Wang et al. do not consider on purpose malicious (or "filtered") payload insertion inside the NIDS flow data, nor do they consider IP-based attacks.

We treat the NIDSes and the host OSes as black boxes in the following. We consider the IP and TCP data stream pushed to the upper layer; thus, our definitions apply to IP and TCP overlap-based attacks.

Let $P = \{$IPv4, IPv6, TCP$\}$ be the Internet protocol set with a chunking mechanism we target. Let C^p be the set of all possible chunks for protocol $p \in P$.

Definition 1 (p protocol data buffer synchronization).   *Given a chunk sequence $c_f...c_l \in C^p$, with c_f (resp. c_l) the first (resp. last) sent chunk, we say that the NIDS and the supervised host have their p data buffers synchronized if the next upper-layer data streams are the same.*

The p's data buffer desynchronization requires chunks linked with a particular overlap type that the NIDS and the host reassemble differently (see Sect. 4) and carefully crafted overlapping data (see Sect. 2.2). Such data is either an *evasion* or an *insertion* payload depending on the target (i.e., the host or the NIDS). *Evasion and insertion attacks aim to desynchronize NIDS and host p protocol data buffers.*

Definition 2 (Evasion in a data overlap-based context).   *An evasion attack consists of some malicious payload that is not visible in the data analyzed by the NIDS while it is visible on the supervised host's reassembled payload.*

Definition 3 (Insertion in a data overlap-based context).   *Symmetrically, an insertion attack consists of some malicious payload that is visible in the flow data analyzed by the NIDS while it is not on the host's reassembled payload.*

Table 1. Attack types illustrated with Fig. 1 reassembly cases and based on Definitions 2 and 3. - means the implementation *ignores* the flow chunk data.

Attack type	Implem.	Target	Reassembled data	Attack scenario	Works		
					[27]	[33]	Us
Evasion	NIDS		-	E1	✓	✓	✓
	Sup. host	✗	"ATTACK"				
	NIDS		"AT00CK"	E2		✓	✓
	Sup. host	✗	"ATTACK"				
Insertion	NIDS	✗	"ATTACK"	I1	✓		✓
	Sup. host		-				
	NIDS	✗	"ATTACK"	I2			✓
	Sup. host		"AT00CK"				

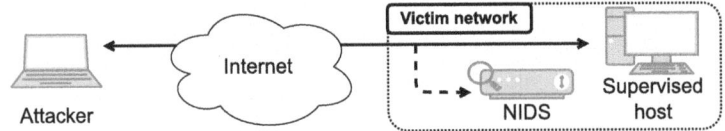

Fig. 2. The considered threat model.

Illustration. We use the overlap chunk sequence introduced in Fig. 1 to illustrate insertion and evasion attack types in Table 1. As we can see, the non-targeted host either reassembles differently (with a benign payload, for example) or completely ignores the chunk sequence. Data buffer desynchronization is one of the most dreaded risks for a NIDS since it eventually allows an attacker to bypass all its security mechanisms. See related CVE 2019-18625, 2019-18792, or 2021-37592, whose scores are high or critical. On the one hand, attackers can use evasions to circumvent the NIDS inspection function. An alert pattern can thus reach the supervised host without the NIDS noticing it, as [27,33] did. An evasion can also impact other NIDS functionalities, such as file or TLS certificate extractions, since a unique (overlapped) bit is sufficient to corrupt them. On the other hand, the insertion attack can alter the NIDS's normal behavior, for instance, by exploiting a known NIDS vulnerability (e.g., 2019-12175, 2023-7242, or 2024-47522 CVEs). Or it can raise false positive alerts, wasting analysts' time.

2.2   Threat Model

The threat model we consider consists of an attacker, a NIDS, and a supervised host, as illustrated in Fig. 2. The NIDS gets a copy of all the network packets the host receives and sends. We also suppose the NIDS is configured such that the supervised host's IP address is associated with the corresponding reassembly policy[2], if the NIDS offers such a feature. The attacker should be able to:

- identify supervised host and NIDS reassembly policies, enabling them to choose a good (i.e., differently reassembled) overlap case candidate.
- craft IP header fields and payload to perform IP fragment-based attack.
- craft TCP header fields and payload to perform TCP segment-based attack.

Gaining Supervised Host and NIDS Reassembly Policy Knowledge. The attacker may want to learn about the monitored host and NIDS reassembly policies to increase the chances of a successful attack. A good approximation to learn about the host reassembly policy is to determine the host OS. Several tools exist to perform OS fingerprinting. Active ones, such as Nmap [2] or Hershel(+) [29], use specifically crafted packet sequences or retransmission times to identify a host OS, while passive tools, such as p0f [35] or nPrintML [17], analyze packet header fields in existing communications. Based on this knowledge and considering the hypothesis that the NIDS is correctly configured, the attacker can determine the NIDS reassembly policy. Finally, they can craft an overlap chunk sequence based on the results reported in Sect. 4.2 and their objectives.

Crafting the Data Chunks. According to the selected overlap case, the attacker chunks the original malicious packet into several pieces. They must appropriately manipulate the header fields *Fragment Offset* and *More Fragments* (resp. *Sequence Number*) to perform an IP (resp. a TCP) chunk-based attack. But, the other header fields of the crafted chunks must be consistent with the carried data (e.g., correct IP or TCP checksum, correct length). The original payload must be in plaintext[3], and choosing the malicious, (i.e., "ATTACK" in Table 1) payload depends on what the attacker wishes to do. However, there are some constraints on the overlapping data portion which is *not* visible in the non-target flow data, especially concerning the syntactic and semantic correctness of the upper-layer protocols. This correctness depends on the performed attack scenario, i.e., E1, E2, I1, or I2, as described in Table 1.

E1 and I1-Related Constraints. The non-targeted implementation ignores the overlap chunk sequence; thus, no data is pushed to the upper protocol data stream. It means that the overlapping data portion does not matter and, ultimately, *any data* that fits the required length is, in fact, possible. The overlapping data portions can even be the same.

---

[2] Based on the current knowledge, we consider that hosts with more recent OS versions than the ones tested by Novak and Sturges in [24,25] (e.g., Debian 12) should have the latest OS family representative reassembly policy (e.g., *linux* because Linux 2.4 was the latest tested version) associated in the NIDS configuration file.

[3] The attacker may however take advantage of the overlap ambiguity during encryption initialization to corrupt the NIDS processing of TLS certificates or ciphersuites for example.

Table 2. Allen's interval algebra relations and the corresponding meaning in terms of Internet packet sequences.

Relation $\mathcal{R}$	Relation $\mathcal{R}$ inverse	Meaning
X M Y $\overline{}^{\,Y}$ $\underline{}_{\,X}$	X Mi Y $\overline{}^{\,Y}$ $\underline{}_{\,X}$	Meet: in-order (resp. out-of-order) contiguous chunks
X B Y $\overline{}^{\,Y}$ $\underline{}_{\,X}$	X Bi Y $\overline{}^{\,Y}$ $\underline{}_{\,X}$	Before: data hole between one chunk ending byte and the other chunk's payload starting byte
X Eq Y $\overline{}^{\,Y}$ $\underline{}_{\,X}$	-	Equal: complete data overlap with the chunks starting and finishing at the same byte offsets. Data retransmissions are Eq overlaps.
X O Y $\overline{}^{\,Y}$ $\underline{}_{\,X}$	X Oi Y $\overline{}^{\,Y}$ $\underline{}_{\,X}$	Overlap: partial data overlap
X S Y $\overline{}^{\,Y}$ $\underline{}_{\,X}$	X Si Y $\overline{}^{\,Y}$ $\underline{}_{\,X}$	Start: partial data overlap
X D Y $\overline{}^{\,Y}$ $\underline{}_{\,X}$	X Di Y $\overline{}^{\,Y}$ $\underline{}_{\,X}$	During: partial data overlap
X F Y $\overline{}^{\,Y}$ $\underline{}_{\,X}$	X Fi Y $\overline{}^{\,Y}$ $\underline{}_{\,X}$	Finish: partial data overlap

E2 and I2-Related Constraints. The NIDS and the supervised host both push data to the upper-layer protocol. For the "AT00CK" reconstruction to be harmless, it must syntactically and semantically conform with all the upper protocols. In particular, as for any upper-layer checksum: 1) if the checksum is contained within the overlapping portion, then it should be correctly adjusted to fit the "AT00CK" reassembled payload; otherwise, 2) the overlapping data should be adapted to fit the checksum[4]. Any upper-layer syntax or semantic direspect may cause unwanted side effects (e.g., NIDS alerts, NIDS or supervised host failure) that could affect the attacker's stealthiness.

3   Testing Method

This section describes the overall method used to obtain OS and NIDS reassembly policies, which are then compared in order to find insertion and evasion opportunities. First, it introduces the algebra used to model overlapping chunk sequences. Then, the section describes all the test case characteristics. Finally, it details the hosts we use to test NIDSes and OSes.

---

[4] Only two dedicated octets are required to make a payload fit any internet checksum because it is computed with 2-octet words [7]. See Appendix C "Deceiving TCP checksum" of Feng et al. work [15] for a payload crafting example.

3.1 Chunk Sequence Modeling

In the present subsection, we document how a spatio-temporal reasoning algebra can be adapted to model overlapping chunks.

Spatio-Temporal Reasoning. Spatio-temporal reasoning is particularly well suited to model packets of protocols that allow chunking and, thus, overlapping. Indeed, we can associate byte offset with one spatial dimension and arrival time with one temporal dimension. Allen's interval algebra [8] is such a spatio-temporal algebra. It consists of 13 different relations, which are described within the first two columns of Table 2. As we can see, there are four non-overlapping Allen relations, i.e., M, Mi, B, and Bi, and nine overlapping ones, i.e., Eq, O, Oi, S, Si, D, Di, F, and Fi. The rightmost column transposes the relation meaning in terms of Internet packet sequence. While these relations describe the relative byte-wise and time-wise position of two chunks, they do not give any information regarding the chunk contents.

Overlap Test Case Modeling. We use Allen's interval relations (or Allen relations for short) to ensure the exhaustiveness of overlap test cases. In the following, a test case is always time-wisely described. In other words, if $c_1 \mathcal{R} c_2$ with $c_1 \in C^p$ and $c_2 \in C^p$, then $t_1 < t_2$ (i.e., c_1 arrives before c_2).

There are nine overlapping Allen relations; thus, we consider that *exhaustiveness in terms of overlap coverage is reached by testing these nine overlap cases*. Thanks to the modeling, we can now prove that Novak and Sturges [24,25] and Atlasis [9] manually found and tested all the possible overlapping test cases. See Table 8 for the other work transposition into Allen formalism.

3.2 Test Case Characteristics

Table 3 summarizes all the overlap test case characteristics. The chunk payloads of a test case are chosen so that 1) they align with the IP header field's unit, which is 8-byte, 2) no matter which overlapping data is preferred, the higher layer checksum is valid, and, of course, 3) the preferred chunk can be distinguished from the other. Novak introduced these payload patterns in [24]. We also use Novak and Sturges's trick [24,25], which ensures that all the chunks have been received before the chunk sequence reconstruction occurs, with ① and ②. Finally, overlaps can be tested *singly* within nine separate chunk sequences ⓐ as [25,27] did ("individual overlap tests" in [25]). Or, differently, *multiple* overlaps can be tested altogether within a unique chunk sequence, resulting in one reassembly ⓑ (Novak and Sturges name it "model overlap tests" in [25]). This last mode has been the most tested in the related works that targeted the OSes [13,24,25,30]. We specifically use Novak and Sturges' *multiple* chunk sequence. The third column of Table 3 illustrates the O relation's test for some introduced characteristics, with simplified chunk payloads to reduce figure size.

Table 3. The IP and TCP test case characteristics.

Characteristic	Description	Example
Overlap type	Allen relation(s)	O
Chunk payload	AABBCCDD $\rightarrow$ DDCCBBAA ensuring checksum validity	$\downarrow$
Reassembly trigerring	① *IP*: the rightmost finishing and lastly sent fragment has the *More Fragments* (MF) bit unset ② *TCP*: extra segment at the byte-wise beginning of the test case segments	
Mode	ⓐ *single*: overlaps tested individually ⓑ *multiple*: overlaps tested altogether	
Upper-layer service	*IP*: ICMP or ICMPv6 Echo *TCP*: TCP Echo	

3.3   OS and NIDS Host Targets

We perform OS testing through a classical Base-Target architecture. The targeted OSes are varying Vagrant/Virtualbox-based boxes. The testing scripts are all launched from the Base box. As for the NIDSes, they are tested within Docker if an official image exists; otherwise, the tests are performed locally. To deduce NIDS reassembly policies, we alert on 1) the chunk payload patterns introduced in Tables 3 and 2) the upper-layer service (i.e., ICMP for IPv4, ICMPv6 for IPv6, and port 7 for TCP). The following IPv4 entry rules are, for example, used to match on the "AABBCCDD" pattern:

- Suricata and Snort: `alert icmp [192.168.0.1] any -> any any (msg: "AABBCCDD detected"; content:"AABBCCDD"; sid:1; rev:7;)`
- Zeek[5]: `signature ipv4-AABBCCDD { ip-proto == icmp src-ip == 192.168.0.1 payload /.*AABBCCDD.*/ event "AABBCCDD detected"}`

In both cases, Network Interface Controller (NIC) offloading is disabled so as not to interfere with the targeted implementation reassembly. See Sect. 6.1 for more details on OS and NIDS reassembly interferences.

4   Results

In this section, we first describe IPv4, IPv6, and TCP reassembly policies of some OSes. The tested OS versions cover a large spectrum for the last 10 years.

---

[5] Since Zeek's preferred pattern-matching method is scripting, we verified that test case reassemblies are the same across the two matching methods.

Table 4. OS IP reassembly policies. o (resp. n) means that oldest (resp. newest) chunk data is prefered and ∅ means that the OS ignores the overlap. Bold blue means that testing modes are reassembled differently. IPv4 (resp. IPv6) cell backgrounds encodes in consistency with [24] (resp. [13]).

OS kernel	Protocol version	Testing mode	Test case								
			Overlapping relation								
			F	Fi	S	Si	O	Oi	D	Di	Eq
Windows 21h2, 23h2	v4	*multiple*	∅	∅	∅	∅	∅	∅	∅	∅	∅
		single	n	∅	n	o	∅	∅	n	o	n
	v6	*multiple*	∅	∅	∅	∅	∅	∅	∅	∅	∅
		single	n	∅	n	o	∅	∅	n	o	n
Linux 4.9, 6.1	v4	*multiple*	∅	∅	∅	∅	∅	∅	∅	∅	∅
		single	n	∅	n	o	∅	∅	n	o	n
	v6	*multiple*	∅	∅	∅	∅	∅	∅	∅	∅	∅
		single	n	∅	n	o	∅	∅	n	o	n
SunOS 5.11	v4	*multiple*	n	o	o	o	o	o	n	o	o
		single	n	∅	n	o	o	o	n	o	n
	v6	*multiple*	n	o	o	o	o	o	n	o	o
		single	n	∅	n	o	o	o	n	o	n
FreeBSD 10.2, 12.1, 14.2	v4	*multiple*	n	o	o	o	o	n	n	o	o
		single	n	∅	n	o	o	n	n	o	n
	v6	*multiple*	∅	∅	∅	∅	∅	∅	∅	∅	∅
		single	n	∅	n	o	∅	∅	n	o	n

We then verify NIDS reassembly consistency with these OSes. The versions we target are:

- *OS*: Windows 10 (21h2) and 11 (23h2), Debian 9 (Linux 4.9) and 12 (Linux 6.1), FreeBSD 10.2, 12.1 and 14.1, OpenBSD[6] 6.0, 6.9 and 7.6 and, Solaris 11.2 to 11.4 (SunOS 5.11).
- *NIDS*: Suricata v7.0.4, Snort v3.1.83, and Zeek v6.2.0.

4.1   OS Reassembly Policies

This sub-section details IPv4, IPv6, and TCP policies for recent OSes in both *multiple* and *single* testing modes. When relevant, we also compare our findings with the latest related work: IPv4 *multiple* mode testing from [24] findings, IPv6 *multiple* mode from [13], and TCP *multiple* and *single* modes from [25].

---

[6] The tested FreeBSD and OpenBSD OS versions reassemble the same way the overlap test cases; therefore, we only report FreeBSD policies.

IP Protocols. Table 4 reports OS IPv4 and IPv6 test case reassembly policies.

IPv4. Windows and Linux policies have evolved for the *multiple* testing mode since Novak [24]. Both OSes now ignore overlap chunks. The other OSes have not changed their policies. However, the newly tested mode, namely *single*, shows different reassemblies for all the OSes when compared to the *multiple* mode. In total, 6 out of 9 overlapping relations are reassembled differently for Windows and Linux-based OSes, while it accounts for 3 out of 9 for the remaining OSes. We hypothesize that the context introduced by the adjacent chunks used inside *multiple* mode causes the discrepancies between the two testing modes. We thus argue that the *single* mode reassemblies should be used to obtain context-agnostic reassemblies inside the NIDSes. In this testing mode, all the OSes reassemble F, S, Si, D, Di, and Eq relations the same way, never ignoring the test cases. Fi relation is never reassembled, possibly due to the fragment with the MF bit unset's drop. Finally, O and Oi are the only overlap test cases that show different reassemblies depending on the OS.

IPv6. FreeBSD OSes do not reassemble O and Oi Allen relations, which differs from the observed IPv4 behavior. Except for FreeBSD, all OSes reassemble IPv4 and IPv6 overlapping fragments the same way, and thus, the previous paragraph descriptions also apply to IPv6 fragments. The Windows and Linux-based OSes ignore all the overlapping relations in a *multiple* test mode, which is inconsistent with Di Paolo's [13] findings for O, Oi, and Eq relations. Since the OS versions that Di Paolo tested are very close to ours, we hypothesize that the lack of tested relation set exhaustiveness for that mode impacts the extracted reassembly.

TCP. Table 5 describes OS TCP reassembly policies and compares findings with state-of-the-art ones [25]. Windows and FreeBSD reassemble similarly the overlaps across testing modes. The former OS always reassembles with the oldest segment data.f These policies are consistent with [25] description. The latest Linux-based OSes reassemble S relation differently depending on the testing mode, favoring old (resp. new) data for *multiple* (resp. *single*) mode. The SunOS also reassembles the Eq overlap differently, which is consistent with [25] findings.

Takeaways. IPv4, IPv6, and TCP reassembly policies are more complex than described in the state-of-the-art as overlap reassemblies change depending on the test mode. IP policies have evolved; for example, the Windows and Linux families now show the same reassemblies. TCP reassembly policies have been unchanged since 2007, except Linux's. Finally, OSes continue to reassemble some overlap test cases differently by favoring old or new data or ignoring the chunks.

4.2   NIDS/OS Reassembly Consistency

This section compares the NIDS reassembly policies we observed with the ones of OSes (see Sect. 4.1) for IPv4, IPv6, and TCP protocols and the *single* mode.

Table 5. OS TCP reassembly policies. o (resp. n) means that oldest (resp. newest) chunk data is prefered. Bold blue means that testing modes are reassembled differently. Cell backgrounds encodes in consistency with [25].

OS kernel	Testing mode	Test case Overlapping relation								
		F	Fi	S	Si	O	Oi	D	Di	Eq
Windows 21h2, 23h2	any	o	o	o	o	o	o	o	o	o
Linux 4.9, 6.1	multiple	n	o	**o**	o	o	n	n	o	o
	single	n	o	**n**	o	o	n	n	o	o
SunOS 5.11	multiple	n	o	n	o	n	o	n	o	**n**
	single	n	o	n	o	n	o	n	o	**o**
FreeBSD 10.2, 12.1, 14.2	any	n	o	o	o	o	n	n	o	o

Because of space issues, we first check reassembly discrepancies between NIDSes and one OS family, namely FreeBSD. We choose this OS because it offers the most insertion and evasion attack[7] opportunities when configured inside Snort and Suricata using the *bsd* policy. We finish by summarizing results for all the tested OSes and providing metrics on NIDS attack opportunities. Full results can be found in https://gitlab.inria.fr/laubard/dimva_2025_artifacts.

Consistency with FreeBSD OSes. Table 6 gathers IP and TCP NIDS reassembly policies in the *single* mode, which is the easiest an attacker can exploit.

IP. NIDSes are all vulnerable to overlap-based attacks with at least two overlap types, except for Snort with IPv4 chunks. Zeek and Suricata can be subject to IPv4 insertion attack with *Fi* relation and IPv4 evasion or insertion with just *Oi*. Moreover, with only two consistent IPv6 test case reassemblies with the FreeBSD, several more relations can be used to perform insertion or evasion attacks on Zeek. Despite being based on the same *bsd* policy description as [24], Suricata reassembles IPv4 overlaps differently. Snort is, interestingly, perfectly consistent with the OS for IPv4 protocol even though no previous work had described reassemblies for the *single* test mode. Additionally, IPv4 and IPv6 fragments are reassembled similarly by Snort and Suricata. Zeek notably reassembles all the overlaps by favoring the oldest IPv6 fragment data.

---

[7] Based on Sect. 2.1 definitions, if the NIDS and the host reassemble differently a test case by not ignoring it (i.e., E2 and I2 from Table 1), the attacker can either perform an insertion or an evasion. The other attack cases E1 and I1 are straightforward.

Table 6. NIDS reassembly consistency with FreeBSD 10.2, 12.1, 14.1 in a *single* testing mode. o (resp. n) means that oldest (resp. newest) chunk data is prefered and ∅ means the OS ignores the test case. Green (resp. red) means that NIDS reassembly is the same as (resp. different from) FreeBSD.

Protocol	Implementation	Overlapping relation								
		F	Fi	S	Si	O	Oi	D	Di	Eq
IPv4	**FreeBSD**	**n**	∅	**n**	**o**	**o**	**n**	**n**	**o**	**n**
	Suricata-*bsd*	n	o	n	o	o	o	n	o	n
	Snort-*bsd*	n	∅	n	o	o	n	n	o	n
	Zeek	n	o	n	o	o	o	n	o	n
IPv6	**FreeBSD**	**n**	∅	**n**	**o**	∅	∅	**n**	**o**	**n**
	Suricata-*bsd*	n	o	n	o	o	o	n	o	n
	Snort-*bsd*	n	∅	n	o	o	n	n	o	n
	Zeek	o	o	o	o	o	o	o	o	o
TCP	**FreeBSD**	**n**	**o**	**o**	**o**	**o**	**n**	**n**	**o**	**o**
	Snort-*bsd*	n	o	o	o	o	n	n	o	o
	Suricata-*bsd*	n	o	o	o	o	n	n	o	o
	Zeek	o	o	o	o	o	o	o	o	o

TCP. Snort and Suricata policies are consistent with the tested FreeBSD OSes for all the overlapping test cases. The NIDSes, thus, consistently implemented the reassembly policies that Novak and Sturges described [25]. Zeek, which does not have such a reassembly policy configuration capability, reassembles F, Oi, and D inconsistently, as it always reassembles with the oldest segment's data. An attacker can use these overlaps to perform insertion and evasion.

Consistency with All OSes. Snort reassembles perfectly consistently IPv4 test cases with the BSD and Sun-based OSes and IPv6 test cases with SunOS. We can find at least one IP test case that is reassembled differently for the remaining OSes, i.e., the Linux and Windows ones. Neither Suricata nor Zeek consistently reassembles all IP overlapping test cases with any of the characterized OSes. Therefore, at least an insertion or an evasion attack can target these NIDS-OS couples. Suppose that Snort or Suricata TCP reassembly policy is correctly associated with the host; there is no possible overlap-based attack except in one case: TCP *solaris* policies and SunOS-based OSes with the Eq test case. On the contrary, Zeek reassembles segments consistently with Windows OSes but does not with the remaining OSes. See https://gitlab.inria.fr/laubard/dimva_2025_ artifacts for more details.

Table 7. NIDS inconsistencies with OS reassemblies and corresponding attack opportunities for a *single* testing mode. W, L, S, and F respectively correspond to the tested Windows, Linux, SunOS and, FreeBSD/OpenBSD kernels.

Protocol	NIDS	Test case inconsistencies		Tested OS kernels with possible attack	
				Evasion	Insertion
IPv4	Suricata	8	22%	F	W, L, S, F
	Snort	4	11%		W, L
	Zeek	9	25%	F	W, L, S, F
IPv6	Suricata	9	25%		W, L, S, F
	Snort	6	17%		W, L, F
	Zeek	28	78%	W, L, S, F	W, L, S, F
TCP	Suricata	1	3%	S	S
	Snort	1	3%	S	S
	Zeek	11	31%	L, S, F	L, S, F

Table 7 gives more general consistency metrics and related attack opportunities. In particular, Zeek and Suricata reassemble IPv4 overlaps inconsistently for about 20% of them. Snort performs better with 11% inconsistent test cases. Snort and Suricata globally perform better than Zeek for IPv6, with the same or fewer OSes that can be targeted with an insertion or evasion attack. Zeek, which has different IPv4 and IPv6 reassembly policies, performs worse for version 6 (28 inconsistencies) than for version 4 (9 inconsistencies). TCP enables fewer attack opportunities, with only one overlap that Suricata and Snort reassemble incorrectly. Zeek exhibits TCP-based evasion or insertion attacks for 3 OSes out of 4. Since one inconsistency is enough to perform an insertion and/or an evasion, the OSes (i.e., columns 4 and 5 in Table 7) that can be targeted are of importance.

Takeaways. As expected, Snort and Suricata (which allow policy configuration) perform overall better than Zeek (which uses a unique policy). Snort can protect itself completely against IPv4 and IPv6 evasion attacks (for the tested OSes) and almost entirely against TCP segment-based attacks. Because OS reassembly policies evolve and are more complex than initially thought, NIDSes must (continuously) verify the consistency of the implemented policies.

Table 8. Summary regarding overlap-based works. "Unknown" means that there is partial or no information on the covered relations for the work tool's run.

Author	Work	Year	Protocol	Testing mode	Tested Allen relations	Target type
Ptacek et al.	[27]	1998	IPv4/TCP	*single*	*Fi, D*	NIDS/OS
Shankar et al.	[30]	2003	IPv4	*multiple*	*O, Oi, Eq*	OS
			TCP	*multiple*	*O, D*	
Novak et al.	[24]	2005	IPv4	*multiple*	all	OS
	[25]	2007	TCP	*multiple single*		
Atlasis	[9]	2012	IPv6	*Na*	all	OS
Khattak et al.	[19]	2013	IPv4/TCP	*single*	all	CS
Wang et al.	[32]	2017	IPv4/TCP	*single*	*Eq*	CS
Lin et al.	[21]	2024	IPv6	*single*	*Eq*	NIDS/OS
Bock et al.	[10]	2019	IPv4/TCP	*Na*	Unknown	CS
Wang et al.	[33]	2020	TCP	*Na*	*F, D, Oi*	CS/NIDS
	[34]	2021	TCP	*Na*	-	OS
Zhang et al.	[36]	2022	IPv4/TCP	*Na*	*Eq*/Unknown	NIDS
Di Paolo et al.	[13]	2023	IPv6	*multiple*	*O, Oi, Eq*	OS
Us	-	-	**IPv4/IPv6/TCP**	***multiple single***	**all**	**NIDS/OS**

5   Related Works

This section presents the related works that analyzed and described IP and TCP implementation reassembly policies. To ease the comparison with these works, we transpose the covered test cases into Allen's formalism in Table 8. We categorize the works that tested implementation reassembly policies in two families according to the test case generation approach.

5.1   Manually Generated Overlap Cases

In 1998, Ptacek and Newsham [27] first showed that OSes could behave differently when reconstructing overlapping IPv4 and TCP chunks. The reassembly ambiguity it poses for NIDSes opened a new research axis, and several works [9,24,25,30] tried to unveil the IPv4, IPv6, and TCP reassembly policies of OSes. Novak and Sturges's [24,25] works reached exhaustivity for the first time regarding the tested overlap types. More recently, Lin et al. [21] showed that some OS and NIDS do not comply with RFC [12] as regards IPv6 data overlaps (which states that implementations should discard the entire fragment sequence in the presence of any overlap type). However, since they tested a unique overlap type, they may have overestimated OS compliance.

In parallel, other works tried to evade censorship systems (CS) with different elusive packet sequence strategies. In 2013, Khattak et al. [19] described the IPv4 and TCP reassembly policies of the Great Firewall of China (GFW) based on the complete set of overlap relations. Wang et al. [32], which manually designed a unique IPv4 and TCP overlap test case, showed that some middleboxes could interfere with the (original) overlapping fragment sequence by dropping or reassembling it, eliminating, thus, any ambiguity. These works, however, do not consider that the server OSes may have different reassembly policies.

Overall, only 3 of the 8 works were exhaustive in regards to the covered overlap relations. The lack of a unified overlap formalization may explain why most recent works target fewer overlap types than before, as shown in Table 8. Finally, none of the works conducted a complete reassembly discrepancy analysis for NIDS/OS couples.

5.2   Semi-automatically Generated Overlap Cases

Other works focused on chunks overlaps from a different perspective. Bock et al. [10] used a genetic algorithm named Geneva to find packet sequences differently processed between CS and hosts. Theoretically, this algorithm can perform evasion attacks with overlapping IPv4 or TCP chunks as it can modify the corresponding header fields and payload. However, it did not find any such chunk sequences. Zhang et al. [36] derived Geneva to find novel packet sequences that bypass Suricata or Snort. Their tool, StateDiver, found one (quite complex) successful technique using IPv4 fragmentation and TCP segmentation. We suppose that Geneva and StateDiver failed to find more overlap-based techniques because 1) successful evasion attempts drove the tools away from this strategy and/or 2) some tested DPIs and hosts may have had the same reassembly policy. One cannot be sure that all the overlap cases were tested exhaustively.

Di Paolo et al. [13] also used a fuzzing-like approach to verify OS compliance with the IPv6 specification [12] when processing overlapping fragments. They derived the Shankar and Paxson model [30] by permuting and duplicating the chunks, and they found that none of the tested OSes (which were Linux, Windows, or BSD-based) conform.

Wang et al. introduced SymTCP and Themis tools [33,34], which both use symbolic execution to find TCP packet sequences that are processed differently between TCP implementations. SymTCP successfully found a data overlapping strategy to evade Zeek version 2.6. However, while this method could theoretically cover all overlap types, SymTCP cannot find exhaustive overlapping test cases because of its incapability to model retroactive behaviors on data buffers (as rightly explained in [33] section IX). Themis could not find any TCP-based attack strategy based on data overlap ambiguity because all the tested implementations were Linux-based. These implementations may, therefore, have the same TCP reassembly policy. In theory, the Themis tool can show discrepancies in reassembly policies if there are any. However, if none are found, one cannot easily retrieve the reassembly policy. The authors also highlight that adapting the tool to any OS may require quite important efforts.

6   Discussions and Future Works

This section discusses the exploitability of Sect. 4 results and NIDS countermeasures regarding overlap data ambiguity, and provides recommendations.

6.1   Overlap-Based Attack Usability

Relation Differences. An attacker that would like to use overlaps to perform an insertion or an evasion attack may struggle differently depending on the relation. If the goal is an evasion by making the NIDS misassemble the transport header, then the attacker would benefit more from S, Si, or Eq overlaps because they make the chunks start at the same byte offset. If these relations are not used, the attacker may need to add a small chunk on the left, which may be considered as "weird" chunks by NIDSes, especially for IP fragments. Zeek, for instance, considers fragments under 64 bytes as too small, producing a "weird" logging entry. Differently, suppose the attacker aims to make the NIDS hash calculation fail for a given file. Any overlap relation is helpful in that case because one bit flip on the overlapping data portion is enough to change the hash.

Context Importance. The overlap relation reassembly may change depending on the testing mode, as mentioned in Sect. 4. IP testing exhibits many reassembly differences between the modes. We hypothesize this is partly due to the strong impact of the fragment with the *More Fragments* bit unset inside *single* testing mode. If this fragment is dropped, the reassembly conditions are not met, and the test case is ignored. Differently, TCP reassembly policies do not change as much for the *single* testing mode. Beyond testing modes, an extra segment was added before (byte-wise) and sent after (time-wise) the overlapping chunks. In future works, we plan to extend overlap cases' *testing context* (e.g., adding an extra non-overlapping chunk after the overlapping ones). This should give a more complete picture of OS reassembly policies.

Chunk Sequence Alteration Before Reaching NIDS or Supervised Host. Offloaded stacks on NIC might impact the OS and NIDS policies described in Sect. 4. NIDS developers advise configuring NIDS instances so nothing alters the supervised traffic, such as NIC offloading. These recommendations, however, do not guarantee that such alteration does not occur on the supervised hosts themselves. We thus plan to analyze NIC's impact in future works.

[20,32] show that some middleboxes drop or reassemble IPv4 fragments, but what middlebox causes this is unclear (e.g., routers, end host's firewall). We also plan to test these middlebox reassembly policies to clarify this point.

Finally, due to well-known and unwanted transport issues, chunks may be delayed or dropped, changing the original overlap relation(s) between the chunks.

6.2   Reassembly Policy Configurability

Suricata and Snort allow one to configure reassembly policies according to the supervised host OS, while Zeek does not. We analyze and compare configurable and non-configurable reassembly policy costs in the following.

Configurability Cost. Making a NIDS configurable regarding various implementation reassembly policies necessitates several steps. As OS network protocol implementations may evolve over time, checking whether their policies have changed regularly is necessary. NIDSes should be able to easily modify and add reassembly policies as well as extend the mapping between OS versions and reassembly policies. NIDS reassembly policies must be carefully tested to ensure the NIDS reassembles consistently with OSes and that no bug was introduced. Finally, NIDS users must correctly configure their NIDS instance to associate IP addresses with reassembly policies. This configuration task is challenging as an organization's IT infrastructure may rapidly evolve and comprise hundreds (or many more) of supervised hosts. Moreover, IP addresses may be non-static, increasing the human cost of such a configuration even more. This configuration could be painlessly automated through passive OS fingerprinting [17,35] or active fingerprinting [29,30]. As several OSes may be behind an IP address, NIDSes should consider changing the IP address-based reassembly to a flow-based reassembly. We plan to investigate these challenges in future works.

Non-configurability Cost. A NIDS that does not make the reassembly policy configurable must propose another countermeasure to face overlap-based attacks. For example, an alert-based solution is possible and would consist in raising an alert whenever an inconsistent data overlap is detected (Zeek, Suricata and Snort implement such a countermeasure). Several approaches may be adopted depending on whether a chunk sequence with overlapping data is inherently considered malicious[8]. If so, there may not be the need for extra information logging, but if not, such a NIDS must log the beginning and finishing byte offsets as well as data on overlapping portions for further analysis. In any case, reassembled data from these chunk sequences must not be used (e.g., for TLS certificate extraction) because the NIDS would not know the monitored host reassembled payload.

6.3   Recommendations for OSes and NIDSes

Overlap ambiguity is a long-standing problem, as Ptacek and Newsham [27] initially reported 25 years ago. OSes have changed their reassembly policies over

---

[8] John and Olovsson's work [18] analyzed the data consistency of some Eq IPv4 fragment overlaps in 2008. But, to our knowledge, no work has systematically analyzed whether overlapping chunks with inconsistent data are observed in the wild, and if so, inferred the beningness of such chunk sequences. There are, however, benigm reasons for complete or partial overlaps to occur (for example, see [16,31]).

time. However, they still exhibit reassembly diversity. We hypothesize that this diversity is partially caused by the lack of recommendations inside IPv4 and TCP RFCs [7,14]. We recommend that the OSes implement the same policy (e.g., always use original data, ignoring overlapping fragments) so that ambiguities (slowly) disappear with new releases. Until then, NIDSes with configurable policies must propose multiple reassembly policies and, continuously testing their consistencies. The NIDSes must especially implement single mode rassemblies because they best describe OS behaviors independently of the testing context.

7   Ethical Considerations

7.1   Responsible Disclosure

We contacted Suricata, Snort, and Zeek developers about NIDS inconsistencies with respect to the latest Windows, Linux, SunOS, and FreeBSD/OpenBSD overlap reassembly policies. We gave the NIDSes some months to fix the reported issues before submitting the paper. Snort did not respond to the solicitations, and Zeek acknowledged the results. The CVE-2024-32867 [22] was assigned to the Suricata *bsd*-related misassemblies. We also notified Suricata that we found a display bug during the TCP tests. In particular, some overlapping chunk payloads appeared twice in the *payload* field of the *eve.json* file (the main logging file). This, however, does not impact the TCP buffer with which the pattern matching is done. This bug is now fixed.

7.2   Censorship Systems

Improving NIDS security and performance has the side effect of improving censorship systems (CS). Different techniques may be used to elude CS, such as using a VPN [23], encapsulating or mimicking a non-censored protocol traffic [1], inserting a packet that desynchronizes host and censorship-related stateful network traffic analysis tools [10,11,19,20,32,33]. The data overlapping strategy falls into the latest technique. Some works showed that it was possible to use data overlaps to circumvent CS at least until 2017 [19,32], but then, works reported the strategy's unusability [10,33]. Thus, our results should not affect the censorship elusion techniques currently used. Nonetheless, even if this strategy is in use to circumvent censorship systems, we consider that improving defense capabilities outweigh the negative impacts on censorship elusion techniques.

8   Conclusion

In this paper, we adapt well-known *evasion* and *insertion* attack types, refining some specific characteristics related to the overlapping ambiguity. We propose to use Allen's interval algebra-based modeling to describe chunk sequences and ensure the enumeration exhaustiveness of overlap types. This enables us to test OS reassemblies completely regarding overlapping pairs of IPv4 and IPv6 fragments as well as TCP segments. The results show that OS reassembly policies have evolved since the last testing campaigns. Overall, we demonstrate that 9 (resp. 6) out of 12 IP or TCP reassembly policies are *inconsistent* with the tested OSes for Suricata (resp. Snort). Zeek only reassembles consistently with Windows OSes the TCP overlaps. This exposes these NIDSes to insertion and evasion attacks. NIDSes with configurable reassembly policies are less subject to attacks, especially segment-based ones, since TCP policies have changed little. The CVE [22] was assigned to the Suricata *bsd*-related misassemblies we uncovered. Finally, we intend to extend the test context (e.g., multi-chunk overlaps) to completely capture OS reassembly policy complexity as test cases are reassembled differently across testing modes.

Acknowledgments. This work was supported by the French National Cybersecurity Agency, ANSSI.

References

1. Meek pluggable transport. https://support.torproject.org/glossary/meek/
2. Nmap. https://nmap.org/
3. Snort IP reassembly policies. https://snort.org/faq/readme-frag3
4. Snort TCP reassembly policies. https://snort.org/faq/readme-stream5
5. Suricata. https://suricata.io/
6. Suricata reassembly policies. https://docs.suricata.io/en/suricata-7.0.4/configuration/suricata-yaml.html#host-os-policy
7. Internet Protocol. RFC 791 (1981). https://doi.org/10.17487/RFC0791. https://www.rfc-editor.org/info/rfc791
8. Allen, J.F.: Maintaining knowledge about temporal intervals. CACM (1983)
9. Atlasis, A.: Attacking ipv6 implementation using fragmentation. Black Hat (2012)
10. Bock, K., Hughey, G., Qiang, X., Levin, D.: Geneva: evolving censorship evasion strategies. In: ACM CCS (2019)
11. Bock, K., Naval, G., Reese, K., Levin, D.: Even censors have a backup: Examining china's double https censorship middleboxes. In: ACM SIGCOMM (2021)
12. Deering, S., Hinden, R.: RFC 8200: internet protocol, version 6 (ipv6) specification (2017)
13. Di Paolo, E., Bassetti, E., Spognardi, A.: A new model for testing ipv6 fragment handling. In: ESORICS (2023)
14. Eddy, W.: Transmission Control Protocol (TCP). RFC 9293 (2022). https://doi.org/10.17487/RFC9293. https://www.rfc-editor.org/info/rfc9293
15. Feng, X., et al.: PMTUD is not panacea: revisiting IP fragmentation attacks against TCP. In: NDSS (2022)
16. Floyd, S., Mahdavi, J., Mathis, M., Podolsky, M.: RFC2883: an extension to the selective acknowledgement (SACK) option for TCP (2000)
17. Holland, J., Schmitt, P., Feamster, N., Mittal, P.: New directions in automated traffic analysis. In: ACM CCS (2021)
18. John, W., Olovsson, T.: Detection of malicious traffic on back-bone links via packet header analysis. CWIS (2008)
19. Khattak, S., Javed, M., Anderson, P.D., Paxson, V.: Towards illuminating a censorship monitor's model to facilitate evasion. In: FOCI (2013)

20. Li, F., Razaghpanah, A., Kakhki, A.M., Niaki, A.A., Choffnes, D., Gill, P., Mislove, A.: lib● erate,(n) a library for exposing (traffic-classification) rules and avoiding them efficiently. In: ACM IMC (2017)
21. Lin, B., Zhang, L., Guo, Y., Zhang, H., Fang, Y.: Research on security protection evasion mechanism based on ipv6 fragment headers. In: IEEE LCN (2024)
22. NIST: CVE-2024-32867
23. Nobori, D., Shinjo, Y.: VPN gate: a volunteer-organized public VPN relay system with blocking resistance for bypassing government censorship firewalls. In: NSDI (2014)
24. Novak, J.: Target-based fragmentation reassembly (2005)
25. Novak, J., Sturges, S.: Target-based TCP stream reassembly (2007)
26. Paxson, V.: Bro: a system for detecting network intruders in real-time. Elsevier Computer networks
27. Ptacek, T., Newsham, T.: Insertion, evasion, and denial of service: eluding network intrusion detection. Technical report, Secure Networks, Inc. (1998)
28. Roesch, M., et al.: Snort: lightweight intrusion detection for networks (1999)
29. Shamsi, Z., Loguinov, D.: Unsupervised clustering under temporal feature volatility in network stack fingerprinting. In: ACM SIGMETRICS (2016)
30. Shankar, U., Paxson, V.: Active mapping: resisting NIDS evasion without altering traffic. In: SP (2003)
31. Touch, J.: RFC 6864: updated specification of the ipv4 id field (2013)
32. Wang, Z., Cao, Y., Qian, Z., Song, C., Krishnamurthy, S.: Your state is not mine: a closer look at evading stateful internet censorship. In: ACM IMC (2017)
33. Wang, Z., Zhu, S.: SymTCP: eluding stateful deep packet inspection with automated discrepancy discovery. In: NDSS (2020)
34. Wang, Z., et al.: Themis: ambiguity-aware network intrusion detection based on symbolic model comparison. In: MTD (2021)
35. Zalewski, M.: p0f (2014). https://lcamtuf.coredump.cx/p0f3/
36. Zhang, Z., Yuan, B., Yang, K., Zou, D., Jin, H.: Statediver: testing deep packet inspection systems with state-discrepancy guidance. In: ACSAC (2022)
37. Zou, Y.H., Bai, J.J., Zhou, J., Tan, J., Qin, C., Hu, S.M.: {TCP-Fuzz}: detecting memory and semantic bugs in {TCP} stacks with fuzzing. In: ATC (2021)

Poster: On the Usage of Kernel Shadow Stacks for User-Level Programs

Marco Calavaro[✉], Pasquale Caporaso, Luca Capotombolo,
Giuseppe Bianchi, and Francesco Quaglia

University of Rome Tor Vergata, Rome, Italy
marco.calavaro@cnit.it

Abstract. Backward edge Control-Flow Integrity (CFI) has been widely supported via user-space shadow stacks. In this paper we introduce innovative kernel shadow stacks for user-space programs. By placing itself at a higher privilege level, the information kept by the kernel shadow stack can no way be altered by the (attacked) user-level code. We provide the main hints of our implementation of the kernel shadow stack for Linux, and report data related to an assessment we carried out of our proposal.

Keywords: Control flow integrity · ROP countermeasures

1 Introduction

Control Flow Integrity (CFI) is a mechanism designed to safeguard against attacks such as Return-Oriented Programming (ROP) [10]. Several CFI solutions provide the support for user-space shadow stacks (e.g., [1,11–13]) which maintain the original return addresses that could have been tampered in the thread stack area. In this article we explore a Kernel Shadow Stack (KSS) alternative.

KSS keeps the original return addresses into a kernel level memory area, hence making it fully inaccessible (and unmodifiable) by the (attacked) application code. Also, KSS fully works at software level, thus being a more flexible alternative to hardware-based support such as Intel Control-Flow Enforcement Technology (CET) [1]. In fact, hardware-based solutions typically enable detecting stack tampering, with no possibility to identify what portion of code did the stack rewriting operation, while KSS can be configured to also offer this functionality. In particular, KSS can support in-depth analysis of the user-level stack for controlling multiple return addresses just when returning from each function.

In this article we present our design of KSS for the Linux kernel and x86-64 CPUs, although the principles KSS relies on can be exploited with other kinds of operating system kernels and other types of processors. Also, our solution is usable in immediate manner since its kernel level part is fully implemented in a Linux Kernel Module (LKM) and does not require any modification/recompilation of the kernel for being used.

M. Egele et al. (Eds.): DIMVA 2025, LNCS 15748, pp. 237–242, 2025.
https://doi.org/10.1007/978-3-031-97623-0_14

The remainder of this paper is structured as follows. In Sect. 2 we discuss related work. KSS is presented in Sect. 3. Experimental data for its assessment are reported in Sect. 4.

2   Related Work

Shadow stacks safeguard a function's return address that stands on the call stack by concealing it within a separate shadow area which is still present in the user-level part of the address space of the program, hence remaining vulnerable to information disclosure and side-channel attacks, as shown in, e.g., [4,13]. Also, hardware-managed shadow stacks require the explicit usage of services for shadow-stack areas setup [2], which does not favor wide usage, especially with legacy applications. Our KSS proposal addresses these aspects by transparently relocating the shadow stack to kernel memory, and adopting a pure software solution. Additionally, the interaction with the shadow stacks kept at kernel level is reinforced by requiring Ring-0 for any modifications. This approach mitigates information leakage and substantially reduces the user-level attack surface.

The solution in [13] relies on return-ids, which map to actual return addresses. The mapping can be changed at runtime in order to (re)randomize the id that can be used to identify an address, but there is still a time window (between two re-randomizations) where the attacker can exploit the current id values and exchange them in the shadow stack area. As hinted, our KSS works at kernel level, thus preventing any possibility of being tampered by the attacker.

The proposal in [9] presents Hardware-Enforced Kernel Control-Flow Integrity (HEK-CFI), which exploits write-protected pages supported via the Intel CET for protecting kernel side return addresses. Our solution is orthogonal to this proposal since, although we exploit kernel-side operations for managing the shadow stack area, they are exploited for protecting the user-side execution flow.

The work in [5] presents a shadow stack managed through the Memory Protection Unit (MPU) on ARM Cortex-M. MPU faults lead to intercept the accesses to the stack area putting registration and elimination of return addresses under control. This solution is devised for embedded systems with a single protection level, while KSS works with user/kernel separation of protection levels.

3   Kernel Shadow Stack

The KSS architecture comprehends two components: 1) the ELF-L (ELF Loader) component, which instruments code before starting the execution of the target program, and 2) the LKM responsible for implementing shadow stacks in the kernel layer. Both of them work transparently to the application code.

The objective of the instrumentation process is twofold: 1) when passing with a thread on an instrumented CALL/RET instruction, an interaction with the kernel layer needs to be activated, so that the kernel can actually manage the shadow stack, and 2) while instrumenting the program, we rely on the binary rewriting of no more than the number of bytes that originally represent the CALL/RET instruction, hence avoiding the rewriting of other parts of the ELF.

As for CALL instructions, we manage both the ones based on displacement (which have 5-byte size) and indirect ones (which have size from 2 to 6 bytes). Each CALL instruction is instrumented by rewriting its first two bytes with a two-byte `int DISP` instruction, which simply triggers a trap—leading to pass control to the kernel layer—with specification of the index (i.e., `DISP`) of the entry of the Interrupt-Descriptor-Table (IDT) that is involved in the trap. This trap is handled by our LKM module. As for RET instructions, they are either of a single byte in length—this is the case of classical `ret` instructions—or are of three bytes in length, these are `ret imm16` instructions. We manage all the different RET instruction lengths, hence aligning our solution to the management of the minimal size (single-byte) variant of these instructions. In particular, they are instrumented by ELF-L via the rewriting of their first byte (or unique byte if the instruction is `ret` or `retf`), with a single byte illegal-opcode instruction. This way, the passage by a thread on the location where the original RET was placed gives rise to a trap—caused by the illegal-opcode—that passes again control to the LKM part of our architecture.

ELF-L packs information related to each instrumented instruction using an array of entries structured as shown in Listing 1.1.

```
struct instruction_info {
  uint8_t instruction_size;
  unsigned long
       instruction_address;
  uint8_t instruction_type;
  unsigned long
       instruction_param;
};
```

```
struct instrumentation_map_data {
  int call_num;
  int ret_num;
  instruction_info * call_array[];
  instruction_info * ret_array[];
  unsigned long randomization_offset;
};
```

Listing 1.1.   **Listing 1.2.**

The LKM includes an `ioctl(...)` for passing the above mentioned array of instrumented instruction information to the kernel. This takes place through the `instrumentation_map_data` structure shown in Listing 1.2, which is used by the kernel layer to build a hash-map of addresses of `int`-based trampolines (and actual trampoline data) or illegal-opcodes (that substituted RET instructions) associated with the current process. This structure includes the randomization offset selected by ELF-L when loading the program, thus enabling the identification of the compile time position of instrumented CALL/RET instructions. Each hash-map element has an access key made of the user-level value of RIP (instruction pointer) that is observed at kernel level when either the `trap` instruction or the illegal-opcode is passed through by the CPU.

The LKM component also takes control when new threads are activated or completed, in order to setup/remove the per-thread shadow stack area. Furthermore, per-process metadata (like the aforementioned hash-map) is removed when the process terminates. To address this aspect, we apply probes in order to take control in our LKM when: 1) the application generates a new thread, like when using `kernel_clone(...)`, and 2) a thread of the application terminates it execution, in particular when using do_exit(...). However, the actual initial setup of the shadow stack area at kernel level is not done until the created thread actually takes control of the CPU. This is done by applying a probe also to the function that performs context-switching, i.e., pick_next_task(...).

KSS is compact, and each stack element records: 1) a return address to be exploited when a thread passes onto an instrumented application-level RET instruction, and 2) the position (the memory address) where that return address should be actually present on the original user level stack of the thread.

Since the original CALL instruction has been replaced by an int instruction, in our LKM we need to simulate the execution of the CALL. To accomplish this task, the user stack pointer is decremented by 8, and the return address is stored in the corresponding user stack location. The correct return address value is obtained from the instrumentation map created by ELF-L, and then passed to the kernel via the aforementioned ioctl(...). In particular, it is determined on the basis of the original address of the instrumented CALL instruction and its size. Such return address is pre-computed by the kernel layer when the instrumentation map is received via the ioctl(...), and is logged in the kernel-level hash-map entry associated with the CALL instruction, thus making it directly available when simulating the execution of the instrumented CALL instruction.

Later, the pt_regs structure—keeping the snapshot of the CPU registers to be restored when returning to user space—is modified by modifying the RIP register value, ensuring a correct return of the execution flow to user level. This is still done exploiting the information kept in the hash-map entry associated with the instrumented instruction—we recall it holds the information we described via Listing 1.1. Before handing over control to user space, the LKM executes the same address-saving operation on the kernel shadow stack, also recoding the address of the user stack position where the saved return value is present.

As for instrumented RET instructions, whose first (or unique) initial byte has been substituted with an illegal-opcode, our LKM intercepts their execution via binary patching the original handler—the "Invalid Instruction" one associated with the 6-th entry of the IDT—and installing a wrapper. Once the LKM has control, it checks via the kernel-level hash-map of instrumented RET instructions if the management of the kernel-level shadow stack needs to be carried out. In the positive case, a pop operation is performed on the user stack, which effectively updates the stack pointer kept in the pt_regs by adding 8. Also, the RIP register kept by pt_regs is restored with the correct value of the return address, which was logged in the kernel-level stack area associated with the thread. Also, if the RET instruction whose execution is intercepted has an additional imm16 parameter, additional bytes are squashed from the stack upon returning.

Before actually performing the aforementioned updates and returning control to user space, KSS performs execution-flow verification actions. In particular, its configuration setup drives how many return addresses are checked by the LKM when executing a single instrumented RET instruction. This enables determining if the function from which we are returning has tampered not just its own return address but some other return address located lower in the user stack

Table 1. Effectiveness against the RecIPE benchmark

Attack Type	target	Total	No KSS	With KSS	Block Perc.
OOBPtrHijack	retaddr	4	4	0	100.0%
BoundOFlow	oldebp	16	16	0	100.0%
BoundOFlow	retaddr	20	20	0	100.0%
PtrHijack	retaddr	4	4	0	100.0%
NBoundOFlow	retaddr	4	4	0	100.0%
Total		48	48	0	100%

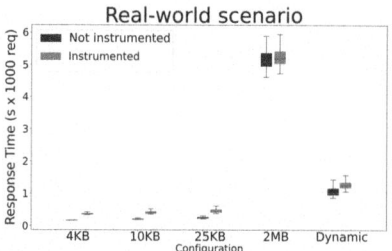

Fig. 1. Response time

area. We denote the number of return addresses that are at most checked as NUM_CHECKS. If at least one of these addresses is tampered: A) a log is produced, with the compile time address of the instrumented RET instruction; this can lead to the identification and analysis of the function that was activated, whose return has observed such tampering; B) the application is simply killed. If we configure NUM_CHECKS to the value 1 and just select mode B) listed above, we make our KSS work similarly to a hardware-managed shadow stack.

4 Experimental Data

We tested KSS on a VMware Esxi Virtual Machine running Debian 12 with Linux Kernel 6.1, configured with 4 vCPUs and 16 GB of RAM. For evaluating its effectiveness, we employed the RecIPE benchmark [8]. It includes 156 attack scenarios designed to standardize the assessment of defense mechanisms against buffer overflow vulnerabilities. We used the 64-bit version of the benchmarks and selected all possible attacks targeting the backward-edge on the stack, resulting in a subset of 48 attacks. We executed each of these attacks twice: one time on a system configuration without KSS and another time on the same system with our solution enabled. When KSS is used, we configured it with NUM_CHECKS set to the value 1 since in this test we are simply identifying the attempt by an application to use a tampered return address at some point in time. The results are reported in Table 1. Without KSS, 48 out of 48 attacks were successful for RecIPE. When KSS was enabled, successful attacks drop to 0. These results demonstrate the effectiveness of our proposal in protecting against backward-edge control-flow attacks—particularly when also considering the impossibility for the user-level code to tamper the kernel-level shadow stack area.

As for the overhead we have assessed KSS using an ample set of benchmarks (e.g. SPEC CPU 2017 [6]) and real-world applications. For space constraints, we report data related to real-world tests with Apache2, where we considered both static Web contents (pre-filled html files) and dynamic contents, generated via the interaction with a backend database. For the latter scenario, we started both Apache2 and the database, i.e. SQLite3 [7], using our ELF-L module.

The size of the static pages used reflects the median page weight taken from the last edition of the HTTP archive's Web Almanac [3], which corresponds to 2 Mb. We chose the median as a baseline and added 3 smaller sized pages. Instrumentation of CALL/RET instruction was applied to both Apache2 Web server and SQLite3 database. As for the dynamic content scenario, our tests have been setup for the case of a login operation exploiting PHP scripts for managing the validation of the provided credentials, and the setup of a session token, a common task offered by a wide set of modern Web applications.

The results in Fig. 1, demonstrate that when the median page weight of a typical Web page is used, the overhead of KSS is fully negligible. The highest overhead, slightly below 100%, is observed when the page size in the static case is reduced to less than 10 KB. According to the Web Almanac, such small page sizes account for less than 10% of Web traffic. Consequently, the practical impact of the overhead for end users is expected to remain within acceptable limits. Also, the dynamic content scenario does not show more than 25% overhead.

References

1. Complex shadow-stack updates (inteló control-flow enforcement technology) (2023). https://www.intel.com/content/www/us/en/content-details/785687/complex-shadow-stack-updates-intel-control-flow-enforcement-technology.html
2. Shadow stacks for 64-bit arm systems (2023). https://lwn.net/Articles/940403/
3. HTTP Archive: Web almanac (2022). https://almanac.httparchive.org/en/2022/
4. Burow, N., Zhang, X., Payer, M.: SoK: shining light on shadow stacks. In: 2019 IEEE Symposium on Security and Privacy (SP), pp. 985–999. IEEE (2019)
5. Choi, W., Seo, M., Lee, S., Kang, B.B.: SuM: efficient shadow stack protection on ARM cortex-m. Comput. Secur. **136**, 103568 (2024). https://doi.org/10.1016/J.COSE.2023.103568
6. Standard Performance Evaluation Corporation: SPEC CPU (2017). https://www.spec.org/cpu2017/
7. Hipp, D.R.: Sqlite (2024). https://www.sqlite.org/
8. Jiang, Y., Yap, R.H.C., Liang, Z., Rosier, H.: RecIPE: revisiting the evaluation of memory error defenses. In: Proceedings of the 2022 ACM on Asia Conference on Computer and Communications Security. ACM, New York (2022)
9. Maar, L., Nasahl, P., Mangard, S.: Beyond the edges of kernel control-flow hijacking protection with HEK-CFI. In: Proceedings of the 19th ACM Asia Conference on Computer and Communications Security, ASIA CCS 2024, Singapore, 1–5 July 2024. ACM (2024). https://doi.org/10.1145/3634737.3661135
10. Shacham, H.: Return-oriented programming: exploits without code injection. In: Proceedings of the 14th ACM Conference on Computer and Communications Security (CCS), pp. 272–283. ACM (2007). https://doi.org/10.1145/1315245.1315313
11. Smith, A.: Memory efficient shadow stack designs for real-time embedded systems (2023)
12. Zou, C., Gao, Y., Xue, J.: Practical software-based shadow stacks on x86-64. ACM Trans. Archit. Code Optim. **19**(4) (2022). https://doi.org/10.1145/3556977
13. Zou, C., Xue, J.: Burn after reading: a shadow stack with microsecond-level runtime rerandomization for protecting return addresses. In: 2020 IEEE/ACM 42nd International Conference on Software Engineering (ICSE), pp. 258–270 (2020)

Poster: Referencing Your Privileges – A Data-Only Exploit Technique for the Windows Kernel

Nicola Stauffer● and Gürkan Gür$^{(\boxtimes)}$●

Institute of Computer Science (InIT), Zurich University of Applied Sciences (ZHAW),
Winterthur, Switzerland
staufnic@students.zhaw.ch, gueu@zhaw.ch

Abstract. In this poster, we present an exploit technique using an integer overflow turned out-of-bounds write inside the paged pool to escalate our privileges on the latest Windows 11 version 24H2. We describe our exploitation strategy and demonstrate how we used it to carry out the exploit. We also identify some future technical directions that could improve this exploit technique.

1 Introduction

Microsoft is continuously raising the bar for kernel attacks. For instance, in Windows 11 24H2, they patched the previousMode technique and removed known kernel address leaks [4]. However, there is still work to be done, as a simple overflow in the paged pool can still lead to a privilege escalation, as demonstrated in this work.

In this paper, we present an exploitation strategy and describe step-by-step how we used an integer overflow turned out-of-bounds write inside the paged pool to escalate our privileges on the latest Windows 11 version 24H2[1]. This exploitation strategy was utilized against NPU driver-related bugs on Windows CoPilot+ systems as presented in BlackHat Asia 2025 (see footnote 1). Our instrumental starting point was the method about *CVE-2022-22715 (Windows Dirty Pipe)* described in [2]. However, with Windows 11 version 24H2, the bug around previousMode, which was often used for better read/write primitives, was fixed. Then, as the write primitive used in [2] by *k0shl* had side effects, we had to replace it with a new technique. Essentially, we used a reference counter to gain an arbitrary increment primitive when we duplicate a _TOKEN object. This primitive was then used to enable the SeDebug privilege in the _TOKEN object of

---

[1] This work is based on our presentation "(Mis)adventures with Copilot+: Attacking and Exploiting Windows NPU Drivers" at BlackHat Asia 2025. The related part of BlackHat Asia 2025 was using this specific technique which led to a privilege escalation using CVE-2024-36336 https://nvd.nist.gov/vuln/detail/CVE-2024-36336. This poster is mainly focused on the exploitation strategy. For the related bugs themselves, please refer to our BlackHat Asia 2025 talk.

M. Egele et al. (Eds.): DIMVA 2025, LNCS 15748, pp. 243–249, 2025.
https://doi.org/10.1007/978-3-031-97623-0_15

the current _EPROCESS and to escalate our privileges. As a positive outcome, when one closes the duplicated _TOKEN objects, their privileges will be automatically dropped again.

Overall, our exploitation strategy follows the steps below:

① Create a predictable pool layout (aka *Fengshui*) (Sect. 2)
② Overwrite a _WNF_STATE_DATA object (Sect. 2)
③ Place a _TOKEN object behind it (Sect. 2)
④ Establish arbitrary read (Sect. 3)
⑤ Use arbitrary increment (Sect. 4)
⑥ Profit! (Sect. 5)

In this paper, we describe step-by-step how we came up with this exploitation strategy and how our exploit works, and we also discuss some future technical directions.

2 Fengshui: Shaping the Paged Pool Layout

Our exploit needs a predictable paged pool layout. In that regard, there are a few resources about how the Windows kernel allocator works which can help [1,3]. We mainly utilized [1] for this purpose. The hardest part about the fengshui was choosing the size and which allocator we wanted to use. We use the Variable Size allocator for our exploit, and our fengshui involves the following three types of objects:

1. **The DirectX Escape** which will trigger the overflow—dynamically sized.
2. **_WNF_STATE_DATA** which is a common initial primitive for linear read/write, as detailed in [2]
3. **A _TOKEN object**

Interestingly, the only object we cannot control the size of is the _TOKEN. It is not a fixed-size allocation – the size depends on what is stored inside the _TOKEN->DynamicPart (possibly, Discretionary Access Control List (DACL) related). For our purposes, _TOKEN objects had a size around 0x7c0, so we adjusted our _WNF_STATE_DATA and vulnerable object to the same size.

Then our strategy emerges as the following (also depicted in Fig. 1):

1. Spray enough _WNF_STATE_DATA to fill up all existing holes and enable dynamic look-aside for this allocation size.
2. Free one _WNF_STATE_DATA.
3. Trigger allocation of our vulnerable object and overflow AllocatedSize and parts of DataSize.
4. Use the corrupted _WNF_STATE_DATA to overwrite fields in the next object with desired values.
5. Replace the non-corrupted _WNF_STATE_DATA objects with _TOKEN objects.
6. Gain full control over a _TOKEN object.

Exploiting involves modifying AllocatedSize and DataSize to enable linear out-of-bounds access using NtQueryWnfStateData and NtUpdateWnfStateData.

Fig. 1. Fengshui layout - _WNF_STATE_DATA and _TOKEN object arrangement

3 Arbitrary Read

After a reliable linear read/write primitive is established, we need arbitrary read capability. To establish an arbitrary read we can modify the _TOKEN->BnoIsolationHandlesEntry and point it to a controlled userspace address as explained in [2]. At this userspace address, we can now forge a _SEP_CACHED_HANDLES_ENTRY. The EntryDescriptor.IsolationPrefix.Buffer at offset 0x30 can be pointed to the kernel address we want to read. Moreover, EntryDescriptor.IsolationPrefix.MaximumLength at offset 0x28 defines how much bytes will be read.

3.1 Locating _EPROCESS

We used the same strategy as used by *StarLabs* [5] in their exploit for CVE-2021-31969. We execute the following steps for that purpose:

1. Search the page containing the _TOKEN->SessionObject of our current Token for an allocation with the tag AlIn
2. Read the pointer at offset 0x38 as it contains an IoCompletion object
3. Search the page containing the IoCompletion object for an allocation with the tag EtwR.
4. Read the pointer at offset 0x30 as it contains an _EPROCESS object
5. Iterate over the ActiveProcessLinks linked list of the _EPROCESS to find our own _EPROCESS object by comparing the UniqueProcessId field to our current process id

Why _EPROCESS? The _EPROCESS is utilized because it has again a pointer to the currently used _TOKEN which contains the privileges held by our current process. Consequently, if we can change the fields _TOKEN.Privileges.Present and _TOKEN.Privileges.Enabled in the _TOKEN of our current process, we can enable powerful SE_Privileges. Those privileges allow us to bypass numerous security checks. For example, with SeDebugPrivilege, we can obtain a handle with full rights to the winlogon.exe process and spawn a subprocess with it.

4 Arbitrary Increment

After Microsoft patched the *previousMode* technique with Windows 11 24H2, the arbitrary write technique described in [2] got borderline impossible to use in a simple way. Basically, it impairs the values surrounding the value we write. Previously, this was not a problem when we simply could set the previousMode to 0 and then use NtReadVirtualMemory and NtWriteVirtualMemory to clean up the corrupted values due to our operation. Therefore, we had to find a different way to enable SeDebugPrivilege for our process. As we only need to flip a bit inside the fields _TOKEN.Privileges.Present and _TOKEN.Privileges.Enabled, we realized a stable arbitrary increment would suffice. After checking the _- TOKEN object, it was evident that _TOKEN->BnoIsolationHandlesEntry has the following type:

```
1  struct _SEP_CACHED_HANDLES_ENTRY
2  {
3      struct _RTL_DYNAMIC_HASH_TABLE_ENTRY HashEntry;              //0x0
4      LONGLONG ReferenceCount;                                     //0x18
5      struct _SEP_CACHED_HANDLES_ENTRY_DESCRIPTOR EntryDescriptor; //0x20
6      ULONG HandleCount;                                           //0x38
7      VOID** Handles;                                              //0x40
8  };
```

Fig. 2. Increment when a _TOKEN handle is duplicated

At offset 0x18, there is an ReferenceCount counter, which could be abused as an increment primitive. Because if we could duplicate a _TOKEN and only the field ReferenceCount gets changed, that would give us the necessary capability. As seen in Fig. 2 if we duplicate a _TOKEN and _TOKEN->BnoIsolationHandlesEntry is set and the _TOKEN->BnoIsolationHandlesEntry.ReferenceCount is less than 0x7fffffffffffffff, only the _TOKEN->BnoIsolationHandlesEntry.Referen- ceCount is changed.

The Decrement Operation. The reference count is decremented once a _- TOKEN handle is closed. As long as that value does not get decremented to 0, which would try to free the _SEP_CACHED_HANDLES_ENTRY, the exploit should still be able to go forward.

If we look at the Present and Enabled fields of a process token running in low integrity, their values are between 0 and 0x7fffffffffffffff as seen in Fig. 3, which is required for our new primitive. As a positive side effect, when we do not need SeDebugPrivilege anymore, we can simply close the handles to our duplicated tokens and it will drop our privileges again.

```
1: kd> dx -id 0,0,ffffc10c0e68d040 -r1 (*((ntkrnlmp!_SEP_TOKEN_PRIVILEGES *)0xfffffe48c47990880))
(*((ntkrnlmp!_SEP_TOKEN_PRIVILEGES *)0xfffffe48c47990880))                    [Type: _SEP_TOKEN_PRIVILEGES]
    [+0x000] Present          : 0x602880000 [Type: unsigned __int64]
    [+0x008] Enabled          : 0x800000 [Type: unsigned __int64]
    [+0x010] EnabledByDefault : 0x40800000 [Type: unsigned __int64]
```

Fig. 3. Privilege struct in low integrity process token

Determining the _TOKEN Duplication Count and Where to Point _TOKEN->BnoIsolationHandlesEntry.ReferenceCount. For this purpose, SE_Privileges provides the instrument. Each SE_Privilege has a value which corresponds to the bit offset inside the Present and Enabled fields which must be 1 so the privilege is enabled. There is only one slight problem: we have a byte precision increment, and we need to flip a bit. To minimize the amount of _TOKEN we need to duplicate, we implemented a simple algorithm. The first thing we need to find is which byte we need to increment. We called this offset. We can calculate this by dividing the privilege value by 8 with an integer division.

At this stage, we need to calculate how many times we need to increment our target byte. For that, we just check the last byte of our privilege value and see which bit we need to flip. We named this property amount:

```
 1 typedef struct _PrivOffsets {
 2     uint16_t amount;
 3     uint16_t offset;
 4 } PrivOffsets;
 5
 6 consteval PrivOffsets CalcPrivIncrement(PRIV priv) {
 7     PrivOffsets offset = { 0 };
 8     offset.offset = priv / 8;
 9     offset.amount = 1 <<  ((priv & 0x7));
10     return offset;
11 }
```

5 Increment in Action

Now let's assemble our new primitive and put it to use:

1. We point token->BnoIsolationHandlesEntry (from the token we r/w into) to our process tokens _TOKEN.Privileges.Present - 0x18 + PrivOffsets.offset
2. Duplicate the controlled token for PrivOffsets.amount
3. Point token->BnoIsolationHandlesEntry (from the token we r/w into) to our process tokens _TOKEN.Privileges.Enabled - 0x18 + PrivOffsets.offset
4. Duplicate the controlled token for PrivOffsets.amount
5. Enjoy your SE_Privileges -> time to profit!

The relevant part of the exploit code is shown below (We omitted the write to the kernel memory after setting the `BnoIsolationHandlesEntry` to make the code more concise):

```
 1  HANDLE* buf_pres = (HANDLE*)calloc(privs.amount, sizeof(HANDLE));
 2  if (buf_pres == NULL) { return NULL; }
 3
 4  HANDLE* buf_en = (HANDLE*)calloc(privs.amount, sizeof(HANDLE));
 5  if (buf_en == NULL) { return NULL; }
 6
 7  token->BnoIsolationHandlesEntry = (void*)(process_token_addr + offsetof(_TOKEN,
         Privileges) + offsetof(_SEP_TOKEN_PRIVILEGES, Present) + privs.offset - 0
         x18);
 8  for (int i = 0; i < privs.amount; i++) {
 9      bool bool_result = DuplicateToken(manipulated_token, SecurityAnonymous, &
             buf_pres[i]);
10      if (!bool_result) {
11          printf("Duplicate token failed pres\n");
12      }
13  }
14
15  token->BnoIsolationHandlesEntry = (void*)(process_token_addr + offsetof(_TOKEN,
         Privileges) + offsetof(_SEP_TOKEN_PRIVILEGES, Enabled) + privs.offset - 0
         x18);
16
17  for (int i = 0; i < privs.amount; i++) {
18      bool bool_result = DuplicateToken(manipulated_token, SecurityAnonymous, &
             buf_en[i]);
19      if (!bool_result) {
20          printf("Duplicate token failed pres\n");
21      }
22  }
```

When we finish using our `SE_Privileges`, we can simply close the handles and the privilege will be gone again:

```
 1  BOOL res = true;
 2  for (int i = 0; i < privs.amount; i++) {
 3      res = CloseHandle(buf_en[i]);
 4      res = CloseHandle(buf_pres[i]);
 5  }
 6  free(buf_en);
 7  free(buf_pres);
```

6   Conclusion

We have presented an exploit technique that can be used for paged-pool overflows in Windows 11 24H2. We have built on prior work in [2,5] and implemented a feasible exploit on that latest Windows kernel. This shows while Microsoft is making progress, there is still more work ahead to make the Windows OS more secure. A potential improvement for this exploit technique would be future-proofing, so a mitigation like Supervisor Mode Access Prevention (SMAP) cannot affect it.

References

1. Bayet, C., Fariello, P.: Scoop the Windows 10 pool! In: Symposium sur la sécurité des technologies de l'information et des communications (2020)
2. k0shl: Break me out of sandbox in old pipe: CVE-2022-22715 (Windows dirty pipe) (2022). https://whereisk0shl.top/post/break-me-out-of-sandbox-in-old-pipe-cve-2022-22715-windows-dirty-pipe. Accessed 26 Oct 2024
3. Shafir, Y.: Windows heap-backed pool: the good, the bad, and the encoded. In: Black Hat USA 2021 (2021)
4. Shafir, Y.: KASLR leaks restriction (2024). https://windows-internals.com/kaslr-leaks-restriction/. Accessed 12 Nov 2024
5. Star Labs: Exploitation of a kernel pool overflow from a restrictive chunk size: CVE-2021-31969 (2023). https://starlabs.sg/blog/2023/11-exploitation-of-a-kernel-pool-overflow-from-a-restrictive-chunk-size-cve-2021-31969/. Accessed 26 Oct 2024

Resilient Systems

PackHero: A Scalable Graph-Based Approach for Efficient Packer Identification

Marco Di Gennaro$^{(\boxtimes)}$, Mario D' Onghia, Mario Polino, Stefano Zanero, and Michele Carminati

Dipartimento di Elettronica, Informazione e Bioingegneria, Politecnico di Milano, Milan, Italy
{marco.digennaro,mario.donghia,mario.polino,
stefano.zanero,michele.carminati}@polimi.it

Abstract. Anti-analysis techniques, particularly packing, challenge malware analysts, making packer identification fundamental. Existing packer identifiers have significant limitations: signature-based methods lack flexibility and struggle against dynamic evasion, while Machine Learning approaches require extensive training data, limiting scalability and adaptability. Consequently, achieving accurate and adaptable packer identification remains an open problem. This paper presents Pack-Hero, a scalable and efficient methodology for identifying packers using a novel static approach. PackHero employs a Graph Matching Network and clustering to match and group Call Graphs from programs packed with known packers. We evaluate our approach on a public dataset of malware and benign samples packed with various packers, demonstrating its effectiveness and scalability across varying sample sizes. Pack-Hero achieves a macro-average F1-score of 93.7% with just 10 samples per packer, improving to 98.3% with 100 samples. Notably, PackHero requires fewer samples to achieve stable performance compared to other Machine Learning-based tools. Overall, PackHero matches the performance of State-of-the-art signature-based tools, outperforming them in handling Virtualization-based packers such as Themida/Winlicense, with a recall of 100%.

Keywords: Packer Identification · Graph Similarity · Graph ML

1 Introduction

Code packing, a widely used *anti-analysis technique* [10], affects the performance of both Machine Learning (ML)-based and traditional signature-based malware detection systems [22]. Packers encrypt or compress executable code, rendering static analysis ineffective [34]. At runtime, an *unpacking stub* embedded in the executable restores the original code by decrypting or decompressing it in memory, allowing the program to execute. This ability to bypass static analysis makes packing particularly appealing to malware authors. The prevalence of

M. Egele et al. (Eds.): DIMVA 2025, LNCS 15748, pp. 253–274, 2026.
https://doi.org/10.1007/978-3-031-97623-0_16

packed malware can bias ML-based detectors into flagging all packed executables as malicious [1]. However, this assumption is flawed, as many benign programs (*goodware*) are also packed to protect intellectual property. For instance, Rahbarinia et al. [31] found that 54% of benign programs and 58% of malware samples in their study were processed with known packers, demonstrating a similar distribution.

Recovering packed code might seem feasible through *dynamic analysis*, as executing a packed program can trigger it to unpack itself. However, modern malware increasingly employs dynamic *evasive behaviors* designed to detect analysis environments and prevent the program from exposing its true functionality at runtime [10]. As a result, packed malware with evasive tactics often resists dynamic-based analysis techniques. Additionally, incorporating dynamic analysis mechanisms into commercial AVs presents challenges, such as requiring kernel-level privileges to execute untrusted code [1,9] and introducing significant computational overhead due to the virtualization infrastructure [22]. Alternatively, *static* identification of the specific packer used in a malware sample could allow AVs to retrieve the original code, if possible, by executing a corresponding unpacker when available. Previous works in this area have applied signature-based methods [13] or ML-based algorithms using static features [15,19,33]. While effective for known packers, these approaches demand substantial effort to accommodate new packers or variations of existing ones. This challenge is amplified by the frequent emergence of *custom packers* in novel malware [33], necessitating either extensive manual signature analysis (e.g., with *Detect It Easy* [13]) or complete re-training of ML-based models. Recent work proposed *PackGenome*, a tool that automates YARA rule generation from packed samples to detect packed binaries [16]. While effective on large and heterogeneous datasets, it relies on dynamic analysis, requiring packers to generate custom-packed samples, limiting the integration of newly discovered packers. These limitations motivated our research into new methodologies for code packer identification, focusing on minimizing packer integration effort. The primary challenge lies in achieving a balance between *accuracy*, rapid *adaptability* for integrating newly discovered packers, resilience against dynamic *evasion* techniques, and overall *scalability*.

This paper presents PackHero, a packer identifier that leverages statically extracted Call Graphs from packed programs. PackHero extracts the CG of a given binary and identifies the packer by comparing it with previously labeled graphs in a stored collection. The graph representation is inspired by the work of X. Li et al. [17]. CGs enable a high level of abstraction and reveal that portions of these graphs remain identical or similar for binaries packed with the same packer. To leverage this, we introduce a heuristic to isolate the graph segment corresponding to the unpacking stub, identifying unique patterns shared by binaries processed by the same packer. To solve the graph similarity problem, we use a specific Graph Neural Network (GNN) [11], known as a Graph Matching Network (GMN) [18]. Additionally, PackHero incorporates a hierarchical clustering approach to group similar graphs, enhancing identification accuracy while ensuring constant inference time when integrating new packers.

We evaluate PackHero on a publicly available dataset of packed Windows Portable Executables (PEs) [1], containing both malware and benign samples, repacked with various commercial and free packers, categorized by complexity according to existing taxonomies [34]. PackHero achieves a macro-average F1-score of 93.7% and an accuracy of 98.7% using only 10 programs per packer during configuration. In its best configuration, utilizing 100 samples per packer, it reaches a macro-average F1-score of 98.3% and an accuracy of 99.8%. The scalability of PackHero, supported by its clustering approach, is validated through comparisons with a non-clustering version in terms of both performance and inference calls to the GMN. Once configured, PackHero requires significantly fewer samples to converge and stabilize than existing ML-based tools, needing just 10 samples versus 40 for the best-performing alternative. Moreover, PackHero features a constant integration cost, whereas the integration cost of other ML-based tools increases linearly with the number of packers recognized. PackHero is a robust alternative to signature-based detection tools, achieving performance aligned with SotA tools. Notably, it performs significantly better against virtualization-based [32] packers like Themida/Winlicense, which employ advanced dynamic evasive behaviors that hinder signature extraction in dynamic analysis-based tools. Specifically, PackHero achieves a perfect recall of 100% on this packer, compared to 92% for DIE and 31% for PackGenome.

In summary, the contributions are the following:

- A hybrid ML and graph signature-based approach for packer identification, enabling automatic and scalable integration of both accessible and non-accessible packers (e.g., custom or closed-source) directly from packed programs throughout the tool lifecycle.
- A heuristic to statically extract a Call Graph of an unpacking routine or a part of it from a packed binary.
- A combination of Graph Matching Network (GMN) with hierarchical clustering to enhance the accuracy and reduce the search space of similar graphs.
- We release PackHero's source code[1] for reproducibility.

2   Background and Motivation

Code packing is a widely used anti-analysis technique [24], where packers, acting as third-party software, transform a program's structure and content, recovering the original software at runtime via a *tail jump* to the original entry point. Initially intended for file compression, most packers now aim to obfuscate and hinder program analysis in legitimate and malicious software. Packers are classified by runtime complexity [34] into six types (I–VI), with most common packers falling within types I–III. Another taxonomy focuses on obfuscation methods, distinguishing *compressors*, *crypters*, and *virtualization-based*

---

[1] https://github.com/necst/packhero

packers, such as Themida [32], which translate code into virtual instructions and implement advanced anti-dynamic analysis techniques. In our experiments, we consider packers from types I–III, with Themida representing the VM-based category.

2.1 Packer Identification

Packer identification is a multi-label classification task aimed at determining the specific packer used to compress or obfuscate a program. This capability allows AV tools to statically unpack programs, thereby enhancing malware detection [22]. In contrast, *packer detection* identifies whether a program is packed, often employing static methods such as similarity comparisons [14,28] or entropy analysis [2,12]. However, these methods are less effective against *low-entropy* packers [22]. This paper focuses on packer identification and categorizes existing approaches into two main families: signature-based methods, which rely on manually or automatically generated signatures, and pattern recognition techniques, predominantly driven by ML-based algorithms.

Signature-Based Methods. Packers often leave specific artifacts that can be used to create signature databases. *Detect It Easy (DIE)* is a well-known tool for packer identification via signature matching [13], outperforming tools like PEiD [27] with its open architecture that allows users to add JavaScript-like scripts for packer detection. However, it requires the manual creation of signature scripts for new packers and their variants, making it challenging to integrate new packers, especially with limited analyzable packed samples. A key limitation of signature-based detection is the need to analyze many samples to identify invariant byte sequences that can be used as signatures. To address this, researchers have explored automating the signature extraction process. Raff et al. propose a method to automatically generate YARA rules [30], a format for defining malware characteristics [3]. Nevertheless, code packing can still easily defeat these rules, similar to other signature schemes. To the best of our knowledge, the State-of-the-art (SotA) tool for signature generation in packed programs is PackGenome [16]. Inspired by biological processes, PackGenome identifies significant instructions in the first unpacking layer (the only statically visible one). It uses Intel *Pintool* [21] to monitor packed programs in a controlled environment, recording and labeling instructions that write "unpacked" instructions. By analyzing multiple executions of programs packed with the same packer and applying similarity metrics, PackGenome extracts *packer-specific* "genes" to generate YARA rules. However, this approach relies on dynamic analysis, making it vulnerable to evasive techniques that hinder the extraction of relevant genes, as empirically confirmed in our experimental evaluation (Subsect. 4.5). Accurate packer identification often requires generating a large number of signatures. For instance, PackGenome recommends using the actual packer to create extensive variations of the unpacking stub. However, this approach is impractical in real-world scenarios where malicious software frequently employs custom packers that are inaccessible to analysts. Additionally, the limited availability of samples for such packers makes it infeasible to build a comprehensive signature database.

Table 1. Packer identifiers and compliance with requirements R1–R4.

Work	Analysis Type	Approach	Code	Replicable	R1	R2	R3	R4
PackGenome [16]	Static + Dynamic	Signature	✓	✓	✓	✓		✓
Detect It Easy (DIE) [13]	Static	Signature	✓	✓	✓		✓	✓
Randomness Profiles [33]	Static	ML		✓	✓	✓	✓	
Binary Diffing [15]	Static	ML		✓	✓	✓	✓	
2SPIFF [19]	Static	ML		✓	✓	✓	✓	
CEG [17]	Static	Signature			✓	✓	✓	
PackHero (our approach)	Static	ML + Signature	✓	✓	✓	✓	✓	✓

ML-Based Packer Identification. The second family of identification methods relies on pattern recognition algorithms, particularly Machine Learning techniques. Proposed approaches include constructing randomness profiles for packed samples [33], applying binary diffing [15], extracting features from the topology of CGs [19], and evaluating the similarity of Consistently-Executing Graphs (CEGs) [17]. S. Li et al. observed that while most packers significantly affect binary entropy, individual packers exhibit distinctive randomness patterns [33]. They used sliding windows [8] to build randomness profiles, training a k-nearest neighbor classifier. Kim et al. employed an SVM classifier with binary diffing measures as kernels, achieving the best performance using the *longest common substring* computed from the first 15 bytes at each program's entry point [15]. This method leverages the similarity of initial instructions in unpacking stubs from the same packer, but its effectiveness is lowered by code obfuscation [16,37]. Hao et al. represented packed programs as CGs and trained an SVM classifier using topological features (e.g., entry point indegree) and general file information like size or section count [19]. While we draw on this idea to represent packed programs through CGs, our methodology directly uses graphs, offering better generalization and results. X. Li et al. [17] proposed a similar approach by comparing graphs using a Weisfeler-Lehman shortest path kernel. Instead of employing CGs, they introduce CEGs to simulate static execution points by traversing procedures, locating branch instructions, and forming flow paths. However, they do not address the challenge posed by the increasing number of graphs that must be compared during identification due to the introduction of new packers. To the best of our knowledge, none of these works have released their code. However, three of the four approaches were straightforward to implement, enabling their comparison with our method (results in Subsect. 4.4). The fourth, CEG, relies on a heuristic for graph extraction, making reimplementation challenging without significant assumptions. Therefore, it was excluded from our study.

2.2   Motivation

Given the limitations of current works, we define key requirements for a novel packer identifier. Table 1 shows that existing SotA approaches only partially meet these requirements, highlighting the need for our solution.

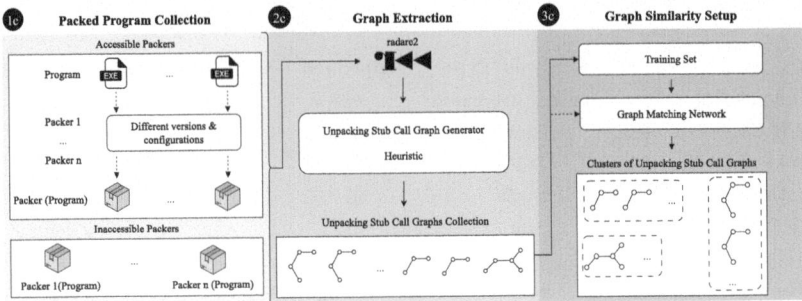

Fig. 1. PackHero Configuration Workflow.

Requirement 1 - High Identification Accuracy. It must achieve high accuracy and low false positives across diverse packer types.

Requirement 2 - Efficient Packer Integration. It must efficiently integrate packers using a limited number of real-world samples, addressing challenges such as inaccessible packers and variations within a single packer family. It should rapidly update its identification capabilities without requiring extensive or frequent retraining. To achieve this, the system should leverage robust algorithms that generalize from existing data, enabling the detection of new variations by integrating them using limited samples with minimal manual intervention.

Requirement 3 - Dynamic Evasive Behavior Management. It must effectively handle evasive behaviors encountered during the collection of wild samples for integration. Wild samples may employ dynamic evasion techniques or be so damaged or corrupted that execution is impossible [35], thereby complicating dynamic signature extraction performed by state-of-the-art tools [16]. To address this challenge, a static analysis-based approach can mitigate these issues by extracting valuable information and artifacts independently of execution.

Requirement 4 - Scalability. It must handle a growing number of packers without performance degradation, integrating new ones seamlessly without major architectural changes or resource demands. In other words, it must handle many/several packer families in parallel.

3 PackHero

PackHero is a *packer identifier* that determines the specific packer used for a given *packed* program. Its approach mirrors the workflow of signature-based detection mechanisms but uses graph "signature" to represent packed programs, with matches determined by similarity rather than exact matching. PackHero leverages a specialized Graph Neural Network (GNN) called Graph Matching Network (GMN) [18]. It operates on *Call Graphs (CGs)* extracted using heuristics. CGs, which represent the invocation relationships between functions in an executable program [6], are chosen for their compact structure and high level of abstraction. Compared to other binary graph representations (e.g., Control-Flow Graph (CFG) and Data Dependence Graph (DDG)), CGs enable an efficient resolution of the similarity problem with GMNs. We divide our approach into two main phases: *configuration* and *inference*.

3.1   Configuration

As depicted in Fig. 1, this phase involves three main step.

1c Collecting Packed Programs. The first step consists of collecting programs for the packers we want to integrate into the tool. It is important to distinguish between an *accessible* and *non-accessible* packer. The former enables the use of the actual code packer to generate packed samples, including all possible versions and configurations. This case is, therefore, ideal. Hence, we consider the "non-accessible packers" scenario to be the general case.

2c Graphs Extraction. PackHero extracts a Call Graph for each collected program. Our implementation relies on radare2 [29] to analyze and extract the CGs. Each vertex of a CG consists of 12 features extracted using radare2 (shown in Table 2). Furthermore, a heuristic designed to filter the unpacking stub part of the CG is applied to simplify the topology of each graph (details in Subsect. 3.3). PackHero collects the generated CGs into a Database (DB) of graphs.

3c Graph Matching Network Training. PackHero identifies intrinsic similarities between extracted CGs using a Graph Matching Network (GMN) [18], a specialized Graph Neural Network (GNN). The GMN processes pairs of graphs and outputs a numeric vector (*embedding*) for each graph. These embeddings result from information propagation between the two graphs, differing from traditional embedding techniques [11] that compute embeddings solely from individual graphs. To train the GMN, we label graph pairs as "similar" if

Table 2. Node features.

Feature	Description
type	Whether it is an internal function, a library imported function, or an entry point containing function
size	The size of the function in bytes
real size	The function size in bytes, including any padding
is pure	Indicates whether the function has any side effects such as modifying external variables or writing to files
calling conventions	The number of calling conventions used by the function
number of basic blocks	The number of basic blocks in the function
number of instructions	The total number of instructions in the function
number of local variables	The number of local variables declared within the function
number of arguments	The number of arguments of the function
edges	The number of edges between basic blocks
indegree	The in-degree of the function in the call graph
outdegree	The in-degree of the function in the call graph

they originate from the same packer and "dissimilar" otherwise. The network is trained to minimize the distance between embeddings of similar graph pairs while maximizing the distance for dissimilar pairs. The loss function is defined as $L(G_1, G_2) = \mathbb{E}_{(G_1, G_2, l)} [\max\{0, \gamma - l(1 - \cos(G_1, G_2))\}]$, where $l \in \{-1, 1\}$ is the label associated with the pair of graphs $< G_1, G_2 >$, γ is a margin parameter and cos is the cosine similarity [38] between the two graphs. In this case, the cos is intended as the cosine similarity between the two embeddings of size 256 extracted from the two graphs via the GMN. Lastly, $\mathbb{E}$ is the empirical risk we want to minimize, which can be done through stochastic gradient descent. The overall GMN design follows the original implementation of the paper that introduced it [18]. Finally, we propose a clustering approach to stabilize the number of matches required to identify each packer and improve the overall performance of the framework. Therefore, we also store the clusters and their respective medoids. Finally, we compute a cluster-specific threshold $t_c = \frac{1}{n^2} \sum_{i=1}^{n} \sum_{j=1}^{n} cos(G_i, G_j) - \sigma$. Namely, the average cosine distance between pairs of graphs belonging to the cluster minus the standard deviation. Such a threshold will then be used at inference time.

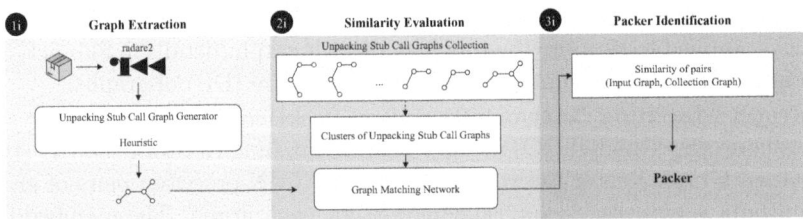

Fig. 2. PackHero Inference Workflow.

3.2 Inference

This phase comprises three steps, depicted in Fig. 2.

1️⃣ Graph Extraction. PackHero must first obtain the CG specific to the unpacking stub extracted through the previously mentioned heuristic.

2️⃣ Similarity Evaluation. The second step consists of evaluating the similarity between the embeddings computed by the GMN for the input graph and the graphs in the DB. Comparing the input graph against all graphs in the DB may be computationally expensive and decrease the general identification performance. Hence, PackHero computes the cosine similarity between the input graph and the medoids associated with each computed cluster to select the "closer" clusters. In other words, each PackHero identification corresponds at least to m GMN inferences, where m is the number of clusters. PackHero selects clusters represented by medoids with a positive cosine similarity with the input graph. Once the clusters are selected, PackHero evaluates the similarity between the input graph and each graph contained in the selected clusters.

Algorithm 1. Extract Unpacking Stub Call Graph

1: $\mathcal{F}$: Set of all functions with call references, $\mathcal{E}$: Set of entry points in the binary
2: **procedure** UNPACKINGSTUBCG($\mathcal{F}, \mathcal{E}$)
3:      $\mathcal{G} \leftarrow$ directed global call graph from $\mathcal{F}$, $C \leftarrow \emptyset$
4:      **for** $e \in \mathcal{E}$ **do**
5:          **if** $\mathcal{G}.hasEdges(e)$ **then**
6:              $C \leftarrow C \cup$ getConnectedComponent($\mathcal{G}, e$)
7:          **end if**
8:      **end for**
9:      $\mathcal{G} \leftarrow C \neq \emptyset$? C : getComponent($\mathcal{G}, \mathcal{E}$) $\cup \mathcal{E}$
10:     **if** $\mathcal{G}.isEmpty()$ **then**
11:         $\mathcal{G} \leftarrow \mathcal{E} \cup$ externalLibraries()
12:     **end if**
13:     **return** $\mathcal{G}$
14: **end procedure**

---

③ Packer Identification. Now, up to m clusters are identified as potential matches, and the similarity between the input graph and all graphs within these m clusters is computed. PackHero identifies the packer with the highest score $s_p := \frac{\sum_{C \in \mathcal{C}_p} \sum_{G_c \in C} \mathbf{1}(\cos(G_{\text{input}}, G_c) \geq t_c)}{\max_{p \in \mathcal{P}} \sum_{C \in \mathcal{C}_p} |C|}$, where G_{input} is the graph extracted from the input program, C_p the set of selected clusters for a packer p, G_c indicates a graph in cluster C, and t_c the threshold for cluster C. $\mathbf{1}(\cos(G_{\text{input}}, G_c) > t_c)$ is a membership function that outputs 1 if the cosine similarity $\cos(G_{\text{input}}, G_c)$ is greater or equal to the threshold t_c, and 0 otherwise. Moreover, $\mathcal{P}$ is the set of all included packers in the selected clusters. Lastly, $\sum_{C \in \mathcal{C}_p} |C|$ is the cardinality of samples in the selected clusters from a packer p. If no cluster is sufficiently "close" to the input CG, PackHero labels the packer as "unknown".

3.3   Extracting the CG of the Unpacking Stub

The Call Graph (CG) is a widely adopted structure [23]. It is also used in security-related tasks such as malware detection [20]. Our approach is based on a principle of "same packer, similar CGs" [19]. To the best of our knowledge, we are the first to exploit this similarity directly. The adoption of CGs offers several advantages. Their structure is straightforward to obtain [6]. Moreover, unlike other binary graph representations, a CG represents the program at a higher level of abstraction. This makes it a compact yet information-rich program representation, which is particularly well-suited for a GMN [18].

To better represent the logic behind a packer, it is necessary to filter the graph to get the unpacking stub. To systematically obtain this filtered CG, we design a heuristic shown in Algorithm 1. The intuitions behind it are: (i) the unpacking stub, or part of it (case of a multilayer packer), must be the first part of the code to be executed, and (ii) except for further obfuscation of the unpacking stub, a part of this routine is always statically visible. Therefore, the heuristic extracts the unpacking stub by exploiting the concept of connected components in undirected graphs, i.e., a subgraph where each pair of nodes is connected via a path [7]. Notice that CGs are directed graphs, but the algorithm requires undirected ones, thereby we convert the CGs into undirected graphs. Given the

Algorithm 2. Packer Call Graphs Clustering

1: DB: a collection of unpacking stub call graphs, $\mathcal{M}$: GMN trained model $\mathcal{P}$: Mapping from graphs
   to their respective packers
2: **procedure** GETCLUSTERING(DB, $\mathcal{M}$, $\mathcal{P}$)
3:     $\mathcal{C} \leftarrow \emptyset$ ▷ Initialize clustering result
4:     **for** each unique packer $p \in \mathcal{P}$ **do**
5:         DB$_p \leftarrow \{G \in \text{DB} \mid \mathcal{P}(G) = p\}$
6:         $\mathcal{D} \leftarrow$ initialize empty distance matrix for DB$_p$
7:         **for** $G_i, G_j \in \text{DB}_p, i \neq j$ **do**
8:             $d \leftarrow$ cosineSimilarity($\mathcal{M}(G_i, G_j)$)
9:             $\mathcal{D}.update(d)$ ▷ Update distance matrix with similarity
10:         **end for**
11:         $\mathcal{C}_p \leftarrow$ hierarchicalClustering($\mathcal{D}$)
12:         **for** $C_j \in \mathcal{C}_p$ **do**
13:             C_j.representative $\leftarrow$ medoid(C_j)
14:         **end for**
15:         $\mathcal{C}.update(\mathcal{C}_p)$
16:     **end for**
17:     **return** $\mathcal{C}$
18: **end procedure**

---

packer could disrupt links between functions, it should create multiple connected components in the CG. Thus, the idea is to extract the connected component containing the program entry point. At the same time, some packers affect the program entry point to make the analysis harder. For instance, analyzing Call Graphs extracted from binaries packed with ASPack [5], we noticed the common part among all the graphs was a second connected component in addition to the single entry function node, which appears to be isolated. Thus, when the entry function is not connected to any other node, a second connected component is maintained in the graph along with the entry function. Otherwise, if the graphs have no edges (UPX [36]), we keep only the program entry functions and any functions from external libraries. As the heuristic suggests, we do not consider a fixed number of functions for each graph. In our experimental evaluation, the average number of functions in the unpacking stubs is ≈ 3.

3.4   Graphs Clustering

Integrating a new packer requires collecting additional graphs. Without clustering, identifying a packer involves matching against *all* graphs in the DB, increasing inference time as new packers are added. To ensure scalability, we introduce a clustering approach that reduces inference time and improves identification performance, as demonstrated in our Experimental Validation. Each cluster contains graphs from only a single packer, allowing the identification of potential sub-groups within the same packer. This *packer unicity* is ensured by constructing the distance matrix in an intra-packer manner, as expressed in Algorithm 2. This approach can mitigate variations in unpacking stubs due to different configurations or versions of the same packer [16]. PackHero employs hierarchical clustering with a single linkage merge criterion, using a distance matrix derived from the trained GMN as input. The silhouette score [38] determines the optimal number of flat clusters for each packer. Finally, PackHero computes a medoid for each cluster, representing the graph with the minimal sum of dissimilarities to all other graphs in the cluster [38].

4   Experimental Validation

We evaluate PackHero through the following four research questions:

RQ1. What is the minimum number of programs required for PackHero's configuration to recognize packers effectively? In addition, once configured, is PackHero able to recognize different packers?

RQ2. How does the clustering-based approach impact PackHero's performance and scalability compared to its non-clustering version?

RQ3. Given an already configured PackHero, how many samples does it require to successfully integrate a new, unseen packer, and how does this integration compare to other ML-based tools?

RQ4. Does PackHero perform better than signature-based tools?

4.1   Experimental Setup

Dataset. We use the *lab* dataset from H. Aghakhani et al. [1], created by repacking Portable Executables (PEs) from benign and malicious Windows x86 software collected from a commercial anti-malware vendor and the EMBER dataset [4]. The samples were repacked using nine widely recognized packers: kkrunchy, MPRESS, Obsidium, PECompact, PELock, Petite, tElock, Themida, and UPX. The dataset includes different packer families. Following a SotA taxonomy [34], it covers Type-I (e.g., UPX), Type-III (e.g., PECompact), and VM-based packers (e.g., Themida). To replicate Experiment II from the original work, we apply the same undersampling strategy, resulting in 15,353 samples per packer. We randomly select 10% of the undersampled dataset while preserving the distribution of malware, benign programs, and packers. This subset, referred to as the *lab-10* dataset, excludes 10 outliers (CGs with more than 500 nodes). In Sect. 4.2, we test PackHero with the *RGD* dataset from PackGenome [16], which consists of three manually constructed programs, compiled from 2–5 lines of C code and packed with several versions and configurations of 20 off-the-shelf packers. To assess transferability, we select only the *RGD* packers also present in *lab-10*.

Hyperparameter Tuning. We optimize the hyperparameters of the GMN by maximizing intra-packet similarity using a grid search approach.

Evaluation Metrics. We evaluate PackHero using metrics [38] such as precision, recall, F1-score, accuracy, False Positive Rate (FPR), and the unknown rate, which indicates how often PackHero fails to recognize a packer. We also measure the *Average Number of Inference Calls*, representing the average calls PackHero makes to the GMN during identification.

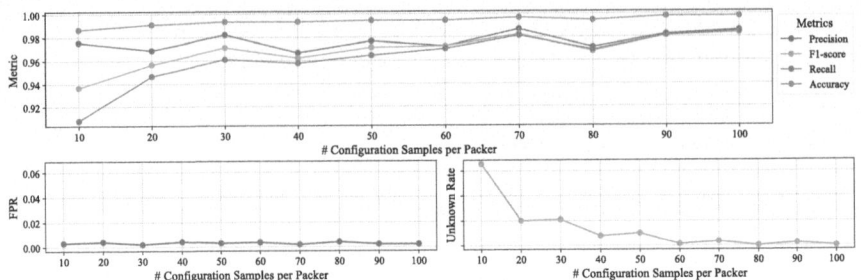

Fig. 3. Metrics trends varying the number of programs to configure PackHero.

4.2 Effective Identification (RQ1)

Configuration. PackHero requires a configuration phase starting with the step of collecting programs. If the packer is accessible, i.e., we can use it to craft new samples, the configuration becomes trivial and we can obtain as many programs as we need. However, if the packer is not accessible, we need to collect packed programs in the wild. These programs can be malware or benign, and it is uncertain whether we can find them in large quantities. Therefore, we test PackHero in a scenario with a limited number of programs, defining the number of programs needed for each packer to achieve a good average performance. From the *lab-10*, we select 100 programs for each packer to configure PackHero, maintaining the original distribution of malware and benign programs. We use the remaining programs for testing. Starting from the 100 programs for each packer, we gradually eliminate 10 programs and create 10 different collections of gradually smaller sizes. We use these 10 collections to configure PackHero. Then, we test our approach, configured with different "training" sizes, using the same test dataset. The metrics we use to evaluate PackHero in this experiment are precision, recall, F1-score, accuracy, FPR, and the unknown rate. Each plot in Fig. 3 shows the macro-average results, i.e., the average results among the 9 packers in the dataset. Looking at the precision, F1-score, recall, and accuracy, the tool does not perform badly even with only 10 programs per packer. Furthermore, starting from 30 samples per packer, PackHero maintains all metrics above 0.96. In addition, the plot shows precision and recall converge in the long run. We also notice that from the configuration of 70 samples for packers, Pack-Hero achieves a good balance between precision and recall, which means a good tradeoff between False Positives (FPs) and False Negatives (FNs). As regards the FPR and the unknown rate trends, they follow all the other metrics. The FPR is overall low and always below 0.00418, i.e., 0.41% of FPs. We also observe higher unknown rates for lower "training sizes", which means PackHero becomes more confident in his choices as the "training" size increases. The plot fluctuations are because the experiment was done with a single run due to the computational cost of training and testing with 10 different configurations. However, a single run places PackHero in a realistic scenario with limited samples and no ability to select the most suitable ones for tool configuration.

Effectiveness on Different Packers. Here, we zoom in on the results obtained from the best configuration in the previous experiment, namely the one with 100 samples for packers. The test set is the remaining part of the dataset. Table 3 shows that PackHero performs very well on each packer in the dataset. We have a near-maximum accuracy in general and equally good results in all other metrics for other packers. The only exception is tElock, which is found to have a higher FPR than the others. This result has chain effects on precision and F1-score. An answer can be found by looking at the clusters' composition. In particular, tElock produces two clusters of 1 and 99 samples. In its current version, PackHero merges single-sample clusters with the nearest one. As a result, for tElock, we obtain a single cluster consisting mostly of similar samples, along with one slightly different sample, which lowers intra-cluster similarity. This leads to a reduced threshold (≈ 0.10 lower than others) and lower confidence in the choice. This observation suggests that treating single-element clusters as outliers and excluding them could enhance PackHero's performance.

Different Versions and Configurations. The *lab-10* dataset includes only one configuration and version per packer. To evaluate transferability, we test PackHero's best configuration (*lab-10*) on the *RGD* dataset [16], which contains multiple versions and configurations for each packer—except for tElock, which is not included in the PackGenome evaluation. Table 4 presents the configurations identified by PackHero for each version in *RGD*. An "identified configuration" occurs when PackHero recognizes all samples, while non-identified configurations show a 0% identification rate, likely due to differences in the unpacking stub. Overall, PackHero generalizes across 16 out of 19 different versions, despite being configured with only a single version and configuration per packer.

Table 3. PackHero performance on lab-10. UR denotes the Unknown Rate.

Packer	#Samples	UR	FPR	Prec	Rec	F1	Acc
kkrunchy	1435	0.0	0.0001	1.00	0.99	1.00	1.00
MPRESS	1435	0.0	0.0018	0.99	0.98	0.98	1.00
Obsidium	1435	0.0	0.0031	0.98	0.98	0.98	0.99
PECompact	1434	0.0	0.0012	0.99	1.00	0.99	1.00
PELock	1435	0.0	0.0002	1.00	0.96	0.98	1.00
Petite	1434	0.0	0.0001	1.00	0.99	0.99	1.00
tElock	1432	0.0	0.0107	0.92	0.98	0.95	0.99
Themida	1433	0.0	0.0007	0.99	1.00	0.99	1.00
UPX	1432	0.0	0.0006	1.00	0.98	0.99	1.00
macro-avg	-	**0.0**	**0.0021**	**0.986**	**0.984**	**0.983**	**0.998**

Answer to RQ1. The number of packed programs required to configure Pack-Hero depends on the desired performance level. With just 10 samples per packer, PackHero achieves a minimum macro-average F1 score of 93.7% and accuracy of 98.7%. Increasing the sample size to 30 can further improve recall and F1-score while maintaining high precision and accuracy. Table 3 demonstrates PackHero's ability to effectively identify multiple packers from different families. Given the dataset's composition, PackHero successfully integrates and recognizes packers of varying complexity in both packed malware and benign programs. Specifically, based on the taxonomy by Ugarte-Pedrero et al. [34], PackHero performs well on Type-I (UPX), VM-based (Themida), and Type-III (PECompact) packers.

4.3   Clustering Effectiveness (RQ2)

We evaluate the impact of the clustering approach on PackHero's performance and scalability. To do this, we replicate the experiment from Sect. 4.2 without the clustering layer: PackHero evaluates similarity with all graphs in the DB and computes packer-specific thresholds instead of cluster-specific ones. Figure 4 illustrates the performance gap between PackHero with and without clustering. This gap is more pronounced for smaller training set sizes and narrows as the training set size increases. Without clustering, the unknown rate consistently drops to 0, as the *unknown* classification (explained in Sect. 3) depends on the similarity step involving clusters' medoids. However, the absence of clustering increases False Positives. These results demonstrate that incorporating clustering into PackHero's workflow is effective and that the chosen medoids are good representatives of subgroups within the same packer.

Table 5 shows the average number of inference calls to the GMN during the identification, i.e., the average number of matched graphs needed to identify a packer from a program. Removing the clustering approach, the similarity with the input graph is evaluated with the entire collection. Thus, the metric is always equal to the size of the entire collection. In contrast, the values we empirically obtain with the use of the clustering approach show are close to the ideal number

Table 4. packhero performance on RGD. A configuration is considered identified if all samples from that configuration are correctly classified.

Packer	Versions (#Identified Configurations/#Configurations)
kkrunchy	v0.23alpha **(0/2)**, v0.23alpha2 **(1/1)**
MPRESS	v1.27 **(1/1)**, v2.18 **(1/1)**, v2.19 **(1/1)**
Obsidium	v1.5 **(7/7)**
PEcompact	v3.02.2 **(18/19)**, v3.11 **(10/12)**
PElock	v1.06 **(0/5)**
Petite	v2.4 **(5/5)**
Themida	v2.37 **(8/9)**, v2.39 (*Winlicense*) **(8/9)**, v3.04 **(0/5)**
UPX	v1.00 **(4/4)**, v1.20 **(8/8)**, v1.25 **(4/4)**, v2.00 **(5/5)**, v3.09 **(8/8)**, v3.96 **(8/8)**

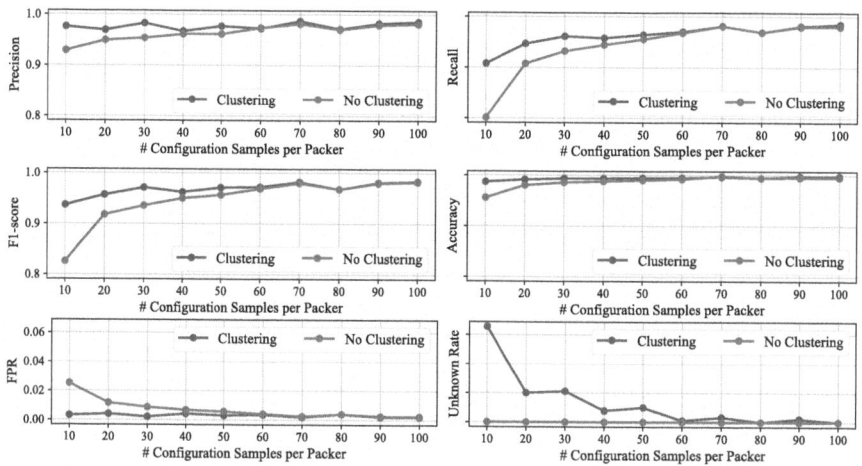

Fig. 4. Comparison between PackHero with and without clustering.

of inference calls to identify the packer. The ideal number of inference calls in the presence of the clustering approach is represented by the sum of m (number of clusters) and the number of programs per packer stored in the DB. Thus, we can state that, involving the clustering approach, the PackHero's number of inference calls does not depend on the number of packers but only on the number of samples for each packer stored in the DB. To further emphasize the significance of the results shown in Table 5, it is important to consider the inference time for our GMN. Indeed, while this network's expressive power surpasses that of its alternatives, it comes with the trade-off of increased temporal complexity.

Table 5. Average number of inference calls made by PackHero to the Graph Matching Network (GMN) during packer identification of a single sample.

Configuration	#Clusters	Clustering		No Clustering
		Ideal Values	**Real Values**	
10	14	24.00	24.14 ± 1.33	90.00 ± 0.00
20	13	33.00	33.02 ± 0.17	180.00 ± 0.00
30	17	47.00	47.03 ± 1.19	270.00 ± 0.00
40	16	56.00	60.59 ± 4.68	360.00 ± 0.00
50	13	63.00	64.31 ± 2.65	450.00 ± 0.00
60	14	74.00	82.68 ± 7.48	540.00 ± 0.00
70	11	81.00	83.71 ± 3.84	630.00 ± 0.00
80	16	96.00	107.54 ± 13.35	720.00 ± 0.00
90	13	103.00	107.30 ± 3.57	810.00 ± 0.00
100	17	117.00	124.47 ± 7.78	900.00 ± 0.00

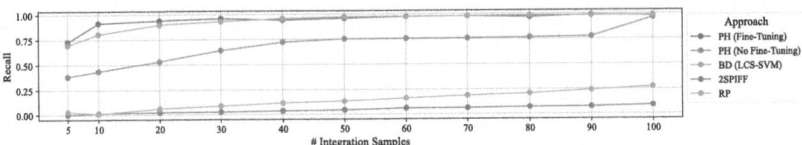

Fig. 5. Recall trend increasing the number of integration samples; PackHero (Fine-Tuned and Not) vs ML-based Tools.

In this experiment, the average single inference time recorded was $1.76ms$. The time difference observed in the tests conducted on 9 packers is not substantial between the version with and without clustering. However, in a scenario with 200 packers and 100 samples in the DB for each packer, the time required for identification would be $\approx 35s$ without clustering but only $21ms$ with it.

Answer to RQ2. The clustering approach improves PackHero's performance, especially with limited samples per packer, while its effectiveness remains unaffected by the number of recognized packers, ensuring scalability.

4.4 Integration of New Packers (RQ3)

As discussed in Subsect. 4.2, to demonstrate the integration is always feasible we have to deal with a scenario in which we need to collect programs in the wild to integrate the new packer. Thus, we have to face the possibility that these programs are not numerous. We have already tested PackHero in a scenario with few samples for its configuration. Here, we aim to evaluate how many samples PackHero needs to "integrate" a new packer with the entire system already configured. The process of integration corresponds to the "configuration" of PackHero described in Subsect. 1. Since the workflow includes a Graph Matching Network (GMN), we can avoid re-training the model from scratch. Indeed, given that we use a Neural Network (NN), it can be updated through fine-tuning, i.e., partially re-training on new samples. At the same time, Pack-Hero may even allow us to avoid fine-tuning the model altogether. Specifically, a new packer can be integrated into PackHero without re-training the GMN, simply by adding its corresponding graphs to the DB. However, this approach is feasible only if the collected graphs for the packer are sufficiently homogeneous. Currently, we assume that manual intervention was previously performed on the packer samples to be integrated, which we assume are always correctly labeled.

We evaluate PackHero with and without fine-tuning the GMN. We train a GMN for each packer, excluding it from the training set, which consists of 100 samples per remaining packer. Then, we integrate samples from the "unseen" packer. In the version without fine-tuning, we add the new packer's graphs directly to the DB. In the fine-tuned version, we fine-tune the GMN using the new graphs before integration. To compare PackHero with SotA tools, we replicate the experiment using three ML-based approaches from Table 1: Randomness Profiles [33], Binary Diffing [15] (best-performing version: LCS-SVM), and

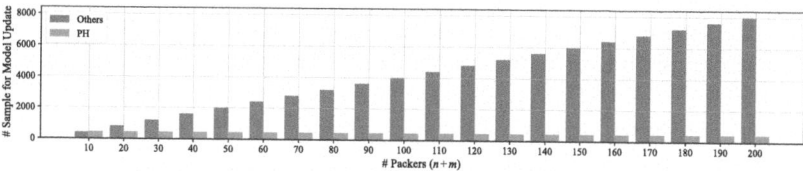

Fig. 6. # Samples involved in the integration ($m = 10$ new packers, $l = 40$ samples per packer, n already integrated packers).

2SPIFF [19]. As implementations of these tools are unavailable, we reimplement them to the best of our ability and validate the implementations by comparing the achieved accuracy on the remaining packers in this experiment. For a fair evaluation, we train them using a stratified 80–20 split of the *lab-10* dataset. Therefore, for each ML-based tool, we train each model by excluding one packer, using 80% of the dataset. This subset includes the 100 samples per packer used to train PackHero. However, none of these approaches utilize a NN. Therefore, we are required to entirely re-train their models each time we want to integrate a new packer. Ultimately, both our method and the other approaches are tested using the test set from the 80-20 split, which corresponds to the stratified 20% of *lab-10*, equal to 307 samples for each packer. Results specific to each packer are obtained by testing the samples from that particular packer on the updated tools. By evaluating separately for each packer, the metric we use is recall, which is equivalent to accuracy in this experimental setup. In Fig. 5, we show the average recall trends for PackHero with and without fine-tuning and compare them against all other ML-based packer identifiers. As the figure shows, 2SPIFF and Randomness Profiles (RP) perform very badly, even in the best configuration. In contrast, both the fine-tuned PackHero and Binary Diffing (BD) with LCS-SVM achieve very good performance and consistency using a small number of samples. Starting from the 40 samples, their performance is aligned, except for small fluctuations due to the single run. However, as the plot shows, PackHero reaches a high recall before BD. Indeed, using 5, 10, 20, and 30 samples to integrate the new packer, PackHero performs better. Similarly, PackHero without fine-tuning also exhibits good performance, although not as good as the other two methods. However, the average performance hides the results for single-packers. Indeed, removing two packers out of nine (Obsidium and Petite) the non-fine-tuned PackHero achieves performance very close to the fine-tuned version and BD starting from 50 integration samples. This result shows that fine-tuning the GMN is unnecessary for low-heterogeneity graphs, saving computation.

Additionally, all ML-based approaches require complete retraining to integrate new packers, a scalability issue PackHero avoids. To demonstrate this, we simulate packer integration during the tool lifecycle. Assuming integration cost depends on the number of samples used, we define the cost function $f(n, m, l)$, where n is the number of known packers, m the new packers, and l the samples per packer. For other tools, $f(n, m, l) = (n + m) \cdot l$, while PackHero's cost is $f(n, m, l) = m \cdot l$, as it only depends on new packers. Figure 6 simulates $m = 10$

and $l = 40$, showing PackHero's constant integration cost, unlike other tools, where cost grows linearly with the number of recognized packers. This result demonstrates that PackHero scales effectively in realistic scenarios where new packers must be integrated over time.

Answer to RQ3. PackHero achieves strong results in integrating a new "unseen" packer with as few as 10 samples. Among 9 tested packers, PackHero outperforms the three other selected ML-based tools for integration sample counts less than 30. Furthermore, a simulation comparing retraining and fine-tuning shows that PackHero's integration cost function remains constant as new packers are added, unlike other ML-based approaches, whose integration costs grow linearly with the number of recognized packers.

4.5   Signature-Based Tool Comparison (RQ4)

In this experiment, we compare the performance of PackHero with State-of-the-art signature-based methods (Detect It Easy (DIE) and PackGenome). DIE [13] is currently the best-performing and most signature-rich packer identifier. Here, we use its latest available version (v3.10). PackGenome [16] is the best tool for automating the extraction of signatures for packer detection, generating YARA [3] rules that can be later used to identify packers. The authors of PackGenome have already compared their framework against DIE, but we include the results for both approaches for completeness. To compare the three frameworks, we use the False Positive Rate (FPR) and recall. We focus on recall because it is the most representative metric when comparing tools designed not just for identification but also for detection (as for DIE and PackGenome). We extract all results from the *lab-10* dataset, removing only the 100 samples for each packer used to configure PackHero. We test DIE using its command-line version, while we test PackGenome by loading the YARA rules provided in the original work. Since DIE does not include PELock signatures and PackGenome's authors did not perform experiments on PELock and tElock, we discard these two packers for this comparison. In Table 6, we show the comparison results.

Table 6. PackHero vs DIE vs PackGenome on lab-10.

		Recall			FPR		
	#Samples	PH	DIE	PG	PH	DIE	PG
kkrunchy	1435	0.99	1.0	1.0	0.0001	0.0	0.0
MPRESS	1435	0.98	1.0	1.0	0.0018	0.0008	0.0034
Obsidium	1435	0.98	1.0	1.0	0.0031	0.0002	0.0000
PECompact	1434	1.0	1.0	1.0	0.0012	0.0	0.0005
Petite	1434	0.99	0.99	0.99	0.0001	0.0	0.0
Themida	1433	1.0	0.92	0.31	0.0007	0.0	0.0499
UPX	1432	0.98	1.0	1.0	0.0006	0.0002	0.0011

Starting with the recall, the three tools perform very similarly, although the two signature-based tools generally exhibit the highest recall. Attention must particularly be directed towards Themida, which poses significant challenges for both DIE and PackGenome, as depicted in the table. An interesting observation is that the matched signatures from DIE are related to the same version of Winlicense/Themida, specifically Themida with Trial/Licensing options [25]. Despite PackGenome including signatures for the same version of this packer, it shows a recall of 0.31. This result motivates the entire work. Indeed, Themida/Winlicense employs advanced (dynamic) evasive behaviors in its packed programs. Furthermore, it is the VM-based packer used during our evaluation. As explained in Sect. 2.1, PackGenome extracts YARA rules by tracing instructions during their execution. Consequently, it is likely to struggle with the evasive behaviors introduced by Themida into the binary during the packing process. Additionally, PackGenome appears to face challenges due to the inherent nature of this packer. Indeed, the result suggests that both the signature itself and its automatic extraction encounter difficulties when dealing with this packer family. Finally, looking at FPR, PackGenome confirms its issues with Themida/Winlicense but shows in-line results for the other packers. PackHero demonstrates a low FPR on average, while DIE generates the fewest FPR.

Answer to RQ4. PackHero matches SotA signature-based tools in accuracy and significantly outperforms them on VM-based packers with advanced evasive behaviors like Themida/Winlicense, demonstrating our approach's effectiveness.

5   Limitations and Future Work

PackHero relies on heuristics to extract filtered CGs. Unpacking stubs play a crucial role in the analysis but other code segments might also contribute to the CG's structure. Even if this work demonstrates that a few statically visible functions are often sufficient to determine the packer's identity, the exact number of functions considered for each CG remains an open aspect. PackHero directly exploits disassembly and function identification, which are inherently challenging problems, especially in the context of malware due to obfuscation techniques, indirect branch resolution, and evasive behaviors. Furthermore, PackHero is potentially vulnerable to adversarial attacks that manipulate the CG to evade identification. An adversary could, for instance, obfuscate the CG by inserting bogus functions, modifying calls, or hiding call targets, significantly complicating packer identification. Additionally, different dynamic evasive behaviors implemented by malware could further impact the accuracy of the extracted CG. Hence, future work may study heuristics to resist adversarial attacks by evaluating CG obfuscation to identify which aspects of our heuristics and features are most susceptible to evasion. In our study, we selected radare2 as the disassembler due to its ease of use and integration. However, recent research has demonstrated that several other open-source disassemblers outperform radare2 performance [26]. This reliance, while currently effective, necessitates further investigation, particularly in scenarios involving adversarial manipulation of the unpacking stub or

CG structure. Finally, PackHero currently focuses on *packer identification* but does not determine whether a sample is packed (*detection*). Preliminary analyses revealed a notable number of False Positives when analyzing non-packed samples, indicating the need for further improvements in this area. Therefore, future work will also address the *packer detection* problem.

6   Conclusion

This paper introduced PackHero, a packer identifier that leverages a heuristic to extract a Call Graph (CG) representing the unpacking routine of a program. Using a clustering approach to enhance performance and reduce the search space, PackHero evaluates the similarity between the extracted CG and labeled CGs stored in a DB, employing a Graph Matching Network (GMN) to compute these similarities and identify the packer. Evaluated on a public dataset of packed benign and malicious programs re-packed multiple times, PackHero meets all key requirements for a novel packer identifier: high accuracy, efficient packer integration, evasive behavior management, and scalability. Relying exclusively on static analysis, PackHero integrates new packers effectively, achieving strong performance with as few as 10 samples, while eliminating the limitations of dynamic analysis, particularly against dynamic evasive behaviors. For some packers, it avoids fine-tuning the GMN, and when fine-tuning is needed, it converges faster than other ML-based tools. Its integration cost remains constant throughout its lifecycle, unlike other methods, where costs grow linearly with the number of packers recognized. PackHero performs comparably to signature-based tools, the current best-performing solutions for packer identification, and significantly outperforms SotA approaches on Themida, a VM-based packer employing advanced dynamic evasive behaviors.

Acknowledgements. This work was partially supported by Project FARE (PNRR M4.C2.1.1 PRIN 2022, Cod. 202225BZJC, CUP D53D23008380006, Avviso D.D 104 02.02.2022) and Project SETA (PNRR M4.C2.1.1 PRIN 2022, Cod. P202233M9Z, CUP F53D23009120001, Avviso D.D 1409 14.09.2022) under the Italian NRRP MUR program, and by Project SERICS (PE00000014) under the MUR National Recovery and Resilience Plan, all funded by the European Union - NextGenerationEU.

References

1. Aghakhani, H., et al.: When malware is packin' heat; limits of machine learning classifiers based on static analysis features. In: Proceedings of Symposium on Network and Distributed System Security (NDSS) (Feb 2020)
2. Al-Anezi, D.M.M.K.: Generic packing detection using several complexity analysis for accurate malware detection. Int. J. Adv. Comput. Sci. Appl. **5**(1) (2014). https://doi.org/10.14569/IJACSA.2014.050102
3. Alvarez, V.M.: Yara. https://virustotal.github.io/yara/ (2024). Accessed 15 Apr 2024

4. Anderson, H.S., Roth, P.: Ember: an open dataset for training static pe malware machine learning models. arXiv preprint arXiv:1804.04637 (2018)
5. ASPack Software: ASPack Software - Application for compression, packing and protection of software. http://www.aspack.com/ (2024). Accessed 15 Apr 2024
6. Callahan, D., Carle, A., Hall, M., Kennedy, K.: Constructing the procedure call multigraph. IEEE Trans. Softw. Eng. **16**(4), 483–487 (1990). https://doi.org/10.1109/32.54302
7. Diestel, R.: Graph Theory. Springer, 5th edn. (2017)
8. Ebringer, T., Sun, L., Boztas, S.: A fast randomness test that preserves local detail. In: Proceedings of the 18th Virus Bulletin International Conference, pp. 34–42. Virus Bulletin Ltd (2008)
9. Egele, M., Scholte, T., Kirda, E., Kruegel, C.: A survey on automated dynamic malware-analysis techniques and tools. ACM Comput. Surv. **44**(2) (mar 2008). https://doi.org/10.1145/2089125.2089126
10. Galloro, N., Polino, M., Carminati, M., Continella, A., Zanero, S.: A systematical and longitudinal study of evasive behaviors in windows malware. Comput. Secur. **113**, 102550 (2022). https://doi.org/10.1016/j.cose.2021.102550
11. Hamilton, W.L.: Graph Representation Learning. Synthesis Lect. Artif. Intell. Mach. Learn. **14**(3), 1–159 (2020), publisher: Morgan and Claypool
12. Hamrock, J., Lyda, R.: Using entropy analysis to find encrypted and packed malware. IEEE Secur. Privacy **5**(02), 40–45 (mar 2007). https://doi.org/10.1109/MSP.2007.48
13. Horsicq: Detect it easy. https://github.com/horsicq/Detect-It-Easy (2024). Accessed 15 Apr 2024
14. Jacob, G., Comparetti, P., Neugschwandtner, M., Kruegel, C., Vigna, G.: A static, packer-agnostic filter to detect similar malware samples. In: International Conference on Detection of Intrusions and Malware, and Vulnerability Assessment, vol. 7591 (01 2010). https://doi.org/10.1007/978-3-642-37300-8_6
15. Kim, Y., Paik, J.Y., Choi, S., Cho, E.S.: Efficient SVM based packer identification with binary diffing measures. In: 2019 IEEE 43rd Annual Computer Software and Applications Conference (COMPSAC). vol. 1, pp. 795–800 (2019).https://doi.org/10.1109/COMPSAC.2019.00117
16. Li, S., et al.: PackGenome: automatically generating robust YARA rules for accurate malware packer detection. In: Proceedings of the 2023 ACM SIGSAC Conference on Computer and Communications Security pp. 3078–3092. CCS '23, Association for Computing Machinery (2023). https://doi.org/10.1145/3576915.3616625
17. Li, X., Shan, Z., Liu, F., Chen, Y., Hou, Y.: A consistently-executing graph-based approach for malware packer identification. IEEE Access **7**, 51620–51629 (2019). https://doi.org/10.1109/ACCESS.2019.2910268
18. Li, Y., Gu, C., Dullien, T., Vinyals, O., Kohli, P.: Graph matching networks for learning the similarity of graph structured objects. In: Chaudhuri, K., Salakhutdinov, R. (eds.) Proceedings of the 36th International Conference on Machine Learning. Proceedings of Machine Learning Research, vol. 97, pp. 3835–3845. PMLR (Jun 2019)
19. Liu, H., Guo, C., Cui, Y., Shen, G., Ping, Y.: 2-SPIFF: a 2-stage packer identification method based on function call graph and file attributes. Appl. Intell. **51**(12), 9038–9053 (2021). https://doi.org/10.1007/s10489-021-02347-w
20. Liu, Z., Wang, R., Japkowicz, N., Gomes, H.M., Peng, B., Zhang, W.: Segdroid: an android malware detection method based on sensitive function call graph learning. Expert Syst. Appl. **235**(C) (Jan 2024). https://doi.org/10.1016/j.eswa.2023.121125

21. Luk, C.K., et al.: Pin: building customized program analysis tools with dynamic instrumentation. SIGPLAN Not. **40**(6), 190–200 (jun 2005). https://doi.org/10.1145/1064978.1065034

22. Mantovani, A., Aonzo, S., Ugarte-Pedrero, X., Merlo, A., Balzarotti, D.: Prevalence and impact of low-entropy packing schemes in the malware ecosystem. In: Proceedings 2020 Network and Distributed System Security Symposium (2020)

23. Muchnick, S.S.: Advanced Compiler Design and Implementation. Morgan Kaufmann, San Francisco, CA (1997)

24. Muralidharan, T., Cohen, A., Gerson, N., Nissim, N.: File packing from the malware perspective: Techniques, analysis approaches, and directions for enhancements. ACM Comput. Surv. **55**(5) (Dec 2022). https://doi.org/10.1145/3530810

25. Oreans Technologies: Winlicense. https://www.oreans.com/WinLicense.php. Accessed 07 Aug 2024

26. Pang, C., et al.: SOK: All you ever wanted to know about x86/x64 binary disassembly but were afraid to ask. In: 2021 IEEE Symposium on Security and Privacy (SP), pp. 833–851 (2021)

27. PEiD: Peid. https://www.aldeid.com/wiki/PEiD (2024). Accessed 15 Apr 2024

28. Perdisci, R., Lanzi, A., Lee, W.: Classification of packed executables for accurate computer virus detection. Pattern Recogn. Lett. **29**(14), 1941–1946 (2008). https://doi.org/10.1016/j.patrec.2008.06.016

29. radare2: radare2: Unix-like reverse engineering framework and command-line tools (2024). https://github.com/radareorg/radare2. Accessed 15 Apr 2024

30. Raff, E., et al.: Automatic yara rule generation using biclustering. In: Proceedings of the 13th ACM Workshop on Artificial Intelligence and Security, pp. 71–82. AISec'20, Association for Computing Machinery, New York, NY, USA (2020). https://doi.org/10.1145/3411508.3421372

31. Rahbarinia, B., Balduzzi, M., Perdisci, R.: Exploring the long tail of (malicious) software downloads. In: 2017 47th Annual IEEE/IFIP International Conference on Dependable Systems and Networks (DSN), pp. 391–402 (2017). https://doi.org/10.1109/DSN.2017.19

32. Rolles, R.: Unpacking virtualization obfuscators. In: Proceedings of the 3rd USENIX Conference on Offensive Technologies, p. 1. WOOT'09, USENIX Association, USA (2009)

33. Sun, L., Versteeg, S., Boztaş, S., Yann, T.: Pattern recognition techniques for the classification of malware packers. In: Steinfeld, R., Hawkes, P. (eds.) Information Security and Privacy, pp. 370–390. Springer, Berlin Heidelberg, Berlin, Heidelberg (2010)

34. Ugarte-Pedrero, X., Balzarotti, D., Santos, I., Bringas, P.G.: SOK: deep packer inspection: a longitudinal study of the complexity of run-time packers. In: 2015 IEEE Symposium on Security and Privacy, pp. 659–673 (2015). https://doi.org/10.1109/SP.2015.46

35. Ugarte-Pedrero, X., Graziano, M., Balzarotti, D.: A close look at a daily dataset of malware samples **22**(1) (Jan 2019)

36. UPX: UPX – the ultimate packer for executables. https://upx.github.io/ (2024). Accessed 15 Apr 2024

37. Yadegari, B., Debray, S.: Symbolic execution of obfuscated code. In: Proceedings of the 22nd ACM SIGSAC Conference on Computer and Communications Security, pp. 732–744. CCS '15, Association for Computing Machinery, New York, NY, USA (2015). https://doi.org/10.1145/2810103.2813663

38. Zaki, M.J., Meira Jr, W.: Data mining and machine learning: fundamental concepts and algorithms. Cambridge University Press, 2 edn. (2020)

A History of GREED: Practical Symbolic Execution for Ethereum Smart Contracts

Nicola Ruaro[⊠], Fabio Gritti, Robert McLaughlin, Dongyu Meng,
Ilya Grishchenko, Christopher Kruegel, and Giovanni Vigna

University of California, Santa Barbara, Santa Barbara, CA, USA
{ruaronicola,degrigis,robert349,dmeng,grishchenko,chris,vigna}@ucsb.edu

Abstract. Smart contracts have transformed blockchain applications, enabling decentralized computation and automated asset management without intermediaries. However, with the growth of decentralized finance, the high financial stakes make smart contract vulnerabilities particularly critical. Because vulnerabilities often go undetected, they lead to substantial losses and diminished trust in blockchain systems.

Symbolic execution has emerged as a powerful technique to uncover subtle vulnerabilities by systematically exploring feasible execution paths. However, most existing symbolic execution tools for smart contracts are tailored to specific vulnerability patterns, making them unsuitable for detecting new types of vulnerabilities. In this paper, we introduce GREED, a highly versatile symbolic execution framework for Ethereum (or EVM-based) smart contracts. GREED features a state-of-the-art symbolic execution engine coupled with a suite of supporting analyses and a modular design that allows security researchers to prototype new analyses rapidly.

To evaluate the effectiveness and extensibility of GREED, we compare it with the state-of-the-art. We first show that GREED can explore significantly more code paths – reaching 84% of all `CALL` statements, as opposed to 9% on average across existing tools. To demonstrate the ease of use (and extensibility) of GREED, we then implement a novel analysis to detect controllable `JUMPI` instructions and evaluate it against all deployed contracts on Ethereum and Binance Smart Chain (BSC), identifying 390 previously unknown vulnerable contracts.

By releasing GREED to the community, we aim to lower the barrier to developing advanced security analyses for smart contracts, empowering security researchers to rapidly prototype new analyses and contribute to a more secure and resilient blockchain ecosystem.

Keywords: Ethereum · Smart Contract · Symbolic Execution

1 Introduction

Ethereum [15] is a global, decentralized blockchain that enables the deployment and execution of decentralized programs (smart contracts). Smart contracts are immutable programs that run on the Ethereum Virtual Machine (EVM) and are executed on demand by blockchain users. Smart contracts have transformed

M. Egele et al. (Eds.): DIMVA 2025, LNCS 15748, pp. 275–296, 2025.
https://doi.org/10.1007/978-3-031-97623-0_17

the way transactions are executed, enabling decentralized applications and automated asset management without intermediaries.

Ethereum (and other blockchains) have witnessed the explosive growth of a new form of blockchain-based finance that is known as decentralized finance (DeFi) – a rich ecosystem of digital currencies, financial tools, and financial services. Because of the exceptionally high stakes involved [12], identifying and fixing vulnerabilities in smart contracts has become critical. Once deployed, smart contracts cannot be easily patched, and exploits can lead to substantial financial damage and loss of trust in blockchain systems [13]. Therefore, rigorous analysis of smart contracts is necessary to ensure their security.

Symbolic execution [1] (SE) has emerged as a powerful technique for smart contract analysis. SE systematically explores a contract in an emulated environment with symbolic variables representing possible (but unknown) inputs. As the execution progresses, the SE system (or engine) tracks the state of the EVM – e.g., program counter, stack, and memory. At specific points in the execution, the engine queries a constraint solver to determine whether a given state is satisfiable – that is, whether each symbolic variable has a feasible concrete solution. When the execution reaches a conditional branch, and both the condition and its negation are satisfiable, the execution path forks, and both branches are explored separately. This enables the generation of concrete inputs that reproduce specific program behaviors, allowing one to uncover subtle bugs that might evade traditional testing methods (e.g., fuzz testing).

Related Work. Over the years, many SE tools have been developed to detect vulnerabilities in smart contracts. Some focus on the formal verification of specific properties [27,29,32,33,36]. For example, VERX [27] uses SE and induction proofs to study safety properties. Others identify known vulnerability patterns [4,10,18,22–24,26,28,31]. For example, TEETHER [23] identifies contracts that leak funds to arbitrary users. While existing tools have shown some success in their respective domains, they suffer from two key limitations: First, the symbolic execution engines of existing tools lack critical analysis features – for instance, a precise memory model – that limit their effectiveness. Second, the architecture of existing tools is typically designed around specific vulnerability patterns, making it challenging to adapt them to new vulnerabilities and extend their capabilities beyond the original scope.

Our Approach. In this paper, we introduce GREED, a highly versatile SE framework designed for the analysis of EVM-based smart contracts. GREED addresses the limitations of existing tools by providing a novel combination of analysis techniques, including both a state-of-the-art SE engine and a suite of supporting analyses. Unlike traditional tools (with a fixed set of predefined analyses), GREED enables security experts to build new analyses tailored to their needs. Our experiments show that GREED's architecture allows for more efficient path exploration – and superior flexibility – without compromising analysis accuracy.

We implemented GREED in approximately 10,000 lines of Python code and released it as an open-source project[1]. GREED has been met with enthusiasm by the community. After the open-source release, the project attracted hundreds of new users (in terms of distinct project downloads, GitHub "stars", and community contributions). We are also aware of several academic institutions and corporations that are either actively using GREED or evaluating it for potential use in their systems.

This paper makes the following contributions:

- We describe GREED, a highly versatile symbolic execution framework designed for EVM-based smart contracts. GREED features a state-of-the-art symbolic execution engine and a novel combination of analysis techniques within a modular and extensible architecture, enabling security experts to tackle complex security challenges.
- We compare GREED against the state-of-the-art and show that it can explore significantly more code paths. GREED outperforms all existing tools, reaching 84% of all CALL statements, compared to 9% across alternatives (on average).
- To demonstrate the ease of adding additional security analysis, we implement a novel checker to detect controllable JUMPI instructions and evaluate it against all contracts in Ethereum and BSC [3], identifying 390 previously unknown vulnerable contracts.

2   Motivation

Existing symbolic execution systems focus on detecting known classes of vulnerabilities. This specialization has led to two main limitations. First, existing systems often forego implementing comprehensive, robust analyses, opting instead for a subset of features tailored to the targeted vulnerabilities. A precise implementation of all analysis features is sometimes unnecessary for individual security analyses. For example, ERC20 tokens rarely interact with external contracts. Thus, a full-fledged cross-contract analysis may be unnecessary for analyzing ERC20 token contracts [19]. Second, in addition to the lack of analysis features, many existing systems lack any underlying static analysis, such as control-flow graph (CFG) recovery. Yet, a balanced integration of static and dynamic analysis is crucial for building sophisticated security tools. The absence of static analyses makes extending and scaling existing systems (for instance, with exploration strategies) inherently challenging. This underscores the necessity for a versatile unified analysis framework that can be repurposed for complex, evolving security analyses.

2.1   Basic Analysis Features

Modern smart contracts frequently use cross-contract interactions, memory operations, and hash functions. Not properly supporting these three features leads to

---

[1] https://github.com/ucsb-seclab/greed.

```solidity
1    pragma solidity ^0.8.0;
2
3    struct Action {
4        address router;
5        bytes data;
6    }
7
8    contract Dispatcher {
9        address router = 0xROUTER;
10
11       function set_router(Action action) public returns (Action) {
12           if (action.router == address(0)) {
13               action.router = router;
14           }
15           return action;
16       }
17   }
18
19   contract Executor {
20       Dispatcher dispatcher = Dispatcher(0xDISPATCHER);
21       mapping(address => uint256) routerCallCounts;
22
23       function execute(Action[] memory actions) public {
24           // Actions (copied in memory) have symbolic offset and size
25           for (uint256 i = 0; i < actions.length; i++) {
26               // Cross-contract call
27               Action memory action = dispatcher.set_router(actions[i]);
28               // Another (controllable) cross-contract call
29               // BUG: ASSUMES ACTION.ROUTER IS ALWAYS SET BY DISPATCHER
30               action.router.call(action.data);
31               // Increment mapping variable
32               routerCallCounts[action.router] += 1;
33           }
34           // Check mapping variable
35           require(routerCallCounts[dispatcher.router] > 1);
36       }
37   }
```

Fig. 1. Simplified Solidity code of the Executor contract. The contract parses a list of provided actions (CALLDATA), interacts with the Dispatcher contract to fetch the **router** address, then interacts with the router and updates the respective interaction counter. RED : requires a precise memory model. YELLOW : requires cross-contract analysis. GREEN : requires a precise SHA model.

significant limitations in the engines' analysis capabilities. For instance, in Fig. 1, we present a contract that – although seemingly simple – cannot be precisely analyzed without implementing the aforementioned analysis features.

Cross-Contract Interactions. Ethereum allows smart contracts to CALL functions of other contracts (Fig. 1: Line 27, Line 30), enhancing modularity and code reuse. However, interactions inherently increase the complexity of smart contracts and can introduce unexpected bugs. For instance, the external contract might operate maliciously and inadvertently change its behavior. Without precise cross-contract analysis, it is impossible to detect vulnerabilities arising from such interactions.

Memory Model. In the EVM, memory is a volatile, mutable storage area that exists only during the execution of a contract function. Any data stored in memory is freed once the execution terminates. Memory is efficient because it avoids the overhead of writing to persistent blockchain storage. This makes it suitable for intermediate calculations, temporary variables, and data manipulation within a function call. Nonetheless, modeling symbolic memory operations is challenging, and existing systems resort to approximations – such as the strategic concretization of symbolic offsets and lengths. When a symbolic memory buffer (for example, the `actions` array on Line 23) is accessed (Line 25), it is undeniably convenient to concretize its length. However, this prevents the system from detecting vulnerabilities that arise from different configurations. For example, the Executor contract reverts unless we provide an array with at least two actions – since the variable `routerCallCounts` is incremented at most once per array element.

Hash Functions. Handling cryptographic hash functions (`SHA`) is crucial due to their pervasive use by dynamic data types – such as arrays and mappings. In Solidity, fixed-size data types have predetermined slots in persistent storage, but dynamic data types grow during execution. To manage this, Solidity computes storage slot offsets dynamically using hash computations: First, all array and mapping variables are assigned a "base slot". Then, the storage slot for an array element with index `i` is calculated as `SHA(base_slot) + i`. Similarly, the storage slot for a mapping element with key `key` is calculated as `SHA(key, base_slot)`. Accurately modeling these hash computations is essential for recognizing data storage patterns (e.g., Line 32 and Line 35) and detecting vulnerabilities related to data access and manipulation.

2.2 Beyond the State-of-the-Art

Robust basic analysis features provide a necessary foundation for smart contract analysis. However, these capabilities alone are insufficient for thoroughly analyzing modern, complex blockchain applications with evolving attack vectors. We argue that it is essential to complement basic analysis features with supporting techniques such as static analysis and exploration strategies. Static analysis techniques – such as control-flow graph recovery and dependency tracking – can isolate critical code regions where vulnerabilities are most likely to reside. Exploration strategies – such as directed search – allow directing the symbolic execution engine toward (previously identified) critical code regions to verify the presence (or absence) of vulnerabilities. Rather than exhaustively exploring all paths, exploration strategies allocate resources to areas with a higher likelihood of revealing subtle bugs, thus addressing long-standing challenges like state explosion. In the following sections, we present our approach to integrating advanced analysis features in our symbolic execution framework.

3 Practical Symbolic Execution with GREED

Figure 2 shows an overview of GREED's architecture. GREED exposes several interfaces that enable both static and dynamic analysis. Initially, the contract

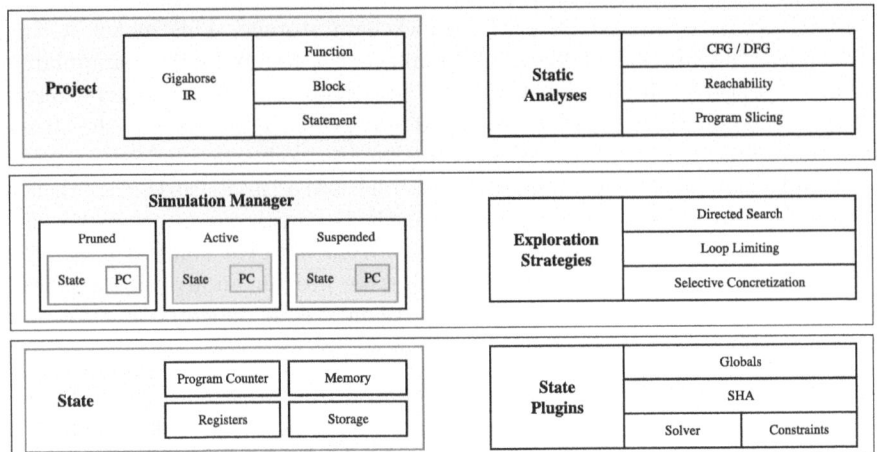

Fig. 2. Overview of GREED. The `project` object exposes static information. The `simulation manager` tracks all execution states and allows one to manipulate them. The `states` store the execution environment and additional context.

is pre-processed using the Gigahorse static analysis framework [20,21]. The contract's intermediate representation (organized in functions, blocks, and statements) is stored in a `project` object. The `project` exposes an interface to all available static analyses (e.g., CFG, Slicing). During execution, the `simulation manager` orchestrates all the execution states, which are organized in "stashes" that indicate whether they are active, pruned, suspended, etc. The `simulation manager` also accepts various `exploration strategies`. At a high level, exploration strategies allow one to programmatically manipulate the execution states and determine which state should be executed next – or which states are uninteresting to explore. Each `state` represents a snapshot of the execution at a specific program location, which stores both the execution environment and additional context. *This is where the basic analysis features live* (see Sect. 2.1). Finally, `state plugins` track additional context (e.g., SHA operations and constraints) that allows for checking the satisfiability of an execution state. The modularity of GREED allows one to easily write new static analyses, exploration strategies, and state plugins – or experiment with different memory models and solvers.

3.1 Static Analysis

GREED operates on the Gigahorse IR, which provides its foundational static analysis capabilities: decompilation, IR lifting, constant folding, basic control-flow and data-flow modeling, and loop analysis. This allows GREED to instead focus on advanced static analyses (e.g., backward and forward program slicing, reachability analysis) and symbolic execution, which are highly valuable for building complex security tools. Below, we discuss some examples of static analyses available in GREED.

Control-Flow Graph (CFG). Gigahorse provides state-of-the-art CFG and call-graph reconstruction for EVM bytecode. This is automatically available in GREED. The CFG encodes control-flow relationships, enabling reasoning about reachability between statements. For instance, this is essential for directing the execution toward a desired statement.

Data-Flow Graph (DFG). Similarly, Gigahorse also provides state-of-the-art DFG reconstruction. The DFG captures data dependencies, allowing one to track how variables are assigned and manipulated throughout the contract.

Reachability. GREED's reachability analysis allows one to automatically determine whether an execution path might exist between two program points. For blocks within the same function, GREED directly analyzes their relationships in the CFG. For blocks in different functions, GREED identifies possible sequences of function calls that connect them. When available, GREED also examines the call stack to identify additional paths that connect the two program points.

Program Slicing. Leveraging the CFG and DFG, GREED can calculate a "slice" of statements that affect (backward) or are affected by (forward) a given variable. For instance, this is essential for implementing under-constrained execution, which enables an approximate but lightweight analysis of local properties.

3.2   Exploration Strategies

Exploration strategies allow for the orchestration of execution states and typically employ a combination of state pruning, prioritization, and manipulation. Pruning allows one to discard states that are unfit for the desired analysis goals. Prioritization allows one to prioritize the exploration of certain states. Manipulation allows one to alter (the execution environment of) certain states. Below, we discuss some examples of exploration strategies available in GREED.

Directed Search. Directed search is an example of an exploration strategy that can leverage both state pruning and prioritization to direct the symbolic execution toward a desired (target) statement. This strategy is supported by a CFG-driven reachability analysis. States closest to the target statement are prioritized. States unfit to reach the target statement are (optionally) discarded. This allows one to focus the analysis on specific execution paths that are relevant to a desired property.

Under-Constrained Search. Under-constrained search allows executing arbitrary program slices by first creating a symbolic state at a specific program location and then manipulating the execution states to manage undefined behavior. First, GREED creates a symbolic execution state at the first program location in the slice. Then, the under-constrained search rewrites all undefined variables to assign them fresh 256-bit symbolic variables. Optionally, the under-constrained search can guide (force) the execution along a predetermined, statically observed path – even if that path is unfeasible in a fully constrained context. This allows one to effectively study the (security) properties of arbitrary program slices without incurring the overhead of fully-constrained symbolic execution.

Loop Limiting. Loop limiting is an essential technique for mitigating state explosion during symbolic execution. In GREED, a counter-based strategy monitors the number of times a given program point is reached. Once a predefined threshold is exceeded, we prune the corresponding execution state. This approach effectively controls redundant loop iterations, ensuring that excessive unrolling does not overwhelm the analysis.

State Monitoring and Rewriting. State rewriting enables the dynamic modification of execution states to incorporate external information – such as concrete execution data, observed blockchain states, or freshly generated symbolic variables. Through this process, one can refine the analysis context to reflect relevant properties or to simulate any desired execution state. For example, a symbolic variable representing an asset's price can be replaced with its actual value retrieved from a live oracle, thereby allowing the analysis to mirror realistic market conditions. Additionally, by coupling state rewriting with state monitoring, GREED can collect valuable metrics (e.g., constraint-solving time) that can be used to identify or prune paths with a desired property – for example, computationally expensive paths.

Selective Concretization. Selective concretization is an example of state manipulation, where a heuristic determines whether any environment variable should be concretized. This is helpful to enforce a specific property ("the value of variable X must be exactly 42 to trigger the vulnerability") or to simplify the analysis when the constraints are too complex (at the cost of possible false negatives).

Classic Prioritization. Depth-first search (DFS) and breadth-first search (BFS) are classic examples of state prioritization. Execution states are never pruned. Instead, a heuristic determines which states should be explored first. In DFS, deep execution states are explored first. In BFS, shallow execution states are explored first. Exhaustive strategies such as BFS or DFS are often impractical for large contracts. In fact, even simple loops or repeated subroutine calls can rapidly inflate the state space. For this reason, exhaustive search strategies are often paired with additional strategies for state pruning.

3.3   Additional Analysis Features and Implementation Details

In the following paragraphs, we discuss important implementation details beyond the analysis features detailed above.

– **Cross-contract interactions:** To handle cross-contract interactions, GREED defaults to concretizing both the target address and the parameters of the CALL instruction. This allows approximating the execution state without incurring the overhead of symbolically executing an external (possibly undetermined) contract. Nonetheless, GREED can be easily configured to support fully symbolic CALL parameters – in fact, this feature (symbolic cross-contract interactions) has been used in the context of other academic works.

- **Memory and storage modeling:** GREED implements a precise memory model [16] that tracks EVM memory as a byte-addressable array supporting symbolic reads, writes, and memcopy-style operations. Our design employs an instantiation-based approach, where memory updates are lazily instantiated (on demand) during reads. To avoid redundant constraint instantiation, we also integrate a caching mechanism such that when a read is performed at a concrete address, the corresponding value (indexed by both the address and read width) is cached. For storage, GREED uses a hybrid model based on array theory, treating storage as an array of 256-bit words keyed by either concrete or symbolic values. Optionally, concrete storage reads (SLOADs) can retrieve actual on-chain data at a specified block number, and our design allows the use of these concrete values in symbolic operations.
- **Hash functions:** We employ a two-phase strategy for handling symbolic hash operations such as SHA. During symbolic execution, SHA instructions are captured as symbolic expressions that record the input parameters (offset, size, and memory contents) in order. When operating in "greedy" mode, GREED first attempts to concretize these parameters. If a unique solution is found, GREED computes its SHA hash value [2] and adds constraints that link the symbolic expression to this concrete value. Otherwise, it instantiates Ackermann constraints [5] to link multiple SHA operations as non-interpretable functions. After execution, a dedicated resolver plugin steps through the observed SHA operations in chronological order and fixes their outcomes by re-evaluating the memory and enforcing the appropriate constraints.
- **Solver integration:** GREED interfaces with an SMT solver – by default, Yices [14] – to query the satisfiability of path constraints. As with most components in our architecture, alternative SMT solvers can be substituted. During development, we evaluated various solvers, such as Z3 [11] and Boolector [8] – and found that Yices consistently offered the best performance.

Finally, as briefly mentioned above, GREED also offers high-level APIs for implementing custom vulnerability checks, exploration strategies (for state pruning, prioritization, and manipulation), and domain-specific analyses, simplifying the development of new smart contract security tools.

4 Evaluation

We evaluate the performance, analysis features, and versatility of GREED through a series of experiments. First, we qualitatively compare its analysis capabilities against existing systems (see Table 1), highlighting comprehensive support for basic analysis features, static analysis, and advanced exploration strategies. Second, we quantitatively compare GREED's targeted exploration capabilities against other state-of-the-art systems. Our results show that GREED reaches significantly more (10x) CALL statements in a sample of (randomly chosen) smart contracts. Third, we study the effect of different configuration settings on the performance of GREED. Finally, to demonstrate the extensibility of GREED, we

Table 1. Comparison of the features of existing systems. ○ Not implemented. ◐ Partially implemented. ● Fully implemented.

Tool	CROSS	MEM	HASH	STATIC	API
OYENTE [24]	○	◐	○	○	○
MAIAN [26]	○	◐	◐	○	○
TEETHER [23]	○	◐	◐	◐	○
MANTICORE [25]	◐	◐	◐	○	○
MYTHRIL [10]	◐	◐	◐	○	○
ETHBMC [18]	●	●	●	○	○
GREED	◐	●	●	●	●

implement a novel analysis to detect controllable JUMPI instructions. GREED identifies 390 previously unknown vulnerable contracts on Ethereum and BSC.

Experimental Setup. For all our experiments, we use a server equipped with 512GB of RAM and dual Intel Xeon Gold 6330 CPUs. We use GNU Parallel [34] to parallelize our tasks, and always limit each task to 5GB of RAM and 60 s of CPU time. We compare against the latest available version of all tools at the time of writing: MAIAN [26] at commit 3965e30, TEETHER [23] at commit 04adf56, MANTICORE [25] at commit 8861005, MYTHRIL [10] at commit 125914a, and ETHBMC [18] at commit e887f33. Notably, integrating MAIAN in our evaluation environment required significant modifications – due to syntax errors, broken dependencies, and missing implementations for several key opcodes. Similarly, we were unable to run OYENTE [24] in our environment, and thus, we have excluded it from our evaluation.

4.1 Analysis Features

In Table 1, we show a comparison between existing systems and GREED, with a focus on basic analysis features (similar to Frank et al. [18]), availability of static analyses, and availability of a high-level API to develop ad hoc static and dynamic analyses. In our comparison, we only include symbolic execution tools that are both publicly available and operate on EVM bytecode – even in the absence of source code.

Cross-Contract Interactions. GREED, MANTICORE, MYTHRIL, and ETHBMC are the only systems that support some form of cross-contract analysis. MANTICORE and MYTHRIL only support CALL instructions with concrete (or concretized) parameters. GREED also supports concrete CALL parameters and handles symbolic parameters through concretization. Nonetheless, GREED can be configured to handle fully symbolic parameters (see Sect. 3.3). ETHBMC supports concrete or fully symbolic CALL parameters.

Memory Model. MAIAN supports symbolic memory reads (not writes). TEETHER, MANTICORE, and MYTHRIL support simple symbolic memory

Table 2. Number (Percentage) of reached `CALL` instructions across different analysis tools. We run each system for 60 s (per contract) to assess its exploration capabilities, highlighting the coverage differences.

	Small	Medium	Large	Total
MAIAN [a]	127 (13%)	20 (2%)	4 (0%)	151 (5%)
teEther	247 (25%)	58 (6%)	1 (0%)	306 (10%)
Manticore	157 (16%)	14 (1%)	2 (0%)	173 (6%)
Mythril	294 (29%)	118 (12%)	12 (1%)	424 (14%)
EthBMC	224 (22%)	87 (9%)	31 (3%)	342 (11%)
Greed	960 (96%)	821 (82%)	745 (75%)	2,526 (84%)

[a] Integrating MAIAN in our evaluation environment required significant modifications.

operations, but must concretize all symbolic memcopy-like operations (e.g., `CALLDATACOPY`). OYENTE does not support any memcopy-like operation. GREED and ETHBMC implement a precise memory model [16] and can handle symbolic memory reads, writes, and memcopy-style operations. As discussed in Sect. 3.3, GREED also implements a caching mechanism to avoid redundant constraint instantiation.

Hash Functions. MAIAN, teETHER, and MYTHRIL support the hashing of memory buffers with fully concrete offsets, lengths, and values. OYENTE does not support symbolic hashing operations, and approximates the result of concrete hashing operations. GREED, MANTICORE, and ETHBMC support the hashing of arbitrary (symbolic or concrete) memory buffers.

Static Analysis. To complement our robust basic analysis features, GREED integrates advanced static analyses that allow us to focus symbolic execution on critical code regions. GREED inherits a number of static analyses from Gigahorse [20,21] (e.g., CFG recovery) and implements additional static analyses (such as program slicing). Among the other tools, only teETHER incorporates static analysis – specifically, CFG recovery and backward slicing.

High-Level APIs. Finally, GREED is the only system that offers a high-level API to develop ad hoc static and dynamic analyses. GREED also offers a number of (built-in) exploration strategies such as directed search, loop limiting, state rewriting, and selective concretization.

4.2 Exploration Capabilities

While existing systems excel at detecting specific vulnerabilities, they prove lacking when evaluated on slightly different tasks. We demonstrate the performance of GREED with a basic code reachability experiment. First, we select a target smart contract with a `CALL` statement x. Then, we alter all existing systems to simply emit a report and terminate when successfully executing (reaching) the chosen statement x. To do this, we leverage the ability of MAIAN, TEETHER, and ETHBMC to detect "prodigal" contracts – i.e., CALL statements with positive Ether value and controllable target address [26]. We (slightly) modified that analysis so that when a CALL statement is reached, instead of verifying the prodigal property, we just check whether the instruction address matches that of the chosen statement x. If so, the analysis simply terminates. Similarly, we modify the execution engine of MANTICORE and MYTHRIL to terminate when executing the chosen statement x. As mentioned above, we were unable to run OYENTE in our environment, and thus, we excluded it from our qualitative evaluation. Finally, we run GREED in its default configuration (with directed search).

We evaluate all tools on a sample of 3,000 (randomly chosen) Ethereum contracts[2] and report our findings in Table 2. In summary, we find that GREED outperforms all existing tools, reaching 84% of all CALLs – whereas others reach 9% on average. We attribute the performance gap observed in related work to a combination of (1) limited basic analysis features and (2) lack of (robust) exploration strategies. In fact, existing systems perform reasonably well on small contracts but struggle to handle the complexity of larger contracts. For example, TEETHER is the only system with a (CFG-driven) exploration strategy, but its CFG recovery often fails on larger contracts. Moreover, we observe that all tools have several failures related to misimplemented instructions and mishandling of external or symbolic data. For example, TEETHER discards any execution path that includes instructions such as RETURNDATACOPY or RETURNDATASIZE, whereas MAIAN fails to model instructions such as SELFBALANCE.

4.3   Ablation Study

In the following paragraphs, we study the effect of different analysis configurations on the performance of GREED. In its default configuration, GREED uses full support for symbolic memory operations (including read, write, and memcopy-like operations), symbolic hash operations, and a directed search strategy that uses prioritization – without pruning.

Directed Search. Table 3 shows the number of reached CALL instructions under different directed search configurations. Disabling pruning (while keeping prioritization active) results in a slight increase in reached targets across all contract sizes (from 958 to 960 for small contracts, from 780 to 821 for medium contracts, and from 708 to 745 for large contracts). We attribute this to imprecisions in the recovered control-flow graph that may incorrectly rule out reachable targets: When this happens, the lack of pruning allows GREED to explore these additional paths. However, this gain comes with an increased memory footprint (rising from an average of 180MB to 260MB per contract). In contrast, disabling prioritization leads to a notable drop in performance, as the execution engine wastes resources exploring paths that are farther from the target state. When

---

[2] We compute the size distribution of all deployed contracts and sample 1,000 small contracts (smallest 25%), 1,000 medium contracts, and 1,000 large contracts (largest 25%) with distinct code.

Table 3. Number of reached `CALL` instructions under different directed search configurations. Disabling pruning yields a slight coverage increase but raises memory usage, whereas disabling prioritization leads to a notable drop in performance – more pronounced in large contracts.

	Prioritization			No Prioritization		
	S	M	L	S	M	L
Pruning	958	780	708	953	745	646
No Pruning	960	821	745	947	513	352

Table 4. Comparison of the number of reached `CALL` instructions under different memory model configurations. Using a concrete memory model results in faster analysis times at the cost of decreased precision. Our caching layer allows for boosting performance without compromising precision.

	Symbolic Memory			Concrete Memory		
	S	M	L	S	M	L
Memory Cache	960	821	745	937	866	757
No Memory Cache	912	707	633	925	745	720

both pruning and prioritization are disabled, the deterioration in performance is even more pronounced, especially for medium and large contracts. Importantly, even in this worst-case configuration, GREED still outperforms all other tools by a wide margin.

Memory Model. We further investigate the impact of our precise symbolic memory model on GREED's performance by replacing it with (gradually) simplified variants – that is, disabling our caching layer and symbolic memory operations. We observe that disabling our caching layer results in a sharp drop in analysis performance across all contract sizes, although this is more evident in larger contracts. As detailed in Table 4, disabling our symbolic memory model (and instead using a concrete one) results in a modest overall boost in performance. We observe that, although approximating symbolic memory operations with their concrete counterparts may result in faster analysis times, *this comes at the cost of a much-decreased analysis accuracy.* In fact, we argue that GREED achieves the best analysis results by combining our symbolic memory model with our caching layer: this configuration yields robust performance without compromising analysis accuracy.

Hash Functions. We find that disabling our precise handling of (symbolic) hash operations results in a slight boost in analysis performance (from 960 to 962 for small contracts, from 821 to 824 for medium contracts, and from 745 to 755 for large contracts). Similar to the observations above, while approximating hash operations might result in faster analysis times, this comes at the cost of a much-decreased analysis accuracy.

```
1   # Discard statements with concrete jump destination
2   stmts = set()
3   for s in proj.stmts:
4       if s.op == "JUMPI" and not s.dest_val:
5           stmts.add(s)
6
7   # Analyze each JUMPI statement
8   for s in stmts:
9       # Set up directed symbolic execution
10      simgr.use_strategy(DirectedSearch(s))
11
12      # Explore each state until we reach the target statement
13      for found in simgr.findall():
14          # Jump condition must be satisfied
15          found.solver.add_constraint(Equal(s.cond_val, TRUE))
16
17          # Jump destination must be controllable
18          found.solver.add_constraint(Equal(s.dest_val, ARBITRARY))
19
20          # Check if the state (with the new constraints) is satisfiable
21          if found.solver.is_sat():
22              yield found.solver.eval_memory(found.calldata, CALLDATASIZE)
```

Fig. 3. Simplified Python code for the controllable JUMPI analysis.

Cross-Contract Interactions. Finally, in the context of this experiment, our reachability analysis stops when it encounters an external interaction (CALL). Therefore, any configuration change in our handling of external interactions does not lead to any change in performance.

Overall, our results underscore the importance of incorporating advanced analysis features – such as exploration strategies and a precise symbolic memory model – in GREED. We observe that while disabling pruning can reveal additional reachable targets, prioritization is essential to guide the search efficiently and keep the state space manageable. Similarly, our precise memory model and caching layer enable GREED's accurate analysis of complex memory operations, thus contributing to its overall superior performance.

4.4   Detecting Controllable JUMPIs

In this section, we demonstrate that GREED can be easily tailored to novel security analyses. To this end, we implement a novel analysis to detect controllable JUMPI instructions – i.e., conditional JUMP instructions. A controllable JUMPI allows an attacker to hijack the program counter, and thus take control of the program execution. This vulnerability has been recently reported [37] in a highly profitable MEV bot and could have resulted in hundreds of thousands of US dollars of financial damage. We implement this analysis in 50 lines of Python code. Figure 3 presents the core of our analysis script.

First, we (statically) inspect all contract statements to identify any JUMPI instructions with a non-constant destination (target) addresses. This lightweight analysis reduces the number of contracts in scope from 4.1M (all contracts with distinct bytecode across Ethereum and BSC) to 1,141. We symbolically execute these contracts and use directed search to reach the target statement. We add additional constraints to enforce that (1) the guarding condition for the JUMPI instruction is satisfied, and (2) the JUMPI destination is controllable. If our engine reaches the JUMPI instruction and the two aforementioned constraints are satisfied, we synthesize a concrete attack and verify it against a private fork of the respective chain. We evaluate our analysis on all deployed contracts in Ethereum and BSC and identify 134 and 256 previously unknown vulnerabilities, respectively, as well as one known vulnerability [37]. We manually confirmed that 130 of the 134 Ethereum contracts are still vulnerable at the time of writing (block 22,279,016). Three of the contracts were vulnerable in the past but have since been destructed and redeployed. One of the contracts contains an invalid JUMPI destination derived from a memory operation that does not appear to be controllable. We confirmed that all 256 BSC contracts are still vulnerable at the time of writing (block 48,398,024). We reported all issues to the Cybersecurity and Infrastructure Security Agency [9].

5   Case Studies

In this section, we illustrate how GREED has been successfully applied to build advanced program analysis systems for Ethereum smart contracts. We focus on two representative case studies: (a) detecting confused deputy vulnerabilities and (b) detecting storage collision vulnerabilities. Both studies leverage GREED's symbolic execution capabilities – augmented with domain-specific rules – to analyze real-world contracts at scale and automatically generate proof-of-concept exploits.

5.1   Detecting Confused Deputy Vulnerabilities

Confused deputy vulnerabilities occur when an attacker hijacks a smart contract's privileged operations via an inter-contract call (e.g., CALL) that is not intended to handle untrusted input. This can lead to unauthorized actions such as transferring assets or modifying critical state variables. For example, the TradingBot contract in Fig. 4 exposes a public execute function that forwards untrusted input directly to any target contract. As a result, an attacker can craft a transaction that redirects this call to the Token contract's transfer function, effectively leveraging the TradingBot's identity (and privileges) to initiate unauthorized asset transfers.

Implementation Overview. While we provide a high-level summary of the approach here, the complete system, JACKAL, is detailed in a separate paper [22]. JACKAL is built on top of GREED's core symbolic execution engine and incorporates several analysis stages tailored to detecting confused deputy vulnerabilities:

– **Confused Contract Discovery.** JACKAL leverages directed symbolic execution to inspect inter-contract calls where untrusted input might influence (control) the target address or function selector. As a result, contracts with controllable CALL instructions are flagged as confused contract "candidates."

```
1   pragma solidity ^0.8.0;
2
3   contract TradingBot {
4       // Public execute function lacking proper access control
5       function execute(address target, bytes calldata data) public {
6           // Forwards untrusted input to the target contract
7           target.call(data);
8       }
9   }
10
11  contract Token {
12      mapping(address => uint256) public balances;
13
14      // The transfer function deducts tokens based on msg.sender
15      function transfer(address recipient, uint256 amount) public {
16          [...]
17      }
18  }
```

Fig. 4. Simplified Solidity code of the TradingBot and Token contracts. The Trading-Bot contract is vulnerable to a confused deputy attack.

- **Target Contract Discovery.** For each confused contract candidate, JACKAL examines historical blockchain transactions to identify interesting external interactions and determines whether such interactions could lead to state modifications (e.g., via SSTORE) that exploit the confused contract's identity. When JACKAL determines that an external interaction could lead to the exploitation of the confused contract's identity, the respective external contract is flagged as a "target" contract.
- **Exploit Generation.** For each target contract, JACKAL leverages GREED to synthesize a transaction that forces the confused contract to invoke sensitive functions in the target contract, thereby demonstrating the exploit. The synthesized transaction is then replayed in a local blockchain simulator to confirm that the attack does not unexpectedly revert.

Through these stages, JACKAL enables end-to-end detection and exploitation of confused deputy vulnerabilities. JACKAL's analysis of over 2.3 million smart contracts identified 529 vulnerable instances and synthesized 31 working end-to-end exploits. All 31 exploits have been manually verified, demonstrating that attackers could potentially compromise digital assets valued at over one million US dollars.

5.2   Detecting Storage Collision Vulnerabilities

Storage collision vulnerabilities arise in proxy-based architectures, where a "proxy" contract delegates calls to separate "logic" contracts via the DELEGATECALL instruction. In this context, although the proxy and logic contracts execute independently, they both share the same underlying persistent storage. As a result, when the two contracts have conflicting interpretations of their storage slots, they might inadvertently overwrite such slots with the wrong value. This allows an attacker to overwrite privileged variables, potentially leading to unauthorized access (privilege escalation) and loss of funds. For example,

```
1   pragma solidity ^0.8.0;
2
3   contract Proxy {
4       // Slot 0 -> implementation
5       address public implementation;
6       fallback() external payable {
7           implementation.delegatecall(msg.data);
8       }
9   }
10
11  contract Implementation {
12      // Slot 0 -> owner (collides with Proxy)
13      address public owner;
14      function setOwner(address _owner) public {
15          owner = _owner;
16      }
17  }
```

Fig. 5. Simplified Solidity code of the Proxy and Logic contracts. The interaction of such contracts results in a storage collision.

in Fig. 5, the `Proxy` contract reserves storage slot zero for its `implementation` variable. Instead, the `Implementation` contract reserves the same storage slot for its `owner` variable. As a result, when the `Proxy` delegates a call (Line 6) to the `Implementation`'s `setOwner` function, the `owner` value overwrites the `implementation` variable in the `Proxy` contract, leading to a storage collision.

Implementation Overview. While we provide a high-level summary of the approach here, the complete system, CRUSH, is presented in a separate paper [28]. CRUSH builds on GREED to automatically detect and exploit storage collision vulnerabilities through the following analysis stages:

- **Component Discovery.** CRUSH analyzes on-chain transactions to identify clusters of contracts – namely, proxies and their corresponding logic contracts – that interact via DELEGATECALL.
- **Collision Discovery.** For each pair of proxy-logic contracts, CRUSH leverages GREED to symbolically execute their bytecode and infer the type of their storage variables. More precisely, after identifying all SLOAD and SSTORE instructions, CRUSH leverages (1) GREED's backward slice analysis to determine how each storage slot is computed and (2) GREED's forward slice analysis to deduce the accessed byte ranges. Then, CRUSH compares the inferred types of the proxy and logic contracts to detect collisions.
- **Exploit Generation.** Once a collision is detected, CRUSH verifies whether an attacker can exploit it by writing to a critical slot in one contract and reading it in another. To do this, CRUSH leverages GREED to synthesize concrete transactions that demonstrate the exploit.

By leveraging GREED's precise modeling of EVM instructions and storage access patterns, CRUSH uncovered critical storage collision vulnerabilities. These vulnerabilities could have led to serious incidents in practice: CRUSH's analysis of over 14 million smart contracts identified 14,891 vulnerable instances and synthesized 956 working end-to-end exploits. All profitable exploits have been manually verified, demonstrating that attackers could potentially compromise digital assets valued at over 6 million US dollars.

6 Discussion and Limitations

GREED inevitably inherits some limitations that arise from our design choices. First, we choose to build GREED directly on top of Gigahorse's IR, rather than extending an existing binaryâĂŞanalysis framework – such as angr [30]. This decision significantly simplifies our modeling of blockchain-specific concepts – e.g., blockchain state, transactions, persistent storage, cross-contract interactions. However, it also implies that sophisticated analyses that already exist in other frameworks, such as taint analysis, are not available out-of-the-box in GREED and must be re-implemented. While this creates unfortunate duplication of effort in the short term, it ultimately enables a more flexible, extensible, and research-friendly framework for smart contract security analysis.

Second, GREED's reliance on Gigahorse's intermediate representation (IR), provides robust static analysis capabilities, but makes GREED's effectiveness partly dependent on Gigahorse's accuracy. For example, inaccuracies such as missing JUMP destinations can cause pruning of paths that are in fact reachable. In Sect. 4.3 we show that this occasionally happens in practice: For some contracts, disabling pruning yields marginal coverage gains at the cost of a sharp increase in memory usage. Although we limit this dependency to well-tested features of Gigahorse (lifting, constant folding, and control-flow analysis), it remains a potential source of inaccuracies.

Other Limitations. Beyond the limitations discussed above, GREED shares modeling limitations common to similar symbolic execution systems. First, our handling of gas costs is deliberately simplified and may potentially miss vulnerabilities that arise from gas-specific behaviors. Second, by default, GREED employs a simplified handling of CALL instructions, which may miss vulnerabilities that require symbolic modeling of contract interactions. Additionally, the blockchain state (e.g., block number, timestamp, difficulty) remains symbolic by default, although one can optionally constraint such a state to actual (concrete) values when needed for more precise analysis. Addressing the aforementioned limitations, including the modeling of gas costs and cross-contract interactions, presents promising avenues for future research.

7 Related Work

Static Analysis. Early research in smart contract security focused on static analysis of the source code. Tools such as SmartCheck [35] and Slither [17] detect common vulnerabilities (e.g., re-entrancy, integer overflows) by scanning Solidity source code using rule-based approaches, offering quick insights to developers. Their availability and ease of use lowered the barrier for preliminary security audits. For example, Slither converts Solidity code into an intermediate representation for detailed data-flow and control-flow analysis, providing both vulnerability detection and potential code optimization insights.

In parallel, other efforts focused on direct analysis of EVM bytecode. Brent et al. proposed Vandal [7] and Ethainter [6], two tools that perform control-flow and data-flow analyses post-compilation, enabling insight even when source code is unavailable. In a similar vein, Grech et al. proposed Gigahorse [20] and Elipmoc [21] – a decompilation framework for EVM bytecode that also provides several rule-based vulnerability analyses. However, these tools often rely on fixed heuristics – such as rigid slicing rules or pattern matching – which may be insufficient to fully capture complex state interactions during execution.

Formal Verification. To provide stronger correctness guarantees, researchers have developed verification frameworks for smart contracts. For instance, Securify [36] operates on EVM bytecode and extracts predicates via a domain-specific language to capture compliance and violation patterns. Similarly, eThor [29] frames safety specifications in terms of reachability and uses an off-the-shelf SMT solver to reason about property violations. On the Solidity side, VerX [27] employs symbolic execution with induction and predicate abstraction to verify safety properties across multiple transactions, while VeriSmart [32] focuses on arithmetic safety through counterexample-guided invariant refinement. Extending these approaches further, Stephens et al. [33] incorporate liveness specifications to broaden the range of verifiable properties. Although these methods promise high-assurance security, they often incur significant engineering overhead, limiting their widespread adoption.

Symbolic Execution. Symbolic execution has emerged as a powerful technique for systematically exploring a contract's execution paths. One of the pioneering systems in this area, Oyente [24], demonstrated that symbolically executing EVM bytecode could effectively uncover vulnerabilities such as re-entrancy and transaction-ordering dependence. Mythril [10] is a symbolic execution-based tool that detects issues including integer overflows, unhandled exceptions, and unprotected self-destruct instructions. Similarly, Teether [23] and Maian [26] also leverage symbolic execution to identify vulnerable states. Manticore [25] and EthBMC [18] further advanced the state-of-the-art by integrating precise memory models and supporting cross-contract analysis. Nonetheless, Manticore does not integrate static analysis techniques – such as control-flow graph recovery or program slicing – limiting its ability to dynamically target critical code regions. Similarly, although EthBMC supports fully symbolic handling of cross-contract calls and a precise memory model, its monolithic design enforces rigid exploration strategies, making it difficult to extend to novel attack vectors.

In contrast to approaches that rely exclusively on static or dynamic methods, our framework GREED integrates static analyses (such as control-flow graph recovery and program slicing) with a flexible suite of symbolic exploration strategies – including directed search, loop limiting, state rewriting, and selective concretization. This unified approach preserves the core advantages of existing systems while adapting more readily to novel attack vectors.

8   Conclusion

We introduce GREED, a versatile open-source symbolic execution framework for EVM-based smart contracts. GREED addresses the limitations of existing tools by providing a novel combination of analysis techniques, including both a state-of-the-art SE engine and a suite of supporting analyses. Our experiments show that GREED reaches significantly more (10x) CALL statements in a sample of (randomly chosen) smart contracts. As a result, GREED enables more efficient path exploration – and superior flexibility – without compromising on the accuracy of the analysis. To demonstrate GREED's flexibility and ease of use, we implement a novel analysis to detect controllable JUMP instructions and evaluate it against all contracts in Ethereum and BSC [3], identifying 390 previously unknown vulnerable contracts. By releasing GREED to the community, we aim to lower the barrier to developing advanced security analyses for smart contracts, empowering security researchers to contribute to a more secure blockchain ecosystem.

Acknowledgments. This material is based upon work supported by NSF under Award No. CNS-2334709. Any opinions, findings, and conclusions or recommendations expressed in this publication are those of the author(s) and do not necessarily reflect the views of the NSF.

References

1. Baldoni, R., Coppa, E., D'elia, D.C., Demetrescu, C., Finocchi, I.: A survey of symbolic execution techniques. ACM Comput. Surv. (CSUR) (2018)
2. Bertoni, G., Daemen, J., Peeters, M., Van Assche, G.: Keccak. In: Annual International Conference on the Theory and Applications of Cryptographic Techniques. Springer (2013)
3. Binance: Binance Smart Chain. https://binance.com/en (2024)
4. Bose, P., Das, D., Chen, Y., Feng, Y., Kruegel, C., Vigna, G.: Sailfish: Vetting smart contract state-inconsistency bugs in seconds. In: 2022 IEEE Symposium on Security and Privacy (SP). IEEE (2022)
5. Bradley, A.R., Manna, Z.: The calculus of computation: decision procedures with applications to verification. Springer Science & Business Media (2007)
6. Brent, L., Grech, N., Lagouvardos, S., Scholz, B., Smaragdakis, Y.: Ethainter: a smart contract security analyzer for composite vulnerabilities. In: Proceedings of the 41st ACM SIGPLAN Conference on Programming Language Design and Implementation (2020)
7. Brent, L., et al.: Vandal: A scalable security analysis framework for smart contracts. arXiv preprint (2018)
8. Brummayer, R., Biere, A.: Boolector: An efficient SMT solver for bit-vectors and arrays. In: Tools and Algorithms for the Construction and Analysis of Systems (TACAS). Springer (2009)
9. CISA: Cybersecurity and Infrastructure Security Agency. https://www.cisa.gov (2024)
10. ConsenSys: Mythril. https://github.com/ConsenSys/mythril (2022)

11. De Moura, L., Bjørner, N.: Z3: An efficient SMT solver. In: International Conference on Tools and Algorithms for the Construction and Analysis of Systems. Springer (2008)
12. DefiLlama: Ethereum. https://defillama.com/chain/Ethereum (2024)
13. DefiLlama: Hacks. https://defillama.com/hacks (2024)
14. Dutertre, B., De Moura, L.: The yices smt solver (2006)
15. Ethereum: Ethereum. https://ethereum.org/en (2024)
16. Falke, S., Merz, F., Sinz, C.: Extending the theory of arrays: memset, memcpy, and beyond. In: Verified Software: Theories, Tools, Experiments (VSTTE). Springer (2014)
17. Feist, J., Grieco, G., Groce, A.: Slither: a static analysis framework for smart contracts. In: 2019 IEEE/ACM 2nd International Workshop on Emerging Trends in Software Engineering for Blockchain (WETSEB). IEEE (2019)
18. Frank, J., Aschermann, C., Holz, T.: ETHBMC: A bounded model checker for smart contracts. In: Proceedings of the 29th USENIX Conference on Security Symposium (2020)
19. Fröwis, M., Fuchs, A., Böhme, R.: Detecting token systems on ethereum. In: Financial Cryptography and Data Security (FC). Springer (2019)
20. Grech, N., Brent, L., Scholz, B., Smaragdakis, Y.: Gigahorse: thorough, declarative decompilation of smart contracts. In: 2019 IEEE/ACM 41st International Conference on Software Engineering (ICSE). IEEE (2019)
21. Grech, N., Lagouvardos, S., Tsatiris, I., Smaragdakis, Y.: Elipmoc: advanced decompilation of Ethereum smart contracts. In: Proceedings of the ACM on Programming Languages (2022)
22. Gritti, F., et al.: Confusum contractum: confused deputy vulnerabilities in ethereum smart contracts. In: 32nd USENIX Security Symposium (USENIX Security 23) (2023)
23. Krupp, J., Rossow, C.: teether: Gnawing at Ethereum to automatically exploit smart contracts. In: 27th USENIX Security Symposium (USENIX Security 18) (2018)
24. Luu, L., Chu, D.H., Olickel, H., Saxena, P., Hobor, A.: Making smart contracts smarter. In: 2016 ACM SIGSAC Conference on Computer and Communications Security (2016)
25. Mossberg, M., et al.: Manticore: A user-friendly symbolic execution framework for binaries and smart contracts. In: 2019 34th IEEE/ACM International Conference on Automated Software Engineering (ASE). IEEE (2019)
26. Nikolić, I., Kolluri, A., Sergey, I., Saxena, P., Hobor, A.: Finding the greedy, prodigal, and suicidal contracts at scale. In: Proceedings of the 34th Annual Computer Security Applications Conference (2018)
27. Permenev, A., Dimitrov, D., Tsankov, P., Drachsler-Cohen, D., Vechev, M.: Verx: safety verification of smart contracts. In: 2020 IEEE Symposium on Security And Privacy (SP). IEEE (2020)
28. Ruaro, N., Gritti, F., McLaughlin, R., Grishchenko, I., Kruegel, C., Vigna, G.: Not your type! detecting storage collision vulnerabilities in Ethereum smart contracts. In: Network and Distributed Systems Security (NDSS) Symposium 2024 (2024)
29. Schneidewind, C., Grishchenko, I., Scherer, M., Maffei, M.: Ethor: Practical and provably sound static analysis of Ethereum smart contracts. In: 2020 ACM SIGSAC Conference on Computer and Communications Security (2020)
30. Shoshitaishvili, Y., Wang, R., et al.: Sok:(state of) the art of war: Offensive techniques in binary analysis. In: 2016 IEEE Symposium on Security and Privacy (SP), IEEE (2016)

31. So, S., Hong, S., Oh, H.: SmarTest: effectively hunting vulnerable transaction sequences in smart contracts through language Model-Guided symbolic execution. In: 30th USENIX Security Symposium (USENIX Security 21) (2021)

32. So, S., Lee, M., Park, J., Lee, H., Oh, H.: Verismart: a highly precise safety verifier for Ethereum smart contracts. In: 2020 IEEE Symposium on Security and Privacy (SP). IEEE (2020)

33. Stephens, J., Ferles, K., Mariano, B., Lahiri, S., Dillig, I.: SmartPulse: automated checking of temporal properties in smart contracts. In: 2021 IEEE Symposium on Security and Privacy (SP). IEEE (2021)

34. Tange, O.: Gnu parallel-the command-line power tool. Usenix Mag (2011)

35. Tikhomirov, S., Voskresenskaya, E., Ivanitskiy, I., Takhaviev, R., Marchenko, E., Alexandrov, Y.: Smartcheck: static analysis of ethereum smart contracts. In: Proceedings of the 1st International Workshop on Emerging Trends in Software Engineering for Blockchain (2018)

36. Tsankov, P., Dan, A., Drachsler-Cohen, D., Gervais, A., Buenzli, F., Vechev, M.: Securify: practical security analysis of smart contracts. In: 2018 ACM SIGSAC Conference on Computer and Communications Security (2018)

37. Zellic: Your Sandwich is My Lunch: How to Drain MEV Contracts V2. https://zellic.io/blog/your-sandwich-is-my-lunch-how-to-drain-mev-contracts-v2 (2023)

FAULTLESS: Flexible and Transparent Fault Protection for Superscalar RISC-V Processors

Moritz Waser[1]($\boxtimes$) , David Schrammel[2] , Robert Schilling[2] , and Stefan Mangard[1]

[1] Graz University of Technology, Graz, Austria
{moritz.waser,stefan.mangard}@tugraz.at
[2] Rivos Inc., Santa Clara, USA
{davidschrammel,rschilling}@rivosinc.com

Abstract. Fault injection (FI) attacks pose a significant threat to the reliability and security of devices. They can cause data or control-flow corruption, leading to system failure or allowing malicious attackers to steal secret data or leak cryptographic keys. To protect against faults, many vendors extend their processors with lockstep capabilities, which require either dedicated hardware duplication or a reconfigurable second core that can act as a shadow core. The former causes a large hardware overhead while the latter requires an inflexible configuration during boot time with additional implications for software design. Software-based fault protection requires recompilation of existing code with custom compilers, which introduces compatibility issues.

This paper presents FAULTLESS: A fault protection mechanism that transparently performs hardware-based instruction duplication and utilizes the existing redundancy in superscalar processors. Contrary to lockstep approaches, our design facilitates a flexible protection approach with marginal hardware overhead that allows developers to toggle the fault protection during runtime, providing a choice between security and performance. The design is fully transparent and compatible with preexisting binaries. We implement our prototype based on the *VeeR EH1* RISC-V processor and show that, when active, our fault protection generates an average performance overhead between 32% and 79%, depending on the hardware configuration. Non-critical applications can deactivate the feature and run without any overheads. On an Artix-7 FPGA, our hardware modifications incur a minimal overhead of 3.5% for LUTs and 2.8% for flip-flops.

Keywords: Fault-Protection · RISC-V · Superscalar CPU

D. Schrammel—The work was done while the author was at Graz University of Technology.

1   Introduction

Faults can have tremendous effects on both the reliability as well as the security of a system. For example, a fault of natural origin, e.g., cosmic radiation, may lead to system failure of an embedded device, causing monetary loss or, even worse, harm to humans. Apart from naturally occurring faults, malicious attackers can perform precise fault injection attacks using power or clock glitches, electromagnetic interference or even lasers. This enables them to escalate privileges [35], bypass security measures [38], or break the security guarantees of ciphers [4]. The traditional threat model for fault injection attacks only considers adversaries with physical access to a device. However, recent research [11,21,33] has shown that faults can even be injected remotely.

Fault protection measures require redundancy, which can be realized in software, hardware, or both [32]. Software-based redundancy, such as instruction duplication, provides hardware portability at the cost of program recompilation and large performance overheads [2,3,6]. As it lacks any consideration for the underlying microarchitecture, it also fails to protect against precise fault injection [18]. Redundancy in hardware can circumvent the performance overhead by increasing the required silicon area and power. A full lockstep design, while providing strong protection, will require more than twice the die area and power [29]. Commodity systems, such as Infineon's TriCore [14] or NXP's S32K3 family [23], give developers the option to either run their system with a multicore configuration or utilize additional cores as secondary checker cores in a lockstep approach. While this provides a tradeoff between additional security and performance, the configuration is set during boot time and cannot be altered at run time. In addition, the duplicated hardware is limited to the CPU. Peripherals like memory require additional fault protection, e.g., through Error Correction Codes (ECCs) such as Hamming codes [13]. Mixed designs achieve a high level of protection with balanced overheads but require an inflexible hardware-software co-design [40]. As embedded devices are limited by area, power, and performance requirements, designers must strike a balance between effective and performant fault protection and implementation costs.

Contribution

In this paper, we introduce FAULTLESS, a fault-protected system design approach for superscalar RISC-V processors that aims for flexibility in terms of performance and application, as well as low hardware overhead. The core idea of the design is a hardware-based, flexible repurposing of a superscalar core's additional execution pipelines, which provide complimentary redundancy as a fault protection measure. FAULTLESS ensures that no instruction can commit a corrupted result to the microarchitectural state while providing a simple, interrupt-based recovery mechanism. The design emphasizes flexibility, as the fault protection can be toggled during runtime, allowing developers to protect important code sections at the cost of a performance overhead while retaining full performance otherwise. This focus stems from the observation that many fault injection attacks in the past specifically targeted security-critical computations like password or signature checks [38]. While large performance overheads are not tolerable in many use cases, dynamically protecting devices from faults during critical operations is an attractive prospect.

We implement and evaluate a prototype of our FAULTLESS design based on the *VeeR EH1* [7] RISC-V core. Furthermore, our security analysis highlights the extensive fault protection of the design. When the fault protection is active, the evaluation shows a performance overhead between 33% and 79%, depending on the hardware configuration. We compare the logic element usage on an Artix-7 FPGA for the unmodified *VeeR EH1* core and our prototype and find that our design uses only 3.5% more LUTs and 2.76% more flip-flops. Finally, we open-source our prototype[1] to facilitate future research.

In summary, our main contributions are as follows:

1. We present FAULTLESS, a system design for superscalar processors that facilitates fault protection with small hardware overhead and flexible performance impact.
2. We detail how our design utilizes existing redundancy to reduce the hardware overhead compared to traditional lockstep designs.
3. We present a proof-of-concept prototype based on the *VeeR EH1* RISC-V core and highlight required hardware changes.
4. We evaluate our prototype to showcase the small hardware overhead of 3.5% for LUTs and 2.76% for flip-flops, as well as the variable performance overhead between 0% and 79%.
5. We extensively analyze the security of our design, underlining all advantages of our design compared to lockstep designs.

2 Background

This section presents fundamental background knowledge on faults, fault attacks and common fault protection measures, which is required for subsequent chapters.

2.1 Fault Attacks

In a fault attack, the adversary induces a glitch into the circuit of a chip and exploits the various effects on the physical level, e.g., transient voltage fluctuations or timing violations [28]. While naturally occurring faults, e.g., caused by exposure to cosmic rays, are mostly of interest for data centers [16] and systems that operate in harsh environments [19], targeted fault attacks are actively used to break the security of embedded devices [24,36]. Exploiting this allows attackers to bypass security measures [35,38], alter the control-flow [11], or leak cryptographic secrets [4,8].

A fault is categorized by its spatial and temporal properties, such as time, duration and location of the fault, as well as its effect. The effect of a fault attack can be described on multiple abstraction layers. On the physical layer, the fault

---

[1] https://extgit.isec.tugraz.at/sesys/faultless.

may charge the gate of a transistor, which leads to a bit-flip in the CMOS logic. For the register transfer layer, this manifests as a computation error. Within the running program, this error can lead to an erroneous comparison, e.g., a wrong branch during a password authentication.

Classic fault attacks require physical access to a device, which allows an attacker to induce electromagnetic pulses, clock glitches, power glitches, or even shoot a laser directly at the decapsulated silicon die [15]. However, works like *Rowhammer.js* [11], *Plundervolt* [21], or *CLKSCREW* [33] presented remote fault attacks mounted solely through software, which further increased the attack surface of fault attacks.

2.2   Fault Protection

Protecting circuits from faults always requires a form of redundancy, which can be realized in software and hardware. The primary goal of fault protection is to detect the occurrence of a fault. Secondary goals are the correction of the fault's effects and pinpointing the fault location.

We distinguish between three common types of redundancy: information, spatial and temporal [32]. Information redundancy, e.g., parity bits, entails adding redundant information to stored or transmitted data. Spatial redundancy describes the physical replication of data or hardware. Temporal redundancy is achieved when a circuit performs a computation multiple times in sequence.

Implementing redundancy in software is attractive because the solution is microarchitecture-agnostic and thus easily applicable to a range of devices. However, software fault protection incurs large performance overheads [2,3,6]. Common software countermeasures include code duplication [34], range checks [27], and signature-based control-flow-integrity schemes [22].

Unlike software countermeasures, redundancy in hardware is tailored for a specific microarchitecture. Hardware fault protection comprises ECCs, e.g., Hamming codes [13], shadow registers [26], and modular redundancy [20]. In addition to redundant hardware, designers can also include measures to shield the device from the cause of a fault or detect physical effects like clock variations or charged particle impacts [37,41].

3   Threat Model

Independent of whether a fault is caused by natural effects or by a malicious attacker, our design goal is to protect processors from single bit-flips in registers and logic lines. We consider protection functional when we can successfully detect any single bit-flip within the processor. We choose a single bit-flip model to showcase the feasibility of our countermeasure. The presented prototype does not protect against multiple bit-flips, but the proposed design facilitates scaling to higher degrees of protection.

Potential adversaries may inject faults anywhere within a CPU, with any methodology, *i.e.*, physical access to the device or remotely. The goals of attackers include crashing systems, or hijacking the control-flow. Both can be achieved

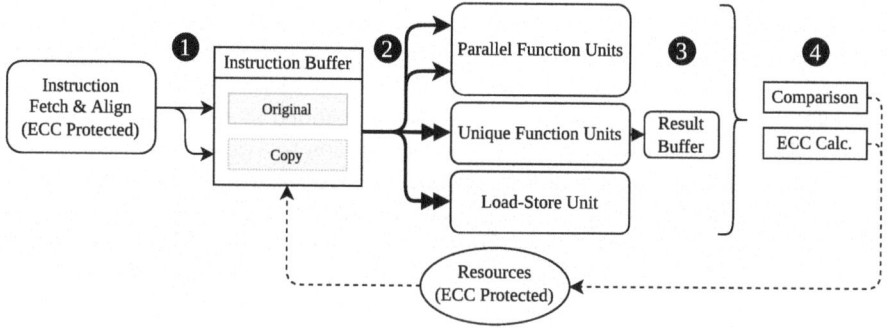

Fig. 1. Concept overview of the FAULTLESS design.

either directly through a corrupted Program Counter (PC) or branch instruction, altered register values or control signals, or by corrupting data that subsequent instructions depend on. Although our design partially protects systems from the effects of permanent, *i.e.*, *stuck-at* faults, we do not generally claim it does. Denial-of-Service (DoS) attacks that aim to delay or block further execution are out of the scope for this work.

4 Design

This section introduces the fundamental concept of the FAULTLESS design and highlights the design rationale. We first present a high-level overview of the concept and then discuss the implications for specific system components in depth.

4.1 Overview

To achieve extensive fault protection for a given system, *i.e.*, a processor and its peripherals, it is paramount to protect both the individual components as well as the communication between them. FAULTLESS is a full system design that achieves fault protection through a combination of hardware-based instruction duplication and ECCs. Unlike lockstep designs, which duplicate the entire processor, we protect the processor pipeline itself through redundant execution of instructions that keeps the hardware overhead small by exploiting the existing redundancy in a superscalar pipeline. Interactions with peripherals are protected with purposefully placed ECCs. In addition, fault protection is controlled through Control and Status Registers (CSRs), which can be modified by software at any given time. With this, developers have full control over the level of protection and related performance overheads, which allows them to selectively protect critical code sections while causing no overhead for other code.

Figure 1 highlights the main contribution of this work, which lies in the hardware-based instruction duplication. The design focuses on the execution path between the issue stage, where instructions leave the instruction buffer, and the commit stage, where results are evaluated and the architectural state is updated. Our design consists of four distinct changes to a superscalar pipeline. First (❶), we duplicate all instructions as they are added to an instruction buffer. Second (❷), we issue both instances of an instruction to available function units. The issuing strategy depends on the availability and presence of redundant function units. Double-headed arrows represent sequential issuing and, thus, temporal redundancy. Third (❸), for instructions that can only be issued sequentially, we introduce an intermediate result buffer when necessary. Finally (❹), we compare the results of both instances.

To complement the protected processor, we assume that all peripherals are protected by ECC or stronger means of protection.

Other processor components, like branch prediction logic, are of little concern to fault protection, as any corrupted state will be eventually corrected by other parts of the microarchitecture. In the case of branch prediction, a corrupted prediction will always be compared to the result of the actual computation of the branch condition at a later point. The point of duplication depends on the microarchitecture, but it must ensure continuous protection. When instructions are protected by ECC as they are fetched, the duplication must happen simultaneously with the ECC decoding and possible error correction. As they pass through the pipeline, duplicated instructions are protected by either spatial or temporal redundancy, depending on instruction type and availability of function units, e.g., Arithmetic Logic Units (ALUs). To keep the hardware overhead small, we only introduce comparisons of the results and their destinations at the final commit point. Whenever an instruction pair, *i.e.*, the original instruction and its redundant copy, can be issued in parallel, the result can be compared at the final pipeline stage without additional buffering. Instruction pairs that were issued sequentially may require an additional result buffer, which holds the result and target of the first instance until the second instance completes execution. In Out-of-Order (OoO) architectures, this buffering can be optimized through the already existing reorder buffer. To ensure full redundancy, the forwarding paths of all instruction instances must be separated. In case of duplication, one instance is marked with a special copy flag that limits available forwarding paths, such that original instructions can only forward results to other original instructions and copies can only forward to copies, respectively. When the design is scaled up to multiple copies, instruction copies must be assigned an identifier that enables distinction of multiple forwarding paths. The same mechanism also prevents instructions from forwarding results to their own copies when source and destination registers are identical.

4.2 Peripherals

Next to the processor pipeline, all interactions with peripherals and the peripherals themselves, *i.e.*, memory, must be protected to achieve full fault protection. This includes main memory, data and instruction caches, and all additional buffers required for e.g., memory transactions. We assume a system where all

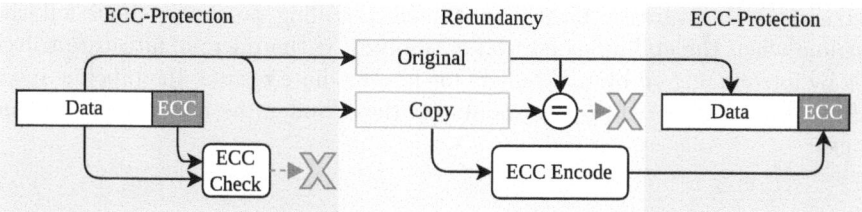

Fig. 2. The transition between ECC-protected domains and the redundancy-protected processor pipeline. The ECC check and the duplication of data happen simultaneously in combinatorial logic. The red cross symbolizes a raised exception due to an ECC check or a failed comparison between duplicated data. (Color figure online)

peripherals are protected by ECCs, meaning all faults that target the peripherals directly are covered.

Implementations must ensure that handshakes between peripherals and the processor provide a seamless transition between ECCs and other forms of redundancy. This is exemplified in Fig. 2. When receiving data from a peripheral, the ECC check and the duplication of the value must happen in combinatorial logic within the same clock cycle. The receiving registers and the ECC decoder must be connected to the same source, such that any fault on the data line is directly detected before the value is latched. Similarly, when sending data to a peripheral, the redundant data must be compared in parallel to the ECC encoding.

4.3   Register Instructions

We consider all instructions that do not interact with the memory interface as register instructions. This includes arithmetics, branches, CSR manipulation, fences, and trap-return operations. For all instructions that are computed by a function unit with several instances, we exploit the existing spatial redundancy. Instructions pairs are issued in parallel and reach their commit point simultaneously, allowing for simple comparison. When instructions require a function unit that only exists once, e.g., a multiplier, we issue them in succession. As we cannot let individual instances of duplicated instructions commit their result, e.g., to the register file, we may require an additional buffer in the commit stage that stores the result of the first instruction for an additional cycle. The requirement depends on the availability of the result within the pipeline and the forwarding capabilities of the given microarchitecture. When the results of all instances of an instruction are available within the pipeline when the first instance reaches the commit point, the buffer is not necessary as the results can be directly compared. OoO processors can optimize this through the reorder buffer. When an additional buffer is necessary, it must be extended with forwarding capabilities, similar to an additional pipeline stage.

For performance optimizations, a pipeline may also include additional stalling points between the issue and commit stages. This complicates forwarding, as instructions that already have passed such a stalling point may have left the pipeline when the stalling condition is resolved. If the microarchitecture solves this by introducing additional buffers for intermediate results, the pipeline must enforce similar forwarding constraints on these buffers as for normal pipeline stages.

4.4 Memory Instructions

To protect loads and stores to storage devices or other memory-mapped peripherals, we must ensure that the values and target addresses are always protected by a form of redundancy for the full transaction. Moreover, duplication of memory instructions requires special attention as they might exhibit non-idempotent properties, *i.e.*, they cause side effects when issued twice. For this reason, we classify all memory-mapped peripherals into two groups: internal and external.

Internal devices include all peripherals that reside on the same SoC and have known behavior in terms of side effects. This encompasses both devices that exhibit strictly idempotent behavior, such as flash or SRAM, as well as non-idempotent behavior, e.g., an interrupt controller.

External peripherals are all devices that are connected to the SoC through a bus. Since external devices are generally unknown from a processor designer's perspective, we assume all such devices as behaving non-idempotent.

For internal peripherals without side effects, we achieve fault protection through instruction duplication, similar to register instructions. Both loads and stores are duplicated and issued sequentially. As described in Sect. 4.2, the microarchitecture must ensure full redundancy for the entirety of the transaction. In this regard, memory interfaces require special assessment, as they often include performance optimizations such as instruction merging for subsequent transactions to the same address.

For internal peripherals with side effects, as well as external peripherals, we require a handshake that transforms the spatial/temporal redundancy within the processor to other forms of protection, e.g., ECCs. The same mechanism as described in Sect. 4.2 can be used to achieve this. As long as the memory instructions are within the processor, we protect them in the same fashion as all other instructions. The issue step, the address computation, and the propagation down the pipeline until the operands are written to a bus buffer or similar, are protected through redundancy. From this point onwards, other forms of redundancy (e.g., ECC) must ensure the integrity of the data. In the case of loads, we require the inverse of the described handshake to protect all values from the moment they enter the processor pipeline. With this, we ensure that all values are always protected by redundancy while avoiding potential issues related to non-idempotent behavior.

4.5   Mode Transition and Recovery

FAULTLESS enables flexible fault protection that can be toggled through software. To control the mode of operation, we add two CSRs, which can be modified by a regular CSR,-write instruction.

First, the u_protectionmode CSR controls if the fault protection using instruction duplication is active or not, *i.e.*, whether instructions get executed redundantly and results are compared. The correct mode of operation must be enforced for all instructions that are fetched after the CSR-write instruction. For a seamless transition between modes, writing to u_protectionmode in the processor's commit stage triggers a full pipeline flush. When enabling fault protection, this prevents all in-flight instructions following the CSR-write from committing their results without being protected by redundancy. After the pipeline flush, the instructions following the CSR-write are re-fetched and executed redundantly. Because the u_protectionmode CSR controls the behavior of the entire pipeline, including result comparisons and forwarding logic, the pipeline must also be flushed when exiting fault-protected mode. This could optionally also be achieved by instrumenting a compiler such that it always places a fence after writes to u_protectionmode.

Second, the u_detectionmode CSR controls how a detected fault within the processor pipeline should be handled. To deal with detected faults, FAULTLESS implements full forward-error-correction. Faults within the pipeline are detected as soon as the affected instruction reaches its commit point. When u_detectionmode is set, a detected fault raises an exception that can be handled in software, which enables further diagnosis. If the CSR is not set, a detected fault triggers a full pipeline flush and rolls back execution to the oldest in-flight instruction. With this, the processor retries executing the same instruction sequence that contained the fault, without requiring any software handling. A detected fault has the highest priority and should override any other microarchitectural effects.

5   Implementation

This section provides details about the prototype implementation of FAULTLESS, which is based on the *VeeR EH1* (formerly known as *SweRV EH1*) RISC-V core that features a 32-bit, superscalar, in-order, dual-issue, 9-stage pipeline with a single privilege level and no virtual memory. The core can be configured to include Tightly Coupled Memory (TCM) that is protected with ECC for both instructions and data.

5.1   Overview

The pipeline consists of three instruction-fetch and align stages, a decoding stage, and five execute stages. A general overview of the application of our FAULTLESS design to the core is given in Fig. 3. Bold arrows symbolize the issue step, which

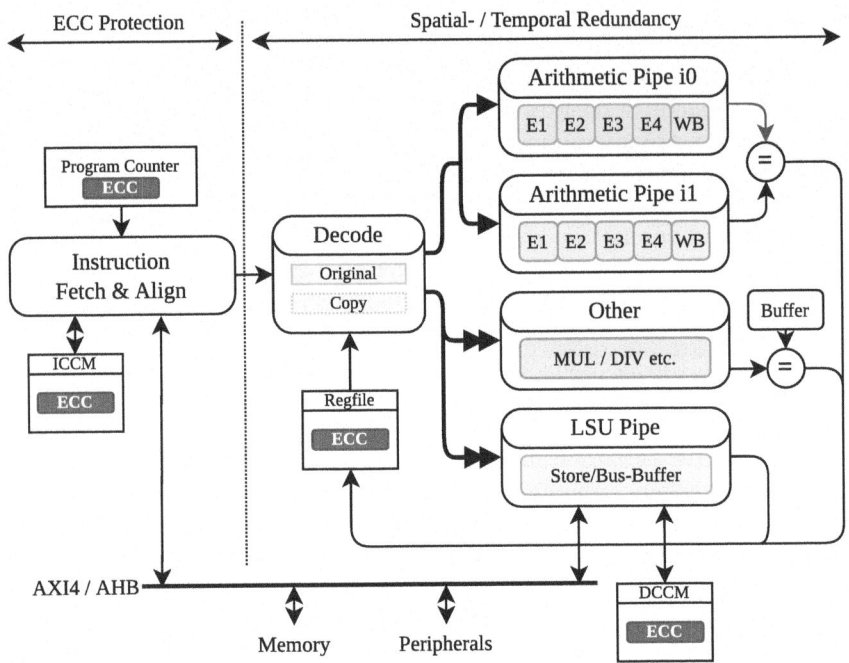

Fig. 3. General overview of our FAULTLESS concept applied to the *VeeR EH1* core.

is handled differently, depending on the instruction type and the availability of function units. Specifically, ALU instructions (including branches) can be issued in parallel, while all other instructions are issued sequentially. All arrows pointing in two directions describe memory interactions, either with TCM or with external devices. All other arrows indicate the path of instructions and their corresponding results within the pipeline. Until instructions reach the decode stage, they are protected by ECC. From the decode stage onward, the duplication provides redundancy to protect the execution.

The decode stage contains an instruction buffer holding up to four instructions, of which two may be issued at every cycle. When ECC-protected instructions reach the decode stage, we ensure continuous protection following Sect. 4.2. The execute stages contain two ALU pipelines, a multiplier, a division unit, and a Load Store Unit (LSU). ALU operations can be either computed immediately or delayed for three cycles for complex forwarding between pipelines.

The following sections highlight our changes to the *VeeR EH1* core by tracing the path of instructions from their duplication to the final commit. Our changes must ensure that instruction pairs are issued together, flow down the pipeline redundantly and commit their result after a comparison confirms that no fault has occurred.

5.2   Issue Step

An instruction buffer within the decode stage is responsible for managing the way individual instructions are issued. All instructions that are written to this buffer are duplicated when fault protection is enabled. The buffer must ensure that all duplicated instructions are issued together, following their designated protection strategy. We modify the issuing behavior such that arithmetics (except multiplication and division), branches, fences, CSR and system instructions will be issued in parallel and are, thus, protected by spatial redundancy. All other instructions are issued sequentially, protecting them with temporal redundancy.

The buffer is implemented as a queue that opportunistically issues the two oldest instructions while receiving new instructions from the align stage. To implement FAULTLESS, we extend the buffer's functionality in two distinct ways. First, we implement instruction duplication by writing incoming instructions to the buffer twice. When doing this, the second instruction is marked as a copy. Second, we modify the issuing behavior such that we ensure that instruction pairs, *i.e.*, the original and its copy, are always issued following the correct redundancy strategy. This is necessary because the buffer could otherwise issue instructions that should be protected with spatial redundancy in parallel to instructions that are issued sequentially. Whenever possible, two different but sequentially issued instructions will be parallelized.

5.3   Execution Pipeline

When instruction pairs are issued, either in parallel or sequentially, they normally receive their operands either from the register file or from subsequent pipeline stages. To improve performance by avoiding stalling conditions, the *VeeR EH1* core can allow instructions to flow down the pipeline with unresolved dependencies that are resolved at later stages. Independent of whether this is the case, we limit the forwarding logic to avoid two newly introduced hazards, as described in Sect. 4.1.

First, instructions must never forward results to their own copy. This would lead to wrong behavior when the destination of a copied instruction is equal to an operand. In the *VeeR EH1* core's pipeline, we stop this from happening for both parallel and sequential instruction pairs.

Second, there must always be two forwarding events for every pair of instructions. One instance of an instruction can never be allowed to forward its result to a pair of subsequent instructions, as this would undermine the redundancy. In the case of the *VeeR EH1* core, this is only a problem because the core has several commit points, which are not all at the final execute stage. If the microarchitecture ensures that all instructions commit their result at the same point, the second condition for the forwarding logic can be dropped, as a fault will always be detected before the corrupted value that was forwarded has any effect.

Memory operations are handled by the LSU, which is responsible for calculating target addresses and delegating requests to their corresponding buffers. The *VeeR EH1* core distinguishes memory accesses between internal targets, e.g., TCM, and external, bus-bound peripherals. We issue all loads and stores redundantly to protect the entire address calculation and the corresponding data until we know whether the address is internal or external. Accesses to TCM are resolved instantly, which allows for these instructions to be resolved within the 5 execute stages without stalling. Since TCM access behaves idempotently, we can protect these instructions with full redundancy, similar to other instructions. When dealing with external targets, the memory request is passed to a bus controller after three stages. We ensure continuous fault protection by implementing a handshake that validates the integrity of the request while it is passed to the bus controller alongside a newly calculated ECC. For loads, the pipeline is stalled at this point until the load is resolved. As soon as the result is available, we read it from the bus controller twice while checking the ECC to again ensure continuous protection.

The overall stalling behavior of the core also requires special attention. In certain conditions, the core can stall only the first three execute stages while the fourth and fifth stages continue execution. To resolve dependencies between the stalled and active stages, after the stalling condition is resolved, the core buffers the results of the final two stages. Since these buffers may also be affected by a fault, we add comparators that verify the integrity of these buffers whenever they hold valid data.

5.4 Commit Points

The *VeeR EH1* core has several commit points that must be extended with comparison logic. All instructions besides branches and bus-bound memory transactions commit their result at the final execute stage. While branches may commit at the final stage, they can also already report a mispredicted branch at the third execute stage, causing a partial pipeline flush. For this reason, the results of both the final and the third stage must be compared for every parallelized instruction pair. Instructions that are issued sequentially also commit at the final stage but require an additional buffer to store the result of the first instance. When the second instance reaches the commit point, the result is compared to the buffer and written to the register file. We also extend the result buffer with forwarding logic to avoid the case where other instructions can only receive a forwarded value from the second instance. In parallel with every result comparison, we calculate a new ECC that is stored alongside the result.

5.5 CSRs and Register File

Following Sect. 4.5, we add two CSRs to the core: u_protectionmode controls the currently active protection mode (either normal or FAULTLESS operation) and u_detectionmode determines, whether a detected fault triggers an exception or a pipeline flush with subsequent re-fetching of the faulted instruction sequence. Writing to u_protectionmode triggers a full pipeline flush to ensure the protection of all following instructions. Since the *VeeR EH1* core by default only includes logic to process CSR instructions in one arithmetic pipeline, we

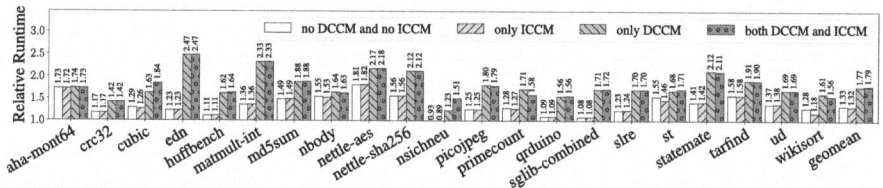

Fig. 4. Relative runtime overhead and geometric mean for all memory configurations for all Embench benchmarks.

extend the second pipeline with the same capabilities so that we can issue CSR operations in parallel. In addition, we add ECC protection for the PC, all CSRs and all general-purpose registers. This requires encoders and decoders for all related read and write ports.

6   Evaluation and Analysis

In the following, we evaluate the performance and hardware overheads of our design using the Embench benchmark suite [10] and FPGA synthesis. In addition, we comprehensively analyze the fault protection of the FAULTLESS design and our prototype.

6.1   Performance

We evaluate the overall performance overhead of FAULTLESS by comparing the runtime (cycles) of the Embench benchmark suite for the base *VeeR EH1* core and our modified design. Our performance evaluation is based on hardware simulation using *Verilator* [39]. As the *VeeR EH1* core is envisioned as an embedded device that runs bare-metal software, we run all benchmarks in this fashion. We found that repeated runs of individual benchmarks show only negligible runtime differences, rendering warmup runs irrelevant. Because multiple runs of the same benchmark provide neither additional information nor follow a normal distribution, we do not report standard deviation. The runtime of any program on the *VeeR EH1* core depends heavily on the configuration regarding TCM usage, both for data (DCCM) and code (ICCM). For this reason, we evaluate performance for all four combinations of TCM usage. When we use the TCM, we preload both code and data of every benchmark into it before starting the simulation. In addition, we also place the top of the stack within the tightly coupled data memory.

Figure 4 shows the individual testbench results for all configurations, as well as the corresponding geometric mean. The biggest overhead of 79% occurs when both data and code are placed within the TCM. Not using the TCM at all or using it only for code leads to an overhead of 32%. The reason why faster data access results in a significant overhead increase is twofold: First, TCM access is instant and does not cause pipeline stalling. Contrary to this, accessing regular memory frequently causes pipeline stalls that mask the additional overhead introduced by our instruction duplication. Second, as TCM access is resolved instantly and requests are sent sequentially, it always causes 100% overhead for memory instructions. In comparison, memory transactions with the bus controller only cause a stall once, as the actual memory access is not performed twice. A noteworthy effect occurred in the *nsichneu* testbench, which performed better when the fault protection was enabled in a configuration without DCCM. When executed without protection, this testbench causes frequent pipeline stalls due to dependencies on loads. The added latency of the instruction duplication allows the pipeline to optimize this, which overall improves the performance.

6.2 Hardware Overhead

The design goal of FAULTLESS is to provide flexible fault protection with negligible hardware overhead. To evaluate this overhead, we synthesize both the original and our modified *VeeR EH1* design for a *Nexys A7* FPGA board and compare the utilization of logic elements. Our modifications do not change the critical path of the core, as the added logic is computed in parallel to much longer combinational paths. The parameters (TCM size, etc.) are identical, so all differences stem from the core itself. Table 1 shows the overhead for both LUTs and flip-flops. Our modifications increase the number of LUTs by 3.5% and the number of flip-flops by 2.76%, underlining the benefit compared to traditional lockstep designs. The largest overhead is located in the decoding and execute stages, which contain the instruction buffer, the register file, and the forwarding logic.

6.3 Security Analysis

The fault protection of our FAULTLESS design is achieved through a combination of different measures: hardware-based instruction duplication for the processor pipeline, ECCs for vulnerable processor state (e.g., general purpose registers or CSRs) and peripherals, and protected communication between components.

Fetched instructions, as well as their related addresses, should always be protected by ECC until they reach the issuing or decoding stages. The PC must also

Table 1. FPGA utilization (LUTs and flip-flops) and relative overhead for all modules of the base *VeeR EH1* and our FAULTLESS design.

	LUTs	Flip-Flops	Combined
Baseline	28978	12293	41271
FAULTLESS	29991	12632	42623
Overhead	3.5%	2.76%	3.28%

be protected through ECC or duplication and comparison. As long as instruction integrity is guaranteed, branch predictors require no modifications, as faults in predictions may cause delays but can never corrupt the system state. Any corrupted branch prediction will be corrected by the actual target computation, while mispredictions that were corrupted into correct predictions will be detected by the redundant execution. When instructions are duplicated as they are placed into a buffer, the ECC decoding must occur in parallel to guarantee continuous redundancy. The protection through duplicated instructions is centered around the comparison that happens at each commit point. This comparison must be as restrictive as possible. Only identical instruction pairs may pass the comparison, which also protects the system from all faults that occur within pipeline registers for control signals. Instructions are considered identical if they are valid, their results and destination match, and one instruction is marked as a copy. Following our threat model, the comparison logic itself does not require additional protection. A bit-flip may occur in either one instruction instance or the comparison logic itself. In the former case, the comparison will detect the fault, while in the latter case, the detection logic may be triggered without any violation of instruction integrity. Both cases lead to a detected fault that can be corrected by reissuing the affected instruction sequence. A corrupted instruction can never pass the comparison logic, as the fault can never corrupt the instructions and the comparison at the same time. The additional result buffer that may be needed for instructions protected through temporal redundancy is also protected by the restrictive comparison. As long as a fault does not turn a valid instruction into an invalid instruction, which would immediately raise an exception, corrupted instructions will not be detected immediately but as soon as they reach their commit point. Propagation of corrupted results within the pipeline is prohibited by the restrictive forwarding logic for original and copied instructions, which have completely separated data paths. When all instructions commit their result at the same point, restricting the forwarding logic is not required since a corrupted instruction will always be detected before a following instruction can commit a corrupted result.

System state such as the PC, the register file, or CSRs must be protected from faults alongside the processor's pipeline. Otherwise, a single bit-flip could corrupt data-dependent control flow or alter microarchitectural behavior, e.g., disable instruction duplication by modifying the u_protectionmode CSR. The PC and register file can be protected with ECCs that are checked at every access. System registers that passively control the processor's functionality require special attention, as checking an ECC upon reading them is not sufficient to prevent faults from silently corrupting system behavior. Thus, critical CSRs, such as our u_protectionmode must be protected with a continuous self-check. This can be achieved using combinatorial ECC checks, which also allow correction. Alternatively, shadow registers that fully duplicate the system's state can be used. This offers detection capabilities that exceed regular ECCs at the cost of higher hardware overhead and, assuming just a single copy of the state, no error correction.

For loads and stores, the degree of fault protection depends on additional measures implemented by the target. Side-effect-free stores to memory-mapped peripherals, like the TCM for the *VeeR EH1* core, can be protected with full redundancy in the entire pipeline. The integrity of loads from such devices depends on the integrity of the data itself, which can be ensured by the use of, e.g., ECC. Implementations must ensure continuous redundancy, as explained in Sect. 4.4. Our instruction duplication protects both the address calculation and the data itself. Bus-bound targets of memory operations that may exhibit side effects require additional protection of the bus protocol and related buffers. Our FAULTLESS design only guarantees correct address calculation and, for stores, data integrity until the data is added to such a buffer. Whether the integrity of a loaded value from such a device is maintained depends on the device. From the moment the value enters the processor pipeline, it is again protected by duplication.

Our design is limited with regard to the amount and type of faults that occur in a system. When multiple transient faults occur simultaneously, the proposed measure cannot guarantee full protection. Similarly, when fault effects are permanent, it depends on the location of the fault whether our design can still function correctly. A single *stuck-at* fault can be corrected when it occurs in a register with ECC , but it can cause permanent erroneous behavior in other places.

7   Related and Future Work

Existing work on fault protection can be roughly categorized into software- and hardware-based designs. The performance and hardware overheads of the discussed designs are collected in Table 2.

The idea of idle execution unit utilization for fault tolerance in software was first mentioned by Schuette et al. [31]. They integrate a software control-flow monitor into idle slots of VLIW instructions without additional hardware modifications. Franklin [9] discusses possible fault targets in processors and mentions redundant execution based on idle function units as a defensive measure.

Table 2. Performance and hardware overheads of related works compared to FAULT-LESS.

		Performance [%]	Hardware [%]
Software-based	HAFT [17]	100	0
	EDDI [25]	13–111	0
	SIMD [5]	364	0
Hardware-based	Lockstep [1]	0	≥100
	[30]	0	75
	SHAKTI-V [12]	25	20
	FAULTLESS (This work)	0–79	3.28

Oh et al. [25] partition general purpose registers into two groups and duplicate instructions to operate on both groups individually. They compare the register sets after each basic block to detect errors. HAFT [17] presents a similar scheme by duplicating data flows and relying on the superscalar pipeline to optimize the execution of two independent threads. Chen et al. [5] utilize SIMD instructions on commodity hardware to duplicate data flow. All of these schemes require custom compilers and exhibit worst-case performance overheads above 100%. In comparison, FAULTLESS is fully compatible with legacy binaries and offers the possibility to flexibly toggle the fault protection and reduce the introduced overhead to zero.

Lockstep designs [1,14,23] achieve comprehensive fault coverage by duplicating a system at a specific granularity, e.g., at package or processor level. This approach causes none or only negligible performance overhead, but requires at least a hardware overhead of 100%. Fault-protected RISC-V designs [12,30] achieve their protection through a combination of ECC and modular redundancy. While this provides reasonable protection, it also causes a significant hardware overhead. Compared to this, our design keeps the hardware overhead minimal while presenting a flexible trade-off between protection and performance.

Potential future work could implement and evaluate our FAULTLESS design on a larger, out-of-order core. Such a core provides higher flexibility and could better optimize dependencies, leading to a smaller performance overhead. The performance implications of adding a data cache to the memory hierarchy could also be examined. Scaling up the duplication to triplication and studying the potential for latency-free fault correction is also possible. This would also provide a possibility to detect permanent faults in specific function units, which could be disabled to ensure correct operation.

8 Conclusion

In this paper, we presented FAULTLESS, a hardware design approach for superscalar processors, which provides fault protection at minimal hardware overhead and dynamic performance overhead. The design performs generic instruction duplication purely in hardware, making it compatible with preexisting binaries. Application developers or the OS can toggle the fault protection, including the performance overhead, during runtime, allowing them to add targeted protection to critical code sections. With this, we present an alternative to lockstep designs that protect processors with a brute-force approach that causes high hardware overheads. Our design achieves fault protection while providing a methodology that can easily be scaled for stronger threat models and larger processors. We implemented and evaluated a proof-of-concept prototype of the design based on the *VeeR EH1* core, showing that extensive fault protection can be achieved with minimal hardware overhead for superscalar processors. Furthermore, we provide a security analysis of our design, highlighting possible fault targets and protective measures used to detect state corruption.

Acknowledgements. We thank the anonymous reviewers for their valuable feedback that improved this work. This project has received funding from the Austrian Research Promotion Agency (FFG) via the AWARE project (FFG grant number 891092). Additional funding was provided by generous gifts from Intel.

References

1. Baleani, M., et al.: Fault-tolerant platforms for automotive safety-critical applications. In: CASES 2003 (2003)
2. Barenghi, A., et al.: Countermeasures against fault attacks on software implemented AES: effectiveness and cost. In: WESS 2010 (2010)
3. Barry, T., et al.: Compilation of a countermeasure against instruction-skip fault attacks. In: CS2@HiPEAC'91 (2016)
4. Boneh, D., et al.: On the importance of checking cryptographic protocols for faults (extended abstract). In: EUROCRYPT 1997 (1997)
5. Chen, Z., et al.: SIMD-based soft error detection. In: CF 2016 (2016)
6. Cojocar, L., et al.: Instruction duplication: leaky and not too fault-tolerant! In: CARDIS 2017 (2017)
7. Digital, W.: RISC-V: high performance embedded SweRV core microarchitecture, performance and CHIPS alliance (2019). https://riscv.org/wp-content/uploads/2019/04/RISC-V_SweRV_Roadshow-.pdf. Accessed 18 Aug 2023
8. Dobraunig, C., et al.: SIFA: exploiting ineffective fault inductions on symmetric cryptography. IACR Trans. Cryptogr. Hardw. Embed. Syst. (2018)
9. Franklin, M.: Incorporating fault tolerance in superscalar processors. In: HIPC 1996 (1996)
10. Free and Open Source Silicon Foundation: Embench: open benchmarks for embedded platforms (nd). https://github.com/embench/embench-iot/. Accessed 13 Jan 2023
11. Gruss, D., et al.: Rowhammer.js: a remote software-induced fault attack in JavaScript. In: DIMVA 2016 (2016)
12. Gupta, S., et al.: SHAKTI-F: a fault tolerant microprocessor architecture. In: ATS 2015 (2015)
13. Hamming, R.W.: Error detecting and error correcting codes. Bell Syst. Tech. J. (1950)
14. Infineon: 32-bit AURIX™TriCore™Microcontroller (nd). https://www.infineon.com/cms/de/product/microcontroller/32-bit-tricore-microcontroller/. Accessed 15 Apr 2024
15. Karaklajic, D., et al.: Hardware designer's guide to fault attacks. IEEE Trans. Very Large Scale Integr. Syst. (2013)
16. Keller, A.M., Wirthlin, M.J.: The impact of terrestrial radiation on FPGAs in data centers. ACM Trans. Reconfigurable Technol. Syst. (2022)
17. Kuvaiskii, D., et al.: HAFT: hardware-assisted fault tolerance. In: EUROSYS 2016 (2016)
18. Laurent, J., et al.: Cross-layer analysis of software fault models and countermeasures against hardware fault attacks in a RISC-V processor. Microprocess. Microsyst. (2019)
19. Luza, L.M., et al.: Impact of atmospheric and space radiation on sensitive electronic devices. In: ETS 2022 (2022)

20. Lyons, R.E., Vanderkulk, W.: The use of triple-modular redundancy to improve computer reliability. IBM J. Res. Dev. (1962)
21. Murdock, K., et al.: Plundervolt: software-based fault injection attacks against intel SGX. In: S&P 2020 (2020)
22. Nicolescu, B., et al.: Software detection mechanisms providing full coverage against single bit-flip faults. IEEE Trans. Nucl. Sci. (2004)
23. NXP: S32K3 Microcontrollers for Automotive General Purpose (nd). https://www.nxp.com/products/processors-and-microcontrollers/s32-automotive-platform/s32k-auto-general-purpose-mcus/s32k3-microcontrollers-for-automotive-general-purpose:S32K3. Accessed 15 Apr 2024
24. O'Flynn, C.: BAM BAM!! on reliability of EMFI for in-situ automotive ECU attacks. IACR Cryptology ePrint Archive (2020)
25. Oh, N., et al.: Error detection by duplicated instructions in super-scalar processors. IEEE Trans. Reliab. (2002)
26. Pattabiraman, K., et al.: Dynamic derivation of application-specific error detectors and their implementation in hardware. In: EDCC 2006 (2006)
27. Rela, M.Z., et al.: Experimental evaluation of the fail-silent behaviour in programs with consistency checks. In: FTCS 1996 (1996)
28. Richter-Brockmann, J., et al.: Revisiting fault adversary models - hardware faults in theory and practice. IEEE Trans. Comput. (2023)
29. Rodrigues, C., et al.: Towards a heterogeneous fault-tolerance architecture based on arm and RISC-V processors. In: IECON 2019 (2019)
30. Santos, D.A., et al.: Characterization of a RISC-V system-on-chip under neutron radiation. In: DTIS 2021 (2021)
31. Schuette, M.A., Shen, J.P.: Exploiting instruction-level resource parallelism for transparent, integrated control-flow monitoring. In: FTCS 1991 (1991)
32. Sorin, D.J.: Fault tolerant computer architecture (2009)
33. Tang, A., et al.: CLKSCREW: exposing the perils of security-oblivious energy management. In: USENIX Security 2017 (2017)
34. Theißing, N., et al.: Comprehensive analysis of software countermeasures against fault attacks. In: DATE 2013 (2013)
35. Timmers, N., et al.: Controlling PC on ARM using fault injection. In: FDTC 2016 (2016)
36. Timmers, N., Mune, C.: Escalating privileges in linux using voltage fault injection. In: FDTC 2017 (2017)
37. Upasani, G., et al.: Framework for economical error recovery in embedded cores. In: IOLTS 2014 (2014)
38. Vasselle, A., et al.: Laser-induced fault injection on smartphone bypassing the secure boot-extended version. IEEE Trans. Comput. (2020)
39. Veripool: Verilator (nd). https://www.veripool.org/verilator/. Accessed 18 Apr 2024
40. Werner, M., et al.: Protecting the control flow of embedded processors against fault attacks. In: CARDIS 2015 (2015)
41. Yuce, B., et al.: FAME: fault-attack aware microprocessor extensions for hardware fault detection and software fault response. In: HASP 2016 (2016)

Poster: Building Confidence in Hardware-Based Ransomware Detection Through Hardware Performance Counter Event Correlation

Ryan Binder[ID], Joshua Byun[ID], Dane Brown[ID], T. Owens Walker III[ID], and Jennie E. Hill[✉][ID]

United States Naval Academy, Annapolis MD, 21402, USA
{m260534,m270870,dabrown,owalker,jehill}@usna.edu

Abstract. Cybercrime is projected to cause over $10 trillion in damages through 2025 and ransomware has increasingly become the weapon of choice for cyber criminals. While hardware performance counters (HPCs) offer promising low-level insights for identifying ransomware behavior, concerns remain about their effectiveness in real-world settings. This work introduces a visualization tool that correlates HPC event data with ransomware execution to identify the most informative counters. Initial testing performed on a real-world dataset collected in a non-virtualized environment demonstrates the tool's potential to enhance early detection and address key concerns surrounding the practical use of HPCs for security applications.

Keywords: Side-channel · Hardware Performance Counters · Ransomware

1 Introduction

The global cost of ransomware, a fast growing type of cyber attack that encrypts a user's critical data and holds it ransom, is predicted to exceed $42 billion USD through 2025 and continue to grow by 30% year-over-year [1]. State of the art ransomware detection methods leverage both software and hardware-based approaches. Software-based detection methods rely on analyzing code behavior, system activity, and file characteristics through techniques such as signature matching, machine learning, and heuristic analysis [14]. In contrast, hardware-based ransomware detection methods analyze the underlying physical behavior of a computing system that may indicate malicious activity, often by monitoring side-channels. One emerging analysis method leverages the microarchitectural side-channel by accessing hardware performance counters (HPCs) embedded in modern CPUs to track metrics such as cache misses, branch mispredictions, and memory access patterns. These metrics can be analyzed to identify deviations from normal application behavior, which may signal ransomware activity [8].

M. Egele et al. (Eds.): DIMVA 2025, LNCS 15748, pp. 316–322, 2025.
https://doi.org/10.1007/978-3-031-97623-0_19

The primary contribution of this work is a visualization tool that facilitates the correlation of HPC event count data with ransomware process execution. This tool enables the evaluation and identification of optimal HPCs for use in the detection of ransomware. Initial testing and analysis of this visualization tool was conducted on a data set of real-world ransomware and benign operations, collected on a non-virtualized system. The remainder of this paper is organized as follows. Section 2 provides background on uses of HPCs for ransomware detection as well as concerns regarding the effectiveness of this method. Section 3 details the process-specific HPC Event Visualization tool developed. The test data collection process, initial results, and a discussion of follow-on research directions are included in Sect. 4.

2   Background: Side-Channels, Hardware Performance Counters, and Ransomware Detection

Side-channels are unintentional sources of information resulting from the physical behavior or microarchitectural implementation of operating computer hardware [8]. The microarchitectural side-channel is a form of information leakage that arises from internal hardware-level optimizations in modern computing systems, such as data values, locations, memory access patterns [7]. For more than a decade, researchers have explored the use of HPCs - special registers built into modern CPUs that track low-level hardware events such as cache hits and misses, branch predictions, and memory accesses - as a microarchitectural side-channel-based malware detection method [12]. Only a handful, however, focus on ransomware, such as RAPPER [2], RanStop [11], DeepWare [10], and HiPeR [3], which use time-series HPC data and machine learning techniques to detect and classify ransomware. Yet, many of these rely on virtualized environments, a limited number of HPCs, or simulated attacks rather than real-world ransomware samples. Startzel et al. analyzed HPCs and specific function calls like those disabling Windows Defender or targeting network resources in leaked source code from actual ransomware campaigns to identify behavioral fingerprints of ransomware variants [13].

A number of researchers have raised concerns regarding the use of HPCs for security applications. Weaver et al. questioned their effectiveness [15], while Dinkarrao et al. demonstrated adversarial attacks reducing accuracy from 80% to under 20% [5]. Zhou et al. identified key issues with HPC-based malware detection, including high overhead, virtualization challenges, and dataset overlap, and found a best-case F1-score of 80.78% with a 15% false positive rate, and ultimately concluded that HPC-based features were not reliably effective for distinguishing malware [16]. This result was later attributed to the PCA technique itself, which was found to disrupt essential correlations within HPC data streams [6]. Further challenges include non-determinism, signal discernibility issues, and performance overheads [9], as well as inaccuracies due to external noise and tool differences [4].

To address these concerns with the practical use of HPCs for security applications, this work provides a tool to researchers to enhance HPC selection and provide confidence in the correlation between identified HPCs and a targeted malicious process execution. Additionally, this work is conducted entirely in non-virtualized environments to avoid the shortcomings of virtualized HPCs. By exploring the correlation of top HPCs for classifying ransomware and benign operations, it builds on previous work [8] to examine how HPC-based detection particularly for identifying ransomware behaviors like encryption may be both practical and effective.

3   Process Visualization Framework

The section describes the primary contribution of this work. The visualization framework facilitates correlation between HPCs and process execution, which is comprised of three components: (1) the development of a snapshot of executing processes, (2) the filtering of HPCs by process, and (3) the presentation of the combined HPC event count data and process execution details on a single timeline plot. Each component is presented and discussed in detail.

Process Snapshot. As Ransomware executes, it often launches multiple process instances as well as external processes to accomplish the desired effect. When using HPCs for side-channel analysis, it is beneficial to know which running processes contributed to the counter metrics and which of those processes are associated with Ransomware execution. To accomplish this, snapshots of the list of running processes were taken at regular intervals during data collection and reduced to only those processes which started after the trial. These snapshots provide valuable correlation information, as parent and child process relationships can be inferred and counter metrics obtained on a per process basis.

Process Filtering. Two types of custom timeline reports were generated using the VTune command line interface. The first included total, system-wide hardware event counts for each combination of operation and HPC collected. The second type focused on process-specific event counts for each HPC collected, along with the corresponding processes that triggered each event during the collection period.

Process Visualization. Using the process filtering reports, time-series plots of total event count were overlaid with process-specific event counts for each

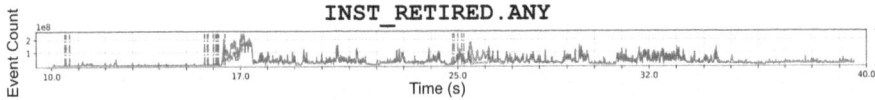

Fig. 1. Process Visualization Example for single PHOBOS ransomware and INST_RETIRED.ANY HPC. The red line indicates instructions specifically retired by the PHOBOS executable, while the blue line shows the total instructions system-wide. (Color figure online)

HPC, creating a visual representation of the portion of the overall event count attributable to the ransomware or benign process of interest as shown in Fig. 1. Additionally, each plot is overlaid with vertical bars indicating the start-time of each instantited processes, obtained from process snapshot results.

4   Initial Results and Analysis

This section details the experimental setup used to generate data for the visualization tool, followed by a presentation and discussion of initial results. It concludes with a brief description of ongoing work and future research directions.

Non-virtualized Experimental Setup. The test system was an Intel Xeon Silver 4208 8-core 2.10 GHz system and Cascade Lake micro-architecture with 32 GB DDR4 memory running Windows 10 Pro Windows Defender disabled. Intel VTune Profiler version 2022.3.0 gathered 90 selected hardware performance metrics. Trials included fifteen ransomware samples from eight leading families representing over 50% of recent threats, as well as 20 benign operations (including 7-Zip AES-256 encryption and SPECworkstation 3.1 benchmarks) to create a realistic test environment. The experimental set-up and test operations were adapted from prior work [8]. Data collection consisted of five trials of each ransomware and benign operation, collected in randomized order at different times of the day to minimize the impact of external factors. A dedicated test script controlled the experimental process, initiating VTune collection of HPCs, triggering the test operation, and logging all processes and parent process IDs at 1 s intervals for the duration of the trial. After each ransomware trial, the VTune results database was safely moved to a firewall-protected data server, and the system was then manually restored to its initial state.

Analysis using Visualization Tool. Figure 2 shows a side-by-side comparison of visualization tool results for a subset of three trials selected for preliminary analysis: PHOBOS ransomware (top), benign 7-Zip encryption (middle), and SPEC Convolution benchmark (bottom). The left plots show counts for a general-purpose HPC, INST_RETIRED.ANY, which counts the total number of instructions retired from the pipeline, while the right plots show a potentially ransomware-indicative counter, L2_RQSTS.CODE_RD_MISS, which counts L2 cache misses when fetching instructions. Each subplot displays a 20-s time series of process-attributed HPC events overlaid on total system-wide counts ransomware in red, benign in green, and total activity in blue. Y-axes are individually scaled for clarity. The INST_RETIRED.ANY results (left) reflect a close correlation between process-specific event count and the total count for all three processes examined while the L2_RQSTS.CODE_RD_MISS results (right) reveal distinguishable bursts uniquely aligned with ransomware execution (top, red). The benign process results for this counter (middle and bottom, green) show little to no correlation with the total event count, highlighting the discriminative potential of L2 cache miss events in detecting ransomware behavior.

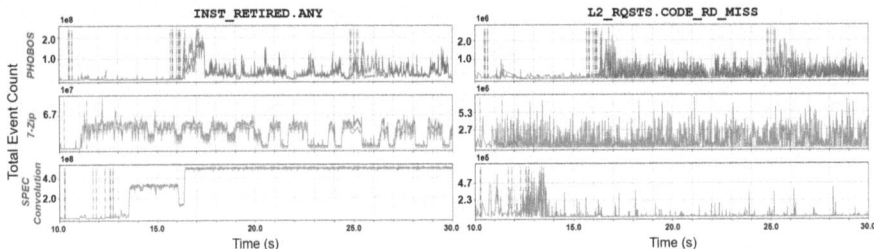

Fig. 2. Vizualization Tool Results for 2 HPCs and 3 operations: PHOBOS ransomware (top), benign 7-Zip encryption (middle), and SPEC Convolution benchmark (bottom). The left side shows event counts for a standard HPC, INST_RETIRED.ANY, while the right side shows event counts for an HPC indicative of ransomware activity (L2_RQSTS.CODE_RD_MISS). Total event count is plotted in blue, while events directly attributed to ransomware processes are overlaid in red, and benign processes are overlaid in green. The y-axis is adjusted for each subplot to maximize the plot height. (Color figure online)

An initial quantitative look at these results aligns with the visual findings. Consider a simple metric, percentage of total HPC counts as defined by process-specific counts divided by total counts for an individual HPC. In the case of L2_RQSTS.CODE_RD_MISS, we see that the percentage of total HPC counts for the PHOBOS ransomware is 50.2% while it is 27.9% and 20.4% for 7-Zip and SPEC Convolution benchmark, respectively. In contrast, this metric is 69.7% for PHOBOS while it rises to 73.8% and 99.2% for 7-Zip and SPEC Convolution benchmark, respectively. While just a single example, this metric highlights the potential to use L2_RQSTS.CODE_RD_MISS as a discriminator for ransomware. It is worth noting that ransomware detection would likely not rely on the use of a single HPC but rather a combination of well-selected HPCs that each, individually, demonstrated high correlation with the target ransomware process.

On-going and Future Work. Ongoing work is focused in two research directions. The first involves the systematic evaluation of the complete list of HPCs for ransomware classification, as identified in prior work and using the associated tools and datasets [8]. The second explores how HPC activity indicates the start of ransomware data encryption on a system. Preliminary work in the latter area has focused on identifying HPCs related to AES instruction executions, such as Intel's AES-NI instruction set and Advanced Vector Extensions (AVX). If these HPCs can be cross-referenced with time of execution for AES instructions, they have potential to be correlated to encryption activity. Additional research could include the application of this tool to an expanded set of ransomware and non-ransomware operations, potentially enabling more precise detection mechanisms. With sufficient data and correlation of the start of encryption-related functions, machine learning techniques could be leveraged to recognize encryption activity based on patterns in HPCs. This information could also potentially be used to properly scope a software reverse engineering effort. An analyst could focus on particular code segments that are known to generate high HPC counts to analyze the most relevant functionality and also confirm the functionality that has been inferred via side-channel analysis.

Acknowledgments. The views expressed in this paper are those of the authors and do not reflect the official policy or position of the U.S. Naval Academy, Department of the Navy, the Department of Defense, or the U.S. Government.

References

1. Cybercrime to cost the world $9.5 trillion in USD annually in 2024. https://www.esentire.com/web-native-pages/cybercrime-to-cost-the-world-9-5-trillion-usd-annually-in-2024. Accessed 1 May 2025
2. Alam, M., Sinha, S., Bhattacharya, S., Dutta, S., Mukhopadhyay, D., Chattopadhyay, A.: Rapper: ransomware prevention via performance counters. arXiv preprint arXiv:2004.01712 (2020)
3. Anand, P.M., Charan, P.S., Shukla, S.K.: Hiper-early detection of a ransomware attack using hardware performance counters. Digital Threats Res. Pract. **4**(3), 1–24 (2023)
4. Das, S., Werner, J., Antonakakis, M., Polychronakis, M., Monrose, F.: SoK: the challenges, pitfalls, and perils of using hardware performance counters for security. In: 2019 IEEE Symposium on Security and Privacy (SP), pp. 20–38. IEEE (2019)
5. Dinakarrao, S.M.P., et al.: Adversarial attack on microarchitectural events based malware detectors. In: Proceedings of the 56th Annual Design Automation Conference 2019, pp. 1–6 (2019)
6. Elnaggar, R., Servadei, L., Mathur, S., Wille, R., Ecker, W., Chakrabarty, K.: Accurate and robust malware detection: running XGBoost on runtime data from performance counters. IEEE Trans. Comput. Aided Des. Integr. Circuits Syst. **41**(7), 2066–2079 (2021)
7. Gruss, D.: Software-based microarchitectural attacks. IT Inf. Technol. **60**(5–6), 335–341 (2018)
8. Hill, J.E., Walker, T.O., Blanco, J.A., Ives, R.W., Rakvic, R., Jacob, B.: Ransomware classification using hardware performance counters on a non-virtualized system. IEEE Access (2024)
9. Mushtaq, M., Benoit, P., Farooq, U.: Challenges of using performance counters in security against side-channel leakage. In: 5th International Conference on Cyber-Technologies and Cyber-Systems (CYBER 2020) (2020)
10. Olani, G., Wu, C.F., Chang, Y.H., Shih, W.K.: DeepWare: imaging performance counters with deep learning to detect ransomware. IEEE Trans. Comput. (2022)
11. Pundir, N., Tehranipoor, M., Rahman, F.: RanStop: a hardware-assisted runtime crypto-ransomware detection technique. arXiv preprint arXiv:2011.12248 (2020)
12. Sayadi, H., He, Z., Makrani, H.M., Homayoun, H.: Intelligent malware detection based on hardware performance counters: a comprehensive survey. In: 2024 25th International Symposium on Quality Electronic Design (ISQED), pp. 1–10. IEEE (2024)
13. Startzel, C., Brown, D., Owens Walker III, T., Hill, J.E.: Identifying ransomware functions through microarchitectural side-channel analysis. In: International Conference on Science of Cyber Security, pp. 19–36. Springer (2024)

14. Urooj, U., Al-rimy, B., Zainal, A., Ghaleb, F.A., Rassam, M.A.: Ransomware detection using the dynamic analysis and machine learning: a survey and research directions. Appl. Sci. **12**(1), 172 (2021)
15. Weaver, V.M., McKee, S.A.: Can hardware performance counters be trusted? In: 2008 IEEE International Symposium on Workload Characterization, pp. 141–150. IEEE (2008)
16. Zhou, B., Gupta, A., Jahanshahi, R., Egele, M., Joshi, A.: A cautionary tale about detecting malware using hardware performance counters and machine learning. IEEE Des. Test **38**(3), 39–50 (2021)

Poster: FEDBLOCKPARADOX - A Framework for Simulating and Securing Decentralized Federated Learning

Gabriele Digregorio$^{(\boxtimes)}$, Francesco Bleggi, Federico Caroli, Michele Carminati, Stefano Zanero, and Stefano Longari

NECSTLab, DEIB, Politecnico di Milano, Milan, Italy
{gabriele.digregorio,michele.carminati,stefano.zanero,
stefano.longari}@polimi.it,
{francesco.bleggi,federico.caroli}@mail.polimi.it

Abstract. A significant body of research in decentralized federated learning focuses on combining the privacy-preserving properties of federated learning with the resilience and transparency offered by blockchain-based systems. While these approaches are promising, they often lack flexible tools to evaluate system robustness under adversarial conditions. To fill this gap, we present FedBlockParadox, a modular framework for modeling and evaluating decentralized federated learning systems built on blockchain technologies, with a focus on resilience against a broad spectrum of adversarial attack scenarios. It supports multiple consensus protocols, validation methods, aggregation strategies, and configurable attack models. By enabling controlled experiments, FedBlockParadox provides a valuable resource for researchers developing secure, decentralized learning solutions. The framework is open-source and built to be extensible by the community.

Keywords: Federated Learning · Blockchain

1 Introduction

In the rapidly evolving landscape of decentralized systems, federated learning holds considerable potential for enhancing privacy and scalability across various applications, fundamentally transforming data management and usage in diverse sectors such as healthcare, finance, and vehicular systems [2,4,14]. However, its integration into real-world scenarios introduces substantial challenges. Traditional federated learning frameworks, which rely on a central coordinating authority, are inherently vulnerable to malicious attacks and present single points of failure [3]. This central dependency not only increases the likelihood of security breaches but also creates potential bottlenecks in data processing and might introduce bias during model aggregation. To address these limitations, recent advancements have explored the elimination of the central orchestrator by developing fully decentralized federated learning systems using blockchain technology [6]. However, while these approaches mitigate certain risks, they also introduce complexities in managing and validating distributed model updates, as well as in accurately assigning the various roles necessary to support the system. These aspects are often overlooked in the current state of the art, with many solutions lacking robust guarantees against sophisticated adversarial attacks. Moreover, the absence of consistent experimental settings makes it difficult to perform meaningful comparisons across architectures. To enable the evaluation, validation, and comparison of different decentralized federated learning systems, we introduce FedBlockParadox, a framework for simulating complex decentralized federated learning environments in a customizable manner. FedBlockParadox supports key features of decentralized federated learning, including various consensus algorithms, validation mechanisms, and heterogeneous data distributions. It is designed to help the research community assess the ability of decentralized systems to withstand known adversarial attacks.

The main contributions of our work are as follows:

- We present FedBlockParadox, a framework for evaluating the resilience of decentralized federated learning approaches, particularly under adversarial conditions often overlooked in existing implementations.
- We make FedBlockParadox highly configurable, supporting a range of validation mechanisms, aggregation techniques, and consensus algorithms. This flexibility enables testing across diverse and realistic scenarios. While the framework includes several integrated algorithms by default, it is intended to be extensible, allowing the community to incorporate and evaluate new, more sophisticated approaches under consistent experimental settings to enable fair comparisons.
- We release FedBlockParadox as open-source software, freely available to the community[1]. Our goal is to support the empirical validation of both existing and future solutions that combine blockchain technology with federated learning.

2   State of the Art

We review key state-of-the-art proposals for decentralized federated learning combined with blockchain technology. Interested readers may refer to additional works summarized in the surveys presented in [10,11,13,17].

In [7], the authors present an implementation of decentralized federated learning that incorporates the Proof of WorkPoW consensus mechanism. However, this approach does not scale well in networks with large numbers of nodes, as each node—regardless of its computational capability—must download all updates from the previous round and perform local aggregation. Furthermore, the system lacks a validation mechanism to detect and discard malicious updates.

---

[1] https://github.com/necst/FedBlockParadox.

The authors of [6] integrate blockchain technology with a novel committee-based consensus mechanism. In each round, a dynamic subset of nodes, referred to as the *committee*, is elected to perform validation. This subset evaluates submitted model updates using local datasets and assigns scores based on validation accuracy. Only updates meeting predefined accuracy thresholds are aggregated into the global model. However, the paper does not explore the effectiveness of alternative validation mechanisms beyond accuracy-based filtering.

In [12], the authors propose an implementation of decentralized federated learning tailored for fog computing environments. The architecture integrates local updates from end devices into a distributed ledger, verified through a Proof of WorkPoW consensus mechanism. A central challenge discussed is the trade-off between privacy protection and system efficiency. Nevertheless, similar to [7], this approach struggles to scale in large networks, as every node must download and locally aggregate all model updates from each round.

The work in [16] introduces a decentralized federated learning architecture that leverages node selection and knowledge distillation to overcome limitations of traditional centralized federated learning. A dynamically selected central node replaces the static server, based on individual node performance and data quality. However, the architecture lacks defenses against malicious behavior such as targeted poisoning or falsified dataset sizes. Moreover, no mechanism is provided to regulate or verify the actions of the elected central node, allowing it to potentially bias the global model without detection.

In [8], a set of nodes replaces the traditional central server, coordinating model updates while preserving system integrity. The aggregation mechanism dynamically adapts to the state of participant data and models by weighting updates based on validation loss and training data volume. A validation step based on accuracy is employed to filter out low-quality or malicious updates. Malicious participants submitting poisoned updates are penalized accordingly.

The authors of [15] introduce a novel consensus protocol termed Proof of FederationPoF, which assigns roles to peers based on their earned stake, accumulated through valid contributions to the global model. This role distribution promotes resilience against centralized control and reduces the likelihood of collusion. The authors also incorporate the Multi-Krum algorithm to detect and reject anomalous updates that deviate significantly from the majority, thereby enhancing robustness against poisoning attacks.

3   FedBlockParadox

We introduce FedBlockParadox, a framework that simulates decentralized federated learning systems under adversarial conditions. Its ultimate goal is to provide a reliable means to test and validate various defenses against common attacks in decentralized federated learning. To support this objective, we release our framework as an open-source tool, encouraging further research and testing of proposed solutions. FedBlockParadox allows customization of key parameters, including consensus algorithms, validation mechanisms, aggregation strategies, and update-sharing methods. The configuration options available at the time of writing are inspired by state-of-the-art implementations in decentralized federated learning systems and detailed below. FedBlockParadox is designed to be extensible, encouraging the integration of additional methods.

General Configurations. These settings define the core parameters of Fed-BlockParadox, including the total number of simulation rounds, the model architecture and initialization, and the number of participating nodes. Moreover, Fed-BlockParadox supports model updates in both weight and gradient form.

Consensus   Algorithms. FedBlockParadox   currently   supports Proof-of-Work (PoW), Proof-of-Stake (PoS), and committee-based consensus, each supporting further customization within the framework.

Validation Algorithms. FedBlockParadox supports a range of validation algorithms. It includes naive strategies such as Pass-Weights and Pass-Gradients, which accept all updates without verification. More advanced methods, such as Global and Local Dataset Validation, evaluate model updates against a shared or local dataset, accepting them only if they meet a predefined accuracy threshold. Finally, Multi-Krum Validation allows validators to assess updates collectively by measuring pairwise distances and filtering out the most divergent ones.

Aggregation Algorithms. FedBlockParadox currently supports three aggregation strategies: FedAvg [9] for model updates in weight form, and Mean and Median Aggregation for model updates in both gradient and weight form.

Malicious Nodes and Attacks. FedBlockParadox allows specific nodes to exhibit malicious behavior. Supported attack types include label flipping, targeted data poisoning, and additive noise. Each attack type is configurable to suit the simulation context and can be tailored to evaluate system robustness under different adversarial scenarios.

Dataset Settings. The simulator supports the use of different datasets. In our initial implementation, we employed the MNIST [1] and CIFAR-10 [5] datasets. Key configuration options include the number of dataset partitions, the percentage of independent and identically distributed IID versus non-IID partitions, and the strategy for assigning partitions to nodes.

4   Preliminary Experiments

We ran a set of experiments to validate our tool and provide baseline performance metrics. The full list of experiments is available in our repository[1], we briefly sum up the results for each attack category in Fig. 1. Tests have been developed varying aggregation methods, validation processes, datasets, architecture, consensus algorithms, and amount of malicious nodes.

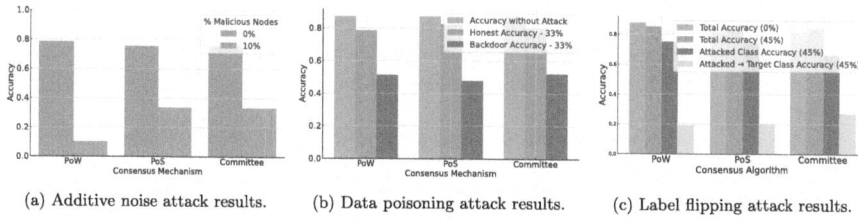

(a) Additive noise attack results.   (b) Data poisoning attack results.   (c) Label flipping attack results.

Fig. 1. Aggregated preliminary results of attacks on the datasets with various parameters, e.g., attacker amounts and aggregation methods.

5   Conclusion

In this work, we presented FedBlockParadox, a simulation framework tailored for modeling decentralized federated learning environments using blockchain-based architectures. Through its modular design and configurability, FedBlockParadox enables experimentation with consensus protocols, validation mechanisms, aggregation techniques, and attack strategies. Our framework addresses critical gaps in the current state of the art by supporting reproducible testing and community-driven extensibility. By releasing FedBlockParadox as open-source software, we encourage the development of new defense mechanisms and validation approaches to advance the field of secure decentralized federated learning.

Future Work. As a research-enabling framework, FedBlockParadox will be employed to empirically and systematically evaluate novel defense strategies and test security assumptions in blockchain-based decentralized federated learning systems proposed in the current state of the art. By providing consistent experimental settings and a standardized evaluation pipeline, it enables meaningful comparisons across architectures and approaches. Our ultimate goal is to assess the real-world impact of blockchain-enabled decentralization on the robustness and security of these systems.

Acknowledgements. This study was carried out within the MICS (Made in Italy – Circular and Sustainable) Extended Partnership and received funding from Next-Generation EU (Italian PNRR – M4 C2, Invest 1.3 – D.D. 1551.11-10-2022, PE00000004). CUP MICS D43C22003120001.

References

1. The MNIST database of handwritten digits. http://yann.lecun.com/exdb/mnist/
2. Boiano, A., et al.: A secure and trustworthy network architecture for federated learning healthcare applications. In: 2024 20th International Conference on Wireless and Mobile Computing, Networking and Communications (WiMob), pp. 124–129 (2024). https://doi.org/10.1109/WiMob61911.2024.10770536
3. Bouacida, N., Mohapatra, P.: Vulnerabilities in federated learning. IEEE Access **9**, 63229–63249 (2021)

4. Digregorio, G., Cainazzo, E., Longari, S., Carminati, M., Zanero, S.: Evaluating the impact of privacy-preserving federated learning on can intrusion detection. In: 2024 IEEE 99th Vehicular Technology Conference (VTC2024-Spring), pp. 1–7. IEEE (2024)
5. Krizhevsky, A., Hinton, G., et al.: Learning multiple layers of features from tiny images (2009)
6. Li, Y., Chen, C., Liu, N., Huang, H., Zheng, Z., Yan, Q.: A blockchain-based decentralized federated learning framework with committee consensus. IEEE Netw. **35**(1), 234–241 (2020)
7. Ma, C.: When federated learning meets blockchain: a new distributed learning paradigm. IEEE Comput. Intell. Mag. **17**(3), 26–33 (2022)
8. Mao, Q., Wang, L., Long, Y., Han, L., Wang, Z., Chen, K.: A blockchain-based framework for federated learning with privacy preservation in power load forecasting. Knowl.-Based Syst. **284**, 111338 (2024)
9. McMahan, B., Moore, E., Ramage, D., Hampson, S., y Arcas, B.A.: Communication-efficient learning of deep networks from decentralized data. In: Artificial Intelligence and Statistics, pp. 1273–1282. PMLR (2017)
10. Moore, E., Imteaj, A., Rezapour, S., Amini, M.H.: A survey on secure and private federated learning using blockchain: theory and application in resource-constrained computing. IEEE Internet Things J. **10**(24), 21942–21958 (2023)
11. Nguyen, D.C., et al.: Federated learning meets blockchain in edge computing: opportunities and challenges. IEEE Internet Things J. **8**(16), 12806–12825 (2021)
12. Qu, Y., et al.: Decentralized privacy using blockchain-enabled federated learning in fog computing. IEEE Internet Things J. **7**(6), 5171–5183 (2020)
13. Qu, Y., Uddin, M.P., Gan, C., Xiang, Y., Gao, L., Yearwood, J.: Blockchain-enabled federated learning: a survey. ACM Comput. Surv. **55**(4), 1–35 (2022)
14. Santos, D.R., et al.: A federated learning platform as a service for advancing stroke management in European clinical centers. In: 2024 IEEE International Conference on E-health Networking, Application Services (HealthCom), pp. 1–7 (2024). https://doi.org/10.1109/HealthCom60970.2024.10880750
15. Shayan, M., Fung, C., Yoon, C.J., Beschastnikh, I.: Biscotti: a blockchain system for private and secure federated learning. IEEE Trans. Parallel Distrib. Syst. **32**(7), 1513–1525 (2020)
16. Zhou, Z., Sun, F., Chen, X., Zhang, D., Han, T., Lan, P.: A decentralized federated learning based on node selection and knowledge distillation. Mathematics **11**(14), 3162 (2023)
17. Zhu, J., Cao, J., Saxena, D., Jiang, S., Ferradi, H.: Blockchain-empowered federated learning: challenges, solutions, and future directions. ACM Comput. Surv. **55**(11), 1–31 (2023)

Author Index

© The Editor(s) (if applicable) and The Author(s), under exclusive license
to Springer Nature Switzerland AG 2025
M. Egele et al. (Eds.): DIMVA 2025, LNCS 15748, pp. 329–331, 2025.
https://doi.org/10.1007/978-3-031-97623-0

The manufacturer's authorised representative in the EU is Springer
Nature Customer Service Centre GmbH, Europaplatz 3, 69115 Heidelberg,
Germany. If you have any concerns regarding our products, please
contact ProductSafety@springernature.com

Printed and bound by CPI Group (UK) Ltd, Croydon, CR0 4YY
29/04/2026
02099461-0011   ·